Coatings: Science and Engineering

Coatings: Science and Engineering

Editor: Falicia Radcliff

NY RESEARCH PRESS

New York

Published by NY Research Press
118-35 Queens Blvd., Suite 400,
Forest Hills, NY 11375, USA
www.nyresearchpress.com

Coatings: Science and Engineering
Edited by Falicia Radcliff

International Standard Book Number: 978-1-63238-669-4 (Hardback)

Cataloging-in-Publication Data

Coatings : science and engineering / edited by Falicia Radcliff.
    p. cm.
Includes bibliographical references and index.
ISBN 978-1-63238-669-4
1. Coatings. 2. Materials. 3. Materials science. I. Radcliff, Falicia.
TA418.9.C57 C63 2019
667.9--dc23

Contents

Preface

A coating is a covering, which is applied on the surface of an object for varied functional or decorative purposes. It can be applied as liquids, solids or gases. Using the techniques of coating, surface properties of the substrate such as corrosion resistance, wear resistance, adhesion or wettability can be changed. It can also provide electrical conductivity or magnetic response in certain cases such as in semiconductor device fabrication. Some of the coating processes and technologies may be grouped into the broad areas of roll-to-roll coating processes, vapor deposition, spraying, physical coating processes, chemical and electrochemical techniques. The aim of this book is to present researches that have transformed coatings technology and aided its advancement. There has been rapid progress in this area and its applications are finding their way across multiple industries. Coherent flow of topics, student-friendly language and extensive use of examples make this book an invaluable source of knowledge.

This book has been the outcome of endless efforts put in by authors and researchers on various issues and topics within the field. The book is a comprehensive collection of significant researches that are addressed in a variety of chapters. It will surely enhance the knowledge of the field among readers across the globe.

It gives us an immense pleasure to thank our researchers and authors for their efforts to submit their piece of writing before the deadlines. Finally in the end, I would like to thank my family and colleagues who have been a great source of inspiration and support.

Editor

Shape Memory and Huge Superelasticity in Ni–Mn–Ga Glass-Coated Fibers

Lei Shao [1], Yangyong Zhao [1], Alejandro Jiménez [2], Manuel Vázquez [2] and Yong Zhang [1,*]

[1]   State Key Laboratory for Advanced Metals and Materials, University of Science and Technology Beijing, No. 30, Xueyuan Road, Beijing 100083, China; shaoleiustb@163.com (L.S.); zhaoyangyongwd@126.com (Y.Z.)

[2]   Institute of Materials Science of Madrid, CSIC, 28049 Madrid, Spain; ajimenezv@icmm.csic.es (A.J.); mvazquez@icmm.csic.es (M.V.)

*   Correspondence: drzhangy@ustb.edu.cn

Academic Editor: Alessandro Lavacchi

Abstract: Ni–Mn–Ga polycrystalline alloy fibers with diameters of 33 μm are reported to exhibit significantly improved ductility and huge superelastic and shape memory strains in comparison to conventional brittle bulk polycrystalline alloys. Particularly, the recoverable strain of the Ni54.9–Mn23.5–Ga21.6 fiber can be as high as 10% at 40 °C. Such optimized behavior has been achieved by a suitable fabrication process via a glass-coating melt spinning method. The superelastic properties at different temperatures and the shape memory effect of Ni54.9–Mn23.5–Ga21.6 fibers were investigated.

Keywords: glass-coated melt spinning; fibers; shape memory; superelasticity; serration behavior

1. Introduction

The shape memory alloy (SMA) is an advanced class of smart materials that exhibit a unique phenomenon. Shape memory and superelasticity effects are believed to be driven by the crystallographically reversible martensitic transformation [1]. The shape memory behavior was first discovered in Au–Cd alloys [2] and afterwards observed in a number of systems, such as Ti–Ni, Cu–Al–Ni [3], Ni–Mn–Ga, and Ni–Co–Mn–In alloys [4–7]. Recently, novel potential systems with SMAs were reported. For example, Fe–Mn–Al–Ni SMA [1] shows a small temperature dependence of the superelastic stress. High-entropy alloy system Ti–Zr–Hf–Co–Ni–Cu [8] shows wide temperature hysteresis of about 90 K at temperatures higher than 400 K of martensitic transformation. Fe–28Ni–17Co–11.5Al–2.5Ta–0.05B SMA [9] exhibits a superelastic strain of more than 13% with a tensile strength above 1 GPa.

SMAs have potential applications in many aspects. The thermo-mechanical properties of a number of SMAs, along with their various applications, have been presented in [10], while the most recent developments are collected in [11]. In their study, Alam et al. found that Fe–Ni–Cu–Al–Ta–B SMA fibers had a 13.5% superelastic strain and a very low austenite finish temperature (−42 °C). Particularly, Ni–Mn–Ga alloys have been extensively explored as ferromagnetic SMAs with a giant magnetic-field-induced strain as large as 9.5% [6,12,13]. Several papers revealed that a high martensitic transformation temperature (up to 350 °C) is observed in given Ni–Mn–Ga alloys with Ni or Mn content higher than the stoichiometric Ni_2MnGa alloy, thus showing interesting potential as high-temperature SMAs [14,15]. Ma et al. [16] reported an over 6.1% shape memory effect (SME) in Ni54–Mn25–Ga21 single crystalline alloy. Meanwhile, a well-pronounced superelastic effect as large as 6% caused by stress-induced martensitic transformation in some high-temperature single crystalline Ni–Mn–Ga alloys were reported by Chernenko et al. [17].

However, it is known that the polycrystalline Ni–Mn–Ga alloys are extremely brittle. Up to now, few studies have been reported in connection with their mechanical behavior, superelasticity, and SME. Li et al. [18] reported that the mechanical and shape memory characteristics of the polycrystalline Ni54–Mn25–Ga21 SMA rods, fabricated by copper mold suction method with high cooling rate, could be improved by grain refinement. However, as the grain size is still as large as 10 to 50 μm, the recovery ratio of the rods is less than 70%, and the maximum shape memory strain is only 4.2%. In-rotating-water-quenching technique was previously used to fabricate a Ni–Mn–Ga fiber about 170 μm in diameter with reduced magnetization and broad martensitic transformation, but the mechanical properties were not greatly improved [19]. Our previous work reported a Ni53.96–Mn24.12–Ga21.92 fiber exhibiting large superelasticity at room temperature [20]. Here, we report the Ni–Mn–Ga fibers, made by glass-coated melt spinning method, which has a higher cooling rate, showing perfect shape memory effect. In addition, the superelastic properties at different temperatures were investigated.

2. Materials and Methods

Glass-coated fibers with different diameters were prepared by glass-coated melt spinning method as shown in Figure 1 [21–23]. An ingot with a nominal composition of Ni53–Mn26–Ga21 was prepared by arc melting the mixture of Ni, Mn, and Ga with a purity higher than 99.9 wt % under a Ti-gettering argon atmosphere. To achieve a homogeneous distribution of elements in the alloys, each alloy was remelted five times. Afterward, the glass-coated melt spinning method, as described elsewhere, was employed to fabricate samples in the shape of fibers [24]. Briefly, the ingot was crushed into pieces, and 3 g of alloy was placed in a borosilicate glass tube with an inner diameter of 8 mm. The heating current are increased to 600 amperes gradually by a high-frequency inductor heater, and the tube and alloy are heated to around 1200 °C under an argon atmosphere. While the alloy melts, the softened glass is drawn down into a fine capillary containing the alloy melt by a glass bar with a super thin tip. The molten alloy inside solidifies rapidly into a metallic fiber by pulling the capillary through cooling water. The diameter of fibers and glass thickness are determined by the rotating speed of the winding wheel. The microstructure of a fiber depends mainly on the cooling rate, which can be controlled by a cooling mechanism when the metal-filled capillary enters water or air on its way to the receiving coil. Finally, the glass coating on the pieces of the fibers for SME and the superelasticity experiments was removed mechanically and carefully.

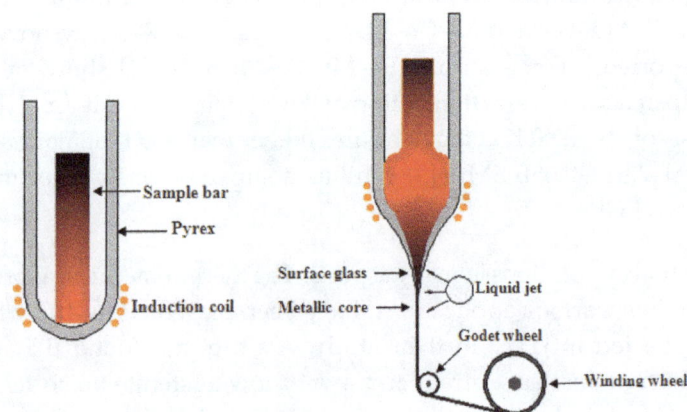

Figure 1. Schematic illustration of the Taylor technology for preparation of fibers.

The surface morphology and the composition were examined by scanning electron microscopy (SEM) and energy dispersive spectrometry (EDS), respectively. In order to analyze the superelastic properties of the fiber, isothermal uniaxial tensile tests were done in a dynamic mechanical analyzer (DMA). Each end of the fiber is mounted in a specially designed paper frame with a rhombic hole at the center by epoxy adhesive. When the testing sample was fixed at the two gripping ends, the paper frame was cut off by the middle. No special surface preparation was conducted on the fibers. The fiber with an initial gauge length of about 5 mm was heated up to 60 °C and kept for 5 min, then cooled to 30 °C, and then subjected to a tensile load–unload cycle with a strain rate of 3.33×10^{-4} s^{-1}. After the tensile test at 30 °C, the fiber was heated up to 40 °C and kept for 5 min, then the loaded–unloaded tensile test was performed at different strains. Similarly, the fiber was heated up to 50 and 60 °C for the same loaded–unloaded tensile test.

The prepared fibers as shown in Figure 2 are shinning and ductile and can be deformed into the shape of the letters U, S, T, and B.

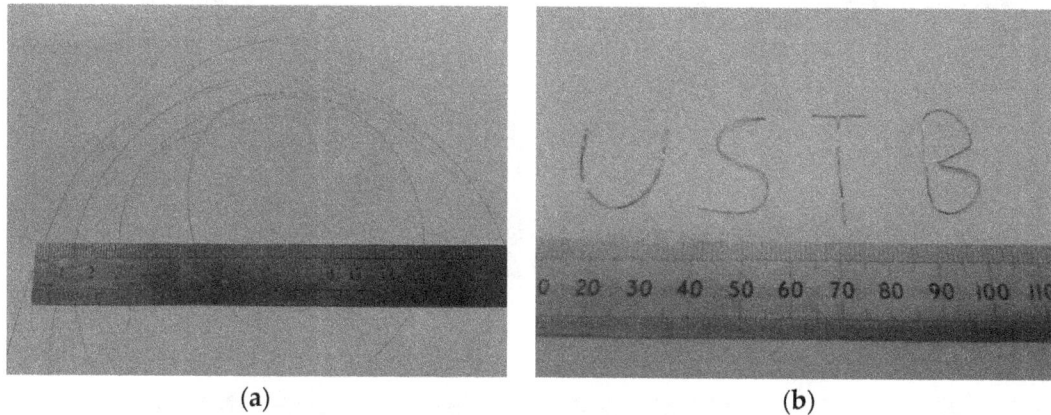

| (a) | (b) |

Figure 2. Pictures of the glass-coated fibers with composition of Ni53–Mn26–Ga21 (**a**) and deformed in to the letters U, S, T, and B with the fibers Ni55–Mn24–Ga21 (**b**).

The shape memory experiments of the fiber were carried out by a dynamic mechanical analyzer (DMA), where the fiber was loaded to 0.05 N (~58 MPa) at room temperature before it was heated to 60 °C. While maintaining this constant load, the fiber was subjected to a thermal cycle across the transformation temperature range from 60 to −70 °C. The stress-assisted SME of the Ni54.9–Mn23.5–Ga21.6 fiber was observed by DMA.

3. Experimental Results and Their Analysis

Figure 3 shows the morphology of surfaces and polished cuts of the fibers for Ni54.9–Mn23.5–Ga21.6 (Figure 3a,b); Ni55.7–Mn23.3–Ga21.0 (Figure 3c,d); Ni56.6–Mn22.4–Ga21.0 (Figure 3e,f). It can be seen that the fibers are very uniform and with a smooth surface, and the grain sizes can be in the sub-micro and micrometer range.

Figure 4a shows the outer appearance of an as-cast fiber with a diameter of 33 μm. The surface is precisely circular, and the diameter of the fiber exhibits a higher uniformity than the fibers prepared by the melt extracted method [25], which demonstrates the suitability of the glass-coated melt spinning method for producing Ni–Mn–Ga fibers. Some cellular grains with a size less than 1 μm could be detected on the surface as shown in Figure 1b, that denotes that the grain size of the fiber is greatly refined by the super cooling rate. The composition of the fiber examined by EDS is Ni54.9–Mn23.5–Ga21.6.

The stress–strain curves for the same fiber tested at different temperatures are shown as Figure 5. A uniaxial tensile strain is applied before unloading at the same rate. Since the fibers show austenite phase at these temperatures, the strain accumulated during loading beyond the elastic regime is caused by the stress-induced martensitic transformation, and this strain almost completely recovers after unloading with the reverse transformation. The fiber shows a perfect superelasticity behavior in a comparatively broader temperature region (from 30 to 60 °C). Differently to the stress–strain curves in Ni–Mn–Ga single crystals, many serration flows occurred during the transformation [18–20,25,26].

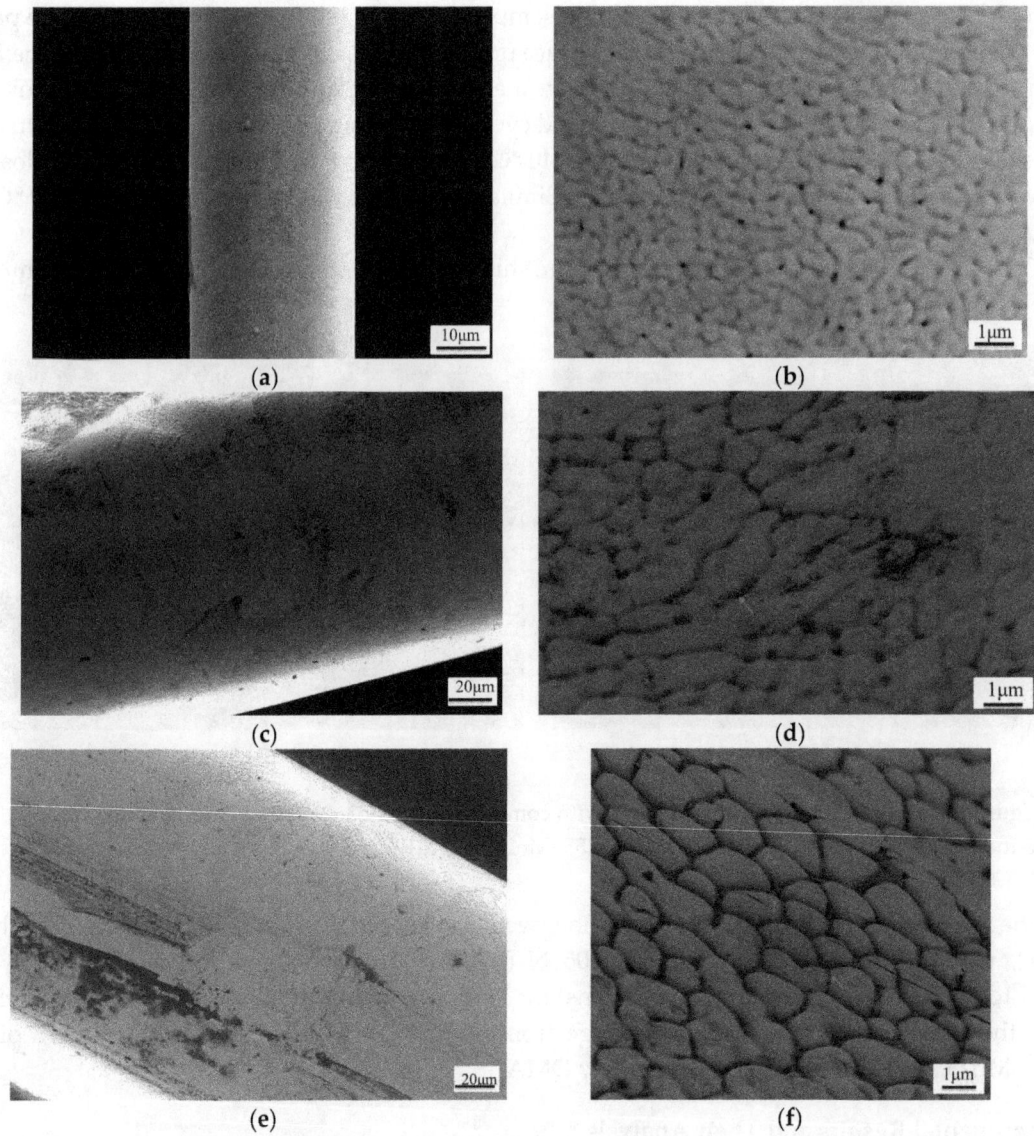

Figure 3. Morphology of surfaces and polished cuts of the fibers. (**a**,**b**) Ni54.9–Mn23.5–Ga21.6; (**c**,**d**) Ni55.7–Mn23.3–Ga21.0; (**e**,**f**) Ni56.6–Mn22.4–Ga21.0.

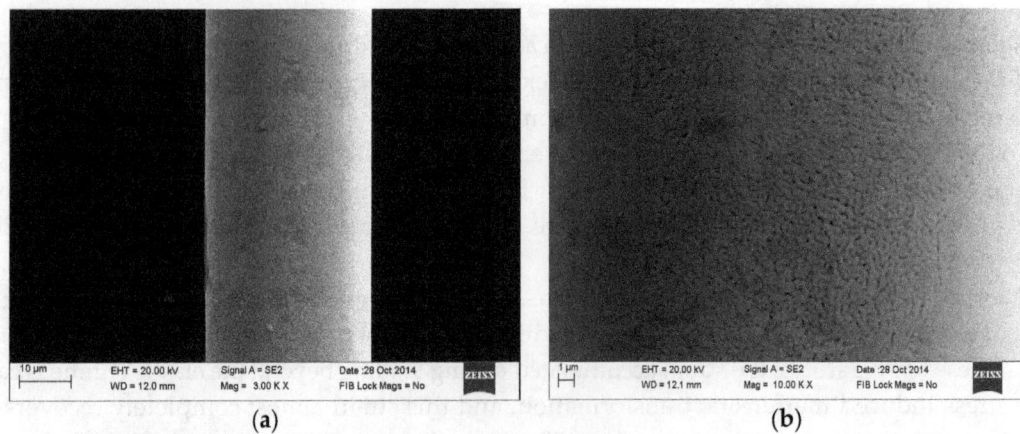

Figure 4. Morphology of the fibers. (**a**) SEM micrograph of the Ni54.9–Mn23.5–Ga21.6 fiber with a diameter of 33 μm; (**b**) magnified image of the surface.

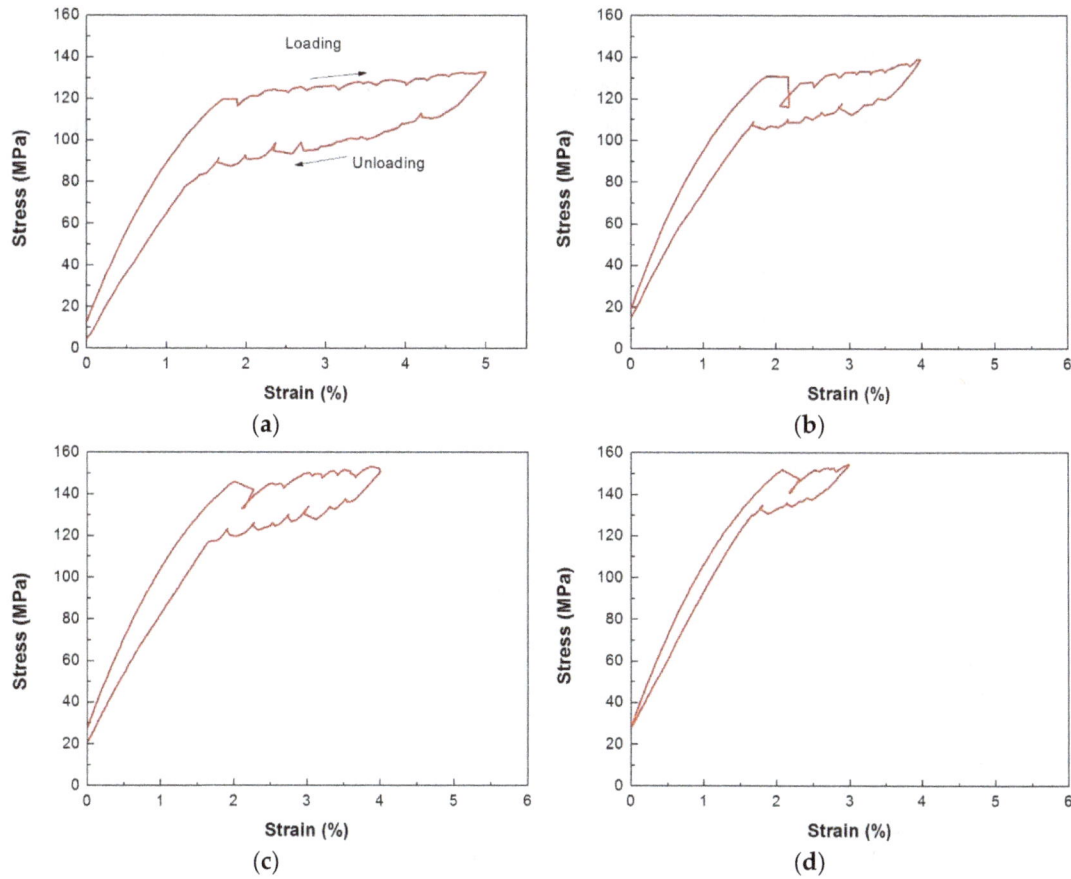

Figure 5. Tensile stress–strain curves of Ni54.9–Mn23.5–Ga21.6 fiber at different temperatures. (**a**) 30 °C; (**b**) 40 °C; (**c**) 50 °C; (**d**) 60 °C.

The critical stress for the stress-induced martensitic (SIM) transformation, σ, which corresponds to the applied stress when the martensitic transformation starts, is plotted against temperature, T, in Figure 6. It increases with increasing temperature in agreement with the Clausius–Clapeyron relationship:

$$\frac{d\sigma}{dT} = -\frac{\Delta S}{\varepsilon \cdot V_m} \tag{1}$$

where ΔS is the molar entropy difference between the parent and martensite phases, ε is the strain caused by the phase transformation, and V_m is the molar volume. The resulting stress–temperature slope, $d\sigma/dT$, is 1.1 MPa/°C.

Table 1 summarizes the comparative stress–temperature slope, $d\sigma/dT$, for the fiber in the present work with that of Ni–Mn–Ga single crystalline and Ni49.9–Mn28.6–Ga21.5 fibers made by melt extraction. As observed, the fiber investigated in this work presents the weakest temperature dependence. This is interpreted to be the reason why the fiber show superelasticity in a wide temperature range from 30 to 60 °C, while the Ni49.9–Mn28.6–Ga21.5 fiber only shows significant superelasticity in a shorter range, i.e., from 28 to 35 °C [25].

On the other hand, the present Ni54.9–Mn23.5–Ga21.6 fiber exhibits repeatable superelastic strains of a magnitude five times higher than polycrystalline the Ni49.9–Mn28.6–Ga21.5 fiber [25]. The superelastic recoverable strain for this fiber is as large as ~10% in Figure 7a, which is also larger than the reported recoverable strain of 6% in the Ni53.1–Mn26.6–Ga20.3 and Ni51.2–Mn31.1–Ga17.7 single crystallines [17] and 5% in the Ni55–Mn20–Ga25 single crystalline [27]. Here, it should be noted that the recoverable strain of the fiber is obtained in tensile mode, while it is in compressive mode for all reported single crystalline systems.

As shown in Figure 7b, the Ni54.9–Mn23.5–Ga21.6 fiber elongates at low temperatures when it undergoes the martensitic transformation, while it contracts at high temperature when it transforms to austenite. The fiber exhibits perfect stress-assisted SME by which the strain caused by the martensitic transformation during cooling is nearly completely recovered by the reverse transformation during heating.

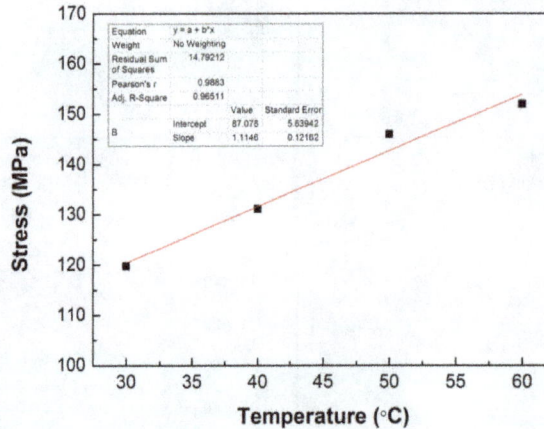

Figure 6. The critical SIM formation stress vs. temperature in the Ni54.9–Mn23.5–Ga21.6 fiber.

Table 1. The stress–temperature slope, $d\sigma/dT$, and recoverable strain of Ni–Mn–Ga single crystalline and fibers made by melt extraction.

Compositions	$d\sigma/dT$ (MPa/°C)	Recoverable Strain (%)
Ni53.1–Mn26.6–Ga20.3	1.9 [001], 3.4 [110]	6%
Ni51.2–Mn31.1–Ga17.7	2.8 [001], 5.2 [110]	6%
Ni55–Mn20–Ga25	2.9 [100]	5%
Ni49.9–Mn28.6–Ga21.5 fiber	8.7	2%
Ni54.9–Mn23.5–Ga21.6 [this work]	1.1	10%

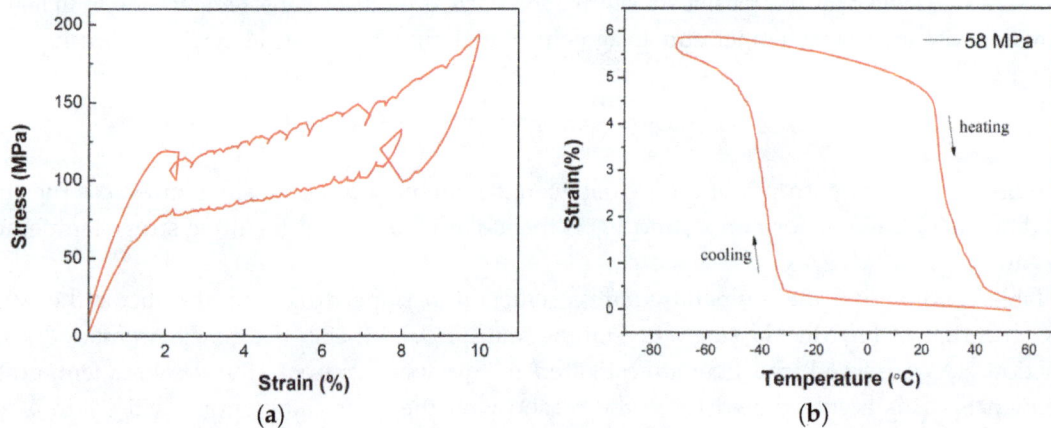

Figure 7. (**a**) Stress–strain curves for Ni54.9–Mn23.5–Ga21.6 fiber unloaded from 10% strain at 40 °C (**b**) Stress-assisted shape memory effect of the Ni54.9–Mn23.5–Ga21.6 fiber under constant tensile load of 0.05 N(~58 MPa).

4. Discussion

Figure 6 and Table 1 show that the fibers presented in this work have the smallest stress–temperature slope (1.1) and the largest recoverable strain (10%). The properties may be due to the high cooling rate of the glass-coated melt-spinning technique. For forming the micro-sized fiber, the cooling rate can be about 10^6 K/s, which will greatly suppress the grain segregations and refine the grains. Moreover, the fibers by this technique are usually with more smooth surface and roundness shape. This is much better than the technique by the melt extraction technique as listed in Table 1 [17]. The properties of the bulk samples and the foam samples are not so good as that by the glass-coated melt-spinning technique, and the reason is believed to be the brittleness of the samples, which has hidden the SIM effect.

From Equation (1), we can see that the critical stress of the SIM is a temperature-dependent parameter. The parameter, $d\sigma/dT$, is a function of S, (the molar entropy difference between the parent and martensite phases), ε, (the strain caused by the phase transformation), and V_m, (the molar volume). The fiber investigated in this work presents the weakest temperature dependence. The lowest value of the $d\sigma/dT = 1.1$ MPa/°C, may be due to the small value of the entropy difference between the austenite and martensite. The high cooling rate by the technique here, which may keep most of the configurations of the father phase, makes the difference between the father phase and son phase smaller.

The serration behavior is a very important phenomena in the materials science field [26,28]. The serration behavior observed in the present results, as shown in Figures 5 and 7a, is interpreted to correlate to the localized martensite phase transformation as shown in Figure 8. Similar results are also reported in [3,20,26].

From Figure 8, when the tensile stress reaches a value, martensite will be locally formed, as shown by the red-colored region in Figure 8. The locally formed martensite phase may induce serration behavior on the stress–strain curve because this SIM may not be a continuous process and may cause a jump in the stress at the macroscale. SIM is very similar to transformation-induced plasticity (TRIP) steels, and the serration behavior has also been reported in other references [20,26].

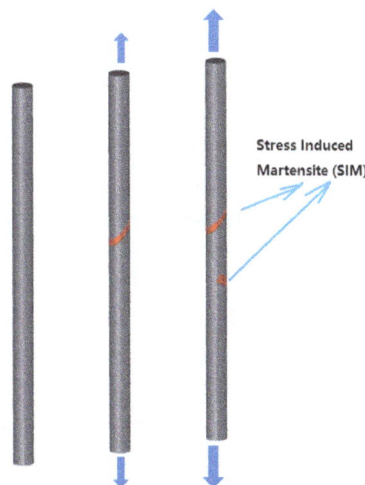

Figure 8. A schematic picture showing the SIM, which results the serration behavior in Figures 4 and 6a.

5. Conclusions

To summarize, Ni54.9–Mn23.5–Ga21.6 fibers with a 33 μm diameter were obtained by a glass-coated melt spinning method. The superelasticity and shape memory characteristics of the polycrystalline Ni–Mn–Ga fiber can be greatly improved by this technique.

The recoverable strain of the Ni54.9–Mn23.5–Ga21.6 fiber is as high as 10% at 40 °C, and the fiber shows superelasticity in a wide temperature range from 30 to 60 °C compared with Ni–Mn–Ga fibers [22]. The mechanism may be due to a different martensite structure.

This achievement was enabled using a glass-coating melt spinning method under suitable fabrication parameters.

Acknowledgments: Y. Zhang would like to acknowledge the financial support of the National Natural Science Foundation of China (NSFC, Granted No. 51471025). Y. Zhao would like to acknowledge the financial support from the doctoral short-term visiting scholarship program by USTB. Research at ICMM/CSIC was supported by the Regional Government of Madrid under Project S2013/MIT-1942850 NANOFRONTMAG-CM. Y. Zhang would like thank the technique help by D.Y. Li.

Author Contributions: L.S. and Y.Z. conceived and designed the experiments; L.S. and Y.Z. performed the experiments; L.S. and Y.Z. analyzed the data; A.J. and M.V. contributed reagents/materials/analysis tools; L.S. and Y.Z. wrote the paper.

Conflicts of Interest: The authors declare no conflict of interest.

References

1. Omori, T.; Ando, K.; Okano, M.; Xu, X.; Tanaka, Y.; Ohnuma, I.; Kainuma, R.; Ishida, K. Superelastic effect in polycrystalline ferrous alloys. *Science* **2011**, *333*, 68–71. [CrossRef] [PubMed]

2. Ölander, A. An electrochemical investigation of solid cadmium-gold alloys. *J. Am. Chem. Soc.* **1932**, *54*, 3819–3833. [CrossRef]

3. Li, D.Y.; Zhang, S.L.; Liao, W.B.; Geng, G.H.; Zhang, Y. Superelasticity of Cu–Ni–Al shape-memory fibers prepared by melt extraction technique. *Int. J. Menerals Metall. Mater.* **2016**, *23*, 928–933. [CrossRef]

4. Otsuka, K.; Ren, X. Physical metallurgy of Ti–Ni-based shape memory alloys. *Prog. Mater. Sci.* **2005**, *50*, 511–678. [CrossRef]

5. Chen, Y.; Zhang, X.; Dunand, D.C.; Schuh, C.A. Shape memory and superelasticity in polycrystalline Cu–Al–Ni microfibers. *Appl. Phys. Lett.* **2009**, *95*, 171906. [CrossRef]

6. Sozinov, A.; Likhachev, A.A.; Lanska, N.; et al. Giant magnetic-field-induced strain in NiMnGa seven-layered martensitic phase. *Appl. Phys. Lett.* **2002**, *80*, 1746–1748. [CrossRef]

7. Kainuma, R.; Imano, Y.; Ito, W.; Sutou, Y.; Morito, H.; Okamoto, S.; Kitakami, O.; Oikawa, K.; Fujita, A.; Kanomata, T.; et al. Magnetic-field-induced shape recovery by reverse phase transformation. *Nature* **2006**, *439*, 957–960. [CrossRef] [PubMed]

8. Firstov, G.S.; Kosorukova, T.A.; Yu, N.K.; Odnosum, V.V. High entropy shape memory alloys. *Mater. Today Proc.* **2015**, *2*, S499–S503. [CrossRef]

9. Tanaka, Y.; Himuro, Y.; Kainuma, R.; Sutou, Y.; Omori, T.; Ishida, K. Ferrous polycrystalline shape-memory alloy showing huge superelasticity. *Science* **2010**, *327*, 1488–1490. [CrossRef] [PubMed]

10. Alam, M.S.; Youssef, M.A.; Nehdi, M. Utilizing shape memory alloys to enhance the performance and safety of civil infrastructure. A review. *Can. J. Civ. Eng.* **2007**, *34*, 1075–1086. [CrossRef]

11. Dezfuli, F.H.; Alam, M.S. Shape memory alloy wire-based smart natural rubber bearing. *Smart Mater. Struct.* **2013**, *22*, 045013. [CrossRef]

12. Ullakko, K.; Huang, J.K.; Kantner, C.; O'Handley, R.C. Large magnetic-field-induced strains in Ni2MnGa single crystals. *Appl. Phys. Lett.* **1996**, *69*, 1966–1968. [CrossRef]

13. Chmielus, M.; Zhang, X.X.; Witherspoon, C.; Dunand, D.C.; Müllner, P. Giant magnetic-field-induced strains in polycrystalline Ni–Mn–Ga foams. *Nat. Mater.* **2009**, *8*, 863–866. [CrossRef] [PubMed]

14. Chernenko, V.A.; Cesari, E.; Kokorin, V.V.; Vitenko, I.N. The development of new ferromagnetic shape memory alloys in Ni–Mn–Ga system. *Scr. Metall. Mater.* **1995**, *33*, 1239–1244. [CrossRef]

15. Jin, X.; Marioni, M.; Bono, D.; Allen, S.M.; O'Handley, R.C. Empirical mapping of Ni–Mn–Ga properties with composition and valence electron concentration. *J. Appl. Phys.* **2002**, *91*, 8222. [CrossRef]

16. Xu, H.B.; Ma, Y.Q.; Jiang, C.B. A high-temperature shape-memory alloy Ni54MnGa21. *Appl. Phys. Lett.* **2003**, *82*, 3206–3208. [CrossRef]

17. Chernenko, V.A.; L'Vov, V.; Pons, J.; Cesari, E. Superelasticity in high-temperature Ni–Mn–Ga alloys. *J. Appl. Phys.* **2003**, *93*, 2394. [CrossRef]

18. Li, Y.; Xin, Y.; Jiang, C.; Xu, H. Shape memory effect of grain refined Ni54Mn25Ga21 alloy with high transformation temperature. *Scr. Mater.* **2004**, *51*, 849–852. [CrossRef]

19. Gómez-Polo, C.; Pérez-Landazábal, J.I.; Recarte, V.; Sánchez-Alarcos, V.; Badini-Confalonieri, G.; Vázquez, M. Ni–Mn–Ga ferromagnetic shape memory wires. *Appl. Phys. Lett.* **2010**, *107*, 123908. [CrossRef]

20. Zhang, Y.; Li, M.; Wang, Y.D.; Lin, J.P.; Dahmen, K.A. Superelasticity and serration behavior in small-sized NiMnGa alloys. *Adv. Eng. Mater.* **2014**, *16*, 955–960. [CrossRef]

21. Zhao, Y.; Hao, H.; Zhang, Y. Preparation and giant magneto-impedance behavior of Co-based amorphous wires. *Intermetallics* **2013**, *42*, 62–67. [CrossRef]

22. Zhang, Y.; Zhao, Y.Y.; Liao, W.B. An Instrument for Continuous Preparing Amorphous Wires. China Patent CN102127720A, 2011.

23. Zhang, Y.; Zhao, Y.Y.; Shao, L. Preparation of a NiMnGa Shape Memory Alloy Wire. China Patent CN105316527A, 2016.

24. Zhao, Y.Y.; Li, H.; Hao, H.Y.; Li, M.; Zhang, Y.; Liaw, P.K. Microwires fabricated by glass-coated melt spinning. *Rev. Sci. Instrum.* **2013**, *84*, 75102. [CrossRef] [PubMed]

25. Qian, M.F.; Zhang, X.X.; Witherspoon, C.; Sun, J.F.; Müllner, P. Superelasticity and shape memory effects in polycrystalline Ni–Mn–Ga microfibers. *J. Alloys Compd.* **2013**, *577*, S296–S299. [CrossRef]

26. Zhang, Y.; Qiao, J.; Liaw, P.K. A brief review of high entropy alloys and serration behavior and flow units. *J. Iron Steel Res. Int.* **2016**, *1*, 2–6. [CrossRef]

27. Pasquale, M.; Sasso, C.P.; Lewis, L.H.; Giudici, L.; Lograsso, T.; Schlagel, D. Magnetostructural transition and magnetocaloric effect in Ni55Mn20Ga25 single crystals. *Phys. Rev. B* **2005**, *72*, 94435. [CrossRef]

28. Li, J.; Yang, X.; Zhu, R.; Zhang, Y. Corrosion and serration behaviors of TiZr0.5NbCr0.5V$_x$Mo$_y$ high entropy alloys in aqueous environments. *Metals* **2014**, *4*, 597–608. [CrossRef]

Hybrid Metaheuristic-Neural Assessment of the Adhesion in Existing Cement Composites

Łukasz Sadowski [1],*, Mehdi Nikoo [2] and Mohammad Nikoo [3]

[1]   Faculty of Civil Engineering, Wrocław University of Science and Technology, Wybrzeże Wyspiańskiego 27, 50-370 Wrocław, Poland
[2]   Young Researchers and Elite Club, Ahvaz Branch, Islamic Azad University, Ahvaz, Iran; sazeh84@yahoo.com
[3]   SAMA Technical and Vocational Training College, Islamic Azad University, Ahvaz Branch, Ahvaz, Iran; m.nikoo2014@gmail.com
*   Correspondence: lukasz.sadowski@pwr.edu.pl

Academic Editor: Paul Lambert

Abstract: The article presents the hybrid metaheuristic-neural assessment of the pull-off adhesion in existing multi-layer cement composites using artificial neural networks (ANNs) and the imperialist competitive algorithm (ICA). The ICA is a metaheuristic algorithm inspired by the human political-social evolution. This method is based solely on the use of ANNs and two non-destructive testing (NDT) methods: the impact-echo method (I-E) and the impulse response method (IR). In this research, the ICA has been used to optimize the weights of the ANN. The combined ICA-ANN model has been compared to the genetic algorithm (GA) and particle swarm optimization (PSO) to evaluate its accuracy. The results showed that the ICA-ANN model outperforms other techniques when testing datasets in terms of both effectiveness and efficiency. As presented in the validation stage, it is possible to reliably map the adhesion level on a tested surface without local damage to the latter.

Keywords: cement mortar; overlay; concrete substrate; interlayer bond; pull-off adhesion; artificial intelligence; metaheuristics; imperialist competitive algorithm; genetic algorithm; particle swarm optimization

---

1. Introduction

Layered cement composites, recently an issue attracting numerous researchers [1–4], usually consist of an overlay placed on an existing concrete substrate. It is always necessary to ensure an appropriate bond between the overlay and existing concrete substrate. The measure of this bond is the value of the pull-off adhesion (f_b), obtained using the pull-off method [5]. This method is time consuming due to the delay time of the curing of the resin used for bonding the steel disc. Moreover, results can be influenced by variation on the rupture surface, the orientation and position of the aggregate onto the disc, the disc material, diameter, and thickness, the backpressure system, and also the speed of load application [6–8]. The tested surface in each of the measuring places is damaged and the efficiency of this method depends on the number of measuring places. The damage has to be repaired after the test. When considering the above, it is advisable to use non-destructive testing (NDT) methods for the purpose of the assessment of the interlayer bond. It has been verified in practice that the impact-echo (I-E) and impulse response (IR) methods may be used individually to make a successful zero-one assessment [9,10]. However, it is not possible to individually use the above-mentioned NDT methods to make a full, reliable assessment of the value of f_b [11–13]. In such a case it is helpful to use artificial neural networks (ANNs) which, in recent years, are increasingly used in civil engineering [14–23].

A new method of identifying f_b by means of the ANN method has recently been proposed [24–29]. The database for this identification was created using 3D morphological parameters, which were evaluated on the existing concrete substrate surface using 3D LASER scanning, and also acoustic parameters obtained with the use of the IR and I-E methods on the surface of the overlay. Various methods have been used, such as radial basis functions (RBF) [24], multi-layer perceptron (MLP) [25–27], and principal component analysis (PCA), in combination with self-organizing feature maps (SOFM) [28].

It is evident that there is no possibility in existing layered cement composites of obtaining 3D morphological parameters of the existing concrete substrate surface. Consequently, it is not possible to adopt the previously-developed method to identify the value of f_b in existing cement composites. Thus, the attempt presented in [29] proved that it is possible to identify the value of f_b between the overlay and existing concrete substrate in existing cement composites on the basis of the acoustic parameters obtained on the overlay surface using the ANN, IR, and I-E methods. The multi-layer perceptron ANN with the gradient descent (MLP-GS) learning algorithm has been found to be useful for this purpose. However, values of determination coefficient R^2 greater than 0.77 were not satisfied. Simple back-propagation (BP) has also been used for optimizing the ANN, which is plagued with inconsistent and unpredictable performances, a slow learning rate, and becoming trapped in local minima [30,31]. Thus, in some applications it is necessary to improve the performance of the ANN with the use of optimization algorithms. The imperialist competitive algorithms (ICA), the genetic algorithm (GA), or particle swarm optimization (PSO) have recently been used for this purpose [32].

The recently developed ICA is a randomized population method inspired by the human political-social evolution [33–37]. It belongs to the metaheuristic group of methods that are expected to become more popular in various engineering applications [38–44]. In the ICA, a number of colonial countries, along with their colonies, try to find a general optimal point in solving the optimization problem. Different methods are then introduced to solve the optimization problems [45,46]. It is worth noting that there have recently been a few attempts to apply the ICA for engineering problems [47–49], e.g., the prediction of soil compaction [50], oil flow rate [51], optimum cost [52] or corrosion current density [53].

With consideration of the above, the article presents the hybrid metaheuristic-neural assessment of the value of the f_b in existing cement composites using the ICA. This method is based solely on the use of ANNs and two NDT methods: I-E and IR. In this research, the ANN was used for prediction and the ICA was used to improve the performance of the ANN. The role of the ICA was to optimize the weights of the ANN. The combined ICA-ANN model has been compared to the GA and PSO in order to evaluate its accuracy.

2. Experimental Setup

As mentioned previously in [29], the NDT tests were carried out on a surface of a two-layer cement composite with dimensions of 2500×2500 mm^2 and with a constant thickness of the overlay equal to 25 mm (Figure 1a,b). The overlay was made of C20/25 class cement mortar with the maximum quartz aggregate grain size equal to 2 mm. The overlay was laid on the existing substrate with a constant thickness of 125 mm made of C30/37 class concrete. In this concrete, the maximum broken basalt aggregate grain size was equal to 8 mm. The surface of the existing concrete substrate was sandblasted. In order to achieve a wider range of adhesion between the overlay and the substrate, half of the surface of the substrate was covered with a bonding agent in the form of concentrate to be diluted with water. Table 1 shows the weight composition of the mixes that were used to make the substrate and the overlay.

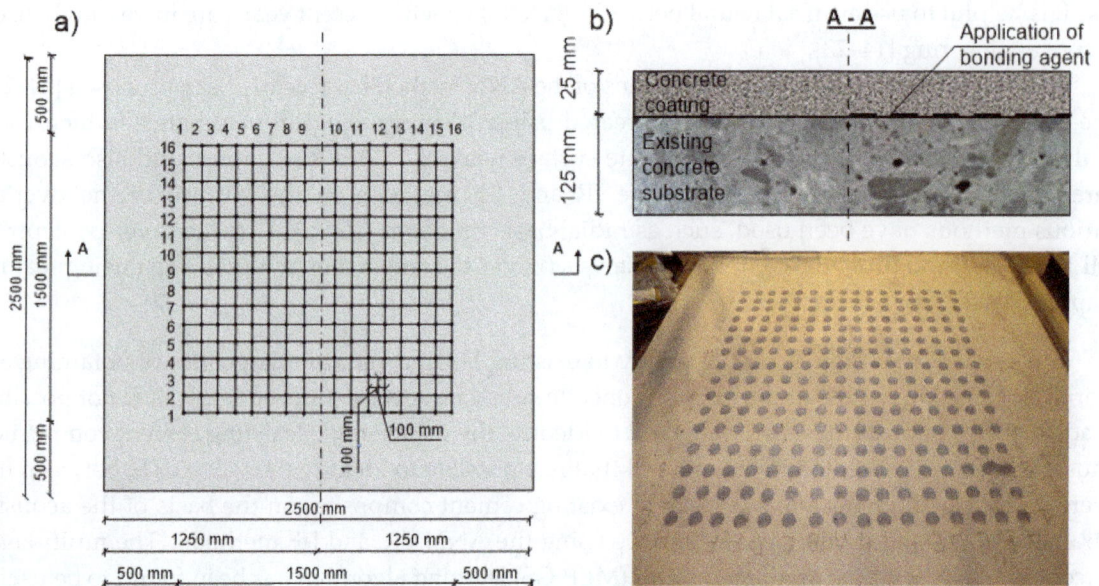

Figure 1. The tested two-layer cement composite: (**a**) arrangement of testing areas on the surface of the overlay; (**b**) cross-section; and (**c**) view of the surface of the overlay after pull-off testing.

Table 1. Weight composition of mixes.

Layer Designation	Portland Cement CEM I 42.5 R	Portland Cement CEM II A-LL 42.5 R	Fly Ash	Water	Fine Aggregate 0–2 mm	Coarse Aggregate 2–8 mm	Plasticizer Visco Flow 6920
	Components of Mixes (kg/m³)						
Overlay	276.0	–	–	138	1599.0	–	–
Substrate	–	352.0	40	165	724.4	1086.6	2.0

The grid of the testing areas was applied on the surface of the overlay 500 mm from its edge (Figure 1a) and the number of testing areas was equal to 256. 90 days after laying the overlay, tests using two acoustic NDT methods were carried out on its surface in 256 designated testing areas (Figure 2a,b).

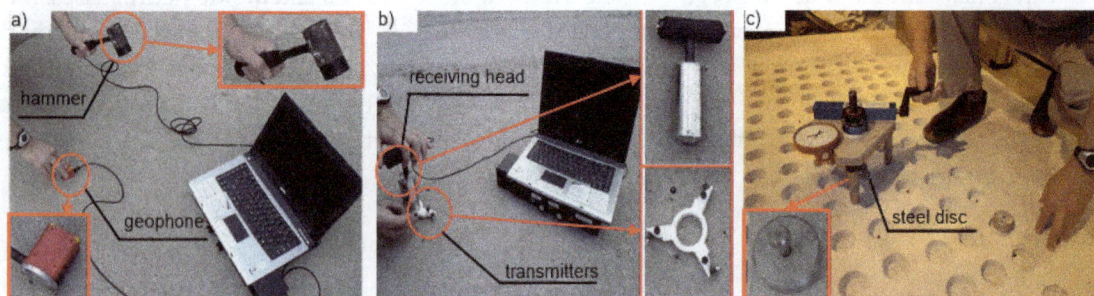

Figure 2. Exemplary view of tests carried out on the surface of the overlay using: (**a**) the IR method; (**b**) the I-E method; and (**c**) the pull-off method.

The values of the following parameters were determined in all these areas using the IR method: average mobility (N_{av}), dynamic stiffness (K_d), mobility slope (M_p/N), and void index (v) according to [54–56]. The value of the frequency of the sound wave reflection from the bottom of a sample (f_T), obtained using the I-E method, was also measured according to [57–59].

After finishing the tests using acoustic NDT methods, tests using the pull-off method were conducted in the same testing areas in order to obtain the real values of the f_b (Figure 2c). The f_b

values were then used as patterns for learning and testing the ANN. The data used for development of the models was obtained from past experiments [29]. In this article, the output value of the pull-off adhesion predicted by the ANN is denoted as $f_{c,b}$. Exemplary data is presented in Table 2.

Table 2. Exemplary data.

| Number of Test | Name of Test Method and Parameters | | | | | |
| | The IR Method | | | | The I-E Method | The Pull-off Method |
	N_{av} (m/s·N)	K_d	M_p/N	v	f_T (kHz)	f_b (MPa)
1	155.000	0.002	0.503	3.149	13.20	1.044
2	128.000	2.000	0.662	3.000	15.63	1.019
3	79.000	2.000	1.413	1.000	12.20	1.070
4	96.000	2.000	0.630	0.709	12.20	0.968
5	80.000	1.000	0.740	1.040	12.60	0.891
6	71.000	0.038	0.531	0.500	5.86	1.248
7	92.000	1.000	0.612	0.629	15.14	1.095
8	89.000	1.000	0.472	1.000	14.65	0.968
9	103.000	1.000	0.661	0.689	12.21	1.070
10	82.000	9.000	0.571	1.000	15.14	0.968
. . .	. . .	. . .	. . .	. . .	. . .	. . .
256	108.065	1.000	1.825	1.000	14.65	0.866

3. Results of Training and Testing

After applying the Chauvenet criterion of the elimination of questionable results and reducing the database to 239 sets of results, the resulting variables were randomly divided into ANN learning, testing, and experimental verification data. Once the data was normalized, out of 239 such sets of results, 70% of the samples (167 patterns) were randomly used for training and 15% of the selected samples (36 patterns) were randomly used to test the ANN. The rest of the samples were randomly used for validation (36 patterns). In order to include all the parameters in a numerical range and make the data dimensionless, the contributing input and output parameters should be normalized prior to the training phase according to Equation (1):

$$x_N = (x - MinX)/(MaxX - MinX) \times 2 - 1 \qquad (1)$$

where x_N is the normalized input and output data, x is the input data, MinX is the minimum of all data, and MaxX is the maximum of all data. Therefore, all of the normalized data is placed in the numerical distance of $[-1,+1]$. The hidden layer node numbers were determined according to [60] by using Equation (2):

$$N_H \leq 2N_I + 1 \qquad (2)$$

where N_H is the maximum number of neurons in the hidden layers and N_I is the number of inputs. Considering that the number of effective inputs obtained is 5, the maximum number of nodes in the hidden layer is equal to 11.

The ICA is used to determine the optimized weight of each of the ANN models. The structure of the ICA-ANN used in the analysis is presented in Figure 3. Table 3 indicates the optimized structure of each model along with features of the ICA. Furthermore, Table 4 indicates the analytical results of training and testing each of the models with optimal structure, which is provided in Table 3. Five statistical parameters of mean error (ME), mean absolute error (MAE), mean squared error (MSE), root mean squared error (RMSE), and mean squared reduced error (MSRE) are presented in Table 5. The performance measurements from all models have been collected and averaged. Due to the randomness of weights and data sampling, each experiment is simulated ten times in order to obtain more reliable results. To determine the performance of the models and to decide on the best model, the MSE test and MSE training criteria obtained from the models are compared with each other and shown in Table 3. According to the indicated results, the 250GEN_5IN model, the weight of which has been optimized by the ICA, has been optimized by 500 countries, 50 empires, and 250 repetitions, and has the best results among the models of its kind.

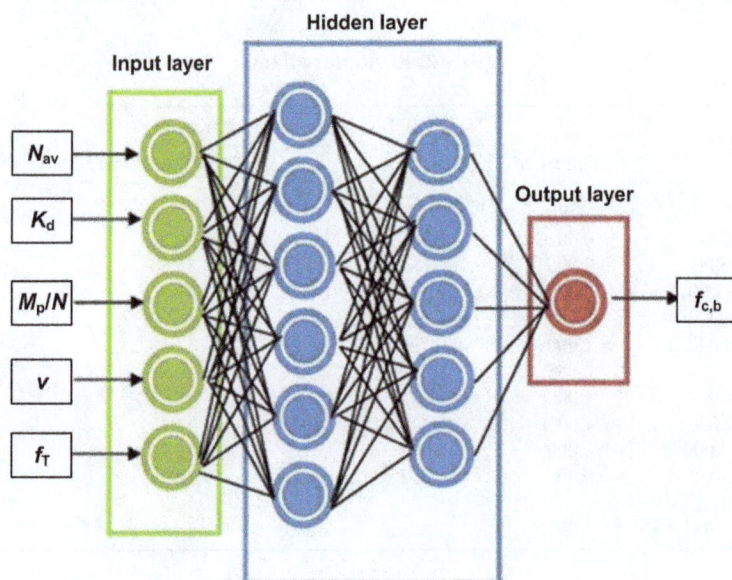

Figure 3. The structure of the ICA-ANN used in the analysis.

Table 3. Optimized structure of the ICA-ANN.

Models Name	ANN Features					Utilized Initialization Parameters in the ICA		
	Number of Input	Number of Output	Number of Hidden Layer	Number of Neurons in Hidden Layer	Transfer Function	Number of Country	Number of Imperialist	Number of Decades
50GEN_5IN	5	1	1	13	satlins	400	40	50
100GEN_5IN	5	1	2	5-5	logsig	300	30	100
150GEN_5IN	5	1	1	9	purelin	450	45	150
200GEN_5IN	5	1	3	6-4-3	satlins	500	50	200
250GEN_5IN	5	1	2	6-5	tansig	500	50	250

Table 4. Results of testing and training of the ICA-ANN.

Model	Testing		Training		Errors		
	Equation	R^2	Equation	R^2	MSE Test	MSE Train	Best Cost
50GEN_5IN	$y = 0.482x - 0.019$	0.215	$y = 0.273x - 0.032$	0.084	0.131	0.152	0.152
100GEN_5IN	$y = 0.476x + 0.134$	0.547	$y = 0.366x + 0.135$	0.456	0.048	0.051	0.051
150GEN_5IN	$y = 0.475x + 0.120$	0.288	$y = 0.380x + 0.115$	0.315	0.082	0.063	0.063
200GEN_5IN	$y = 0.002x - 0.179$	0.001	$y = 0.001x - 0.179$	0.041	0.289	0.226	0.226
250GEN_5IN	$y = 0.802x + 0.041$	0.858	$y = 0.679x + 0.045$	0.844	0.044	0.047	0.047

Table 5. Statistical results of ICA-ANN errors.

Model	ME		MAE		MSE		RMSE		MSRE	
	Train	Test	Train	Test	Train	Test	Train	Test	Train	Test
50GEN_5IN	−0.169	−0.153	0.308	0.270	0.151	0.128	0.389	0.357	1.648	1.346
100GEN_5IN	0.016	−0.001	0.158	0.148	0.051	0.043	0.227	0.208	0.560	0.455
150GEN_5IN	−0.001	−0.012	0.184	0.198	0.064	0.077	0.252	0.278	0.693	0.815
200GEN_5IN	−0.367	−0.438	0.399	0.458	0.226	0.286	0.475	0.535	2.459	3.011
250GEN_5IN	−0.015	−0.010	0.104	0.094	0.017	0.014	0.131	0.117	0.188	0.144

As indicated in Tables 3 and 4, in model 5 the values of the determination coefficient R^2 for the parameter of f_b at the training and testing stages are equal, respectively, to 0.844 and 0.858. The slope of the straight line for this parameter equals 0.679 and 0.802, which represents the accuracy of the model and less modeling error. According to [61], in this model the nonlinear tan-sigmoid function (TANSIG), which shows the minimal error, was considered as the transfer function:

$$Tansig(n) = \frac{2}{1 + e^{-2n}} - 1 \qquad (3)$$

According to Table 5, the coefficients of *ME*, *MAE*, *MSE*, *RMSE*, and *MSRE* in two training and testing stages for the ANN model with a 1-5-6-5 structure, and also the properties of 500 countries, 50 empires, and 250 repetitions, are less than for all of the other models. This represents lower errors of this ANN in comparison to other models. Therefore, the ICA-ANN model with the title of 250GEN_5IN is more accurate than other models of its kind. The result of comparing the selected 250GEN_5IN ICA-ANN model with the observation data is presented in Figure 4. The minimum cost and the mean cost diagrams are shown in Figure 5.

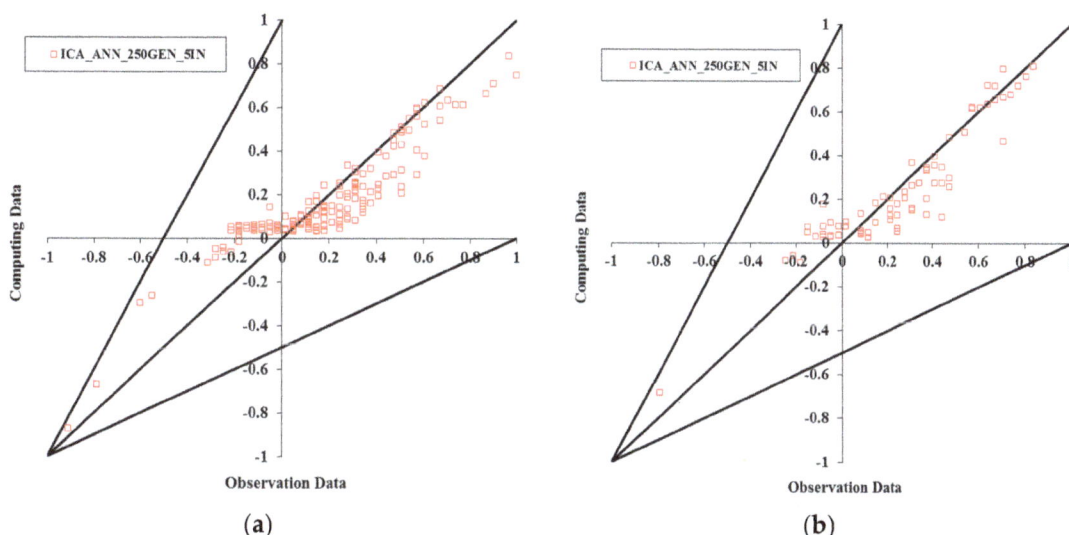

Figure 4. Results of the ICA-ANN for (**a**) training and (**b**) testing.

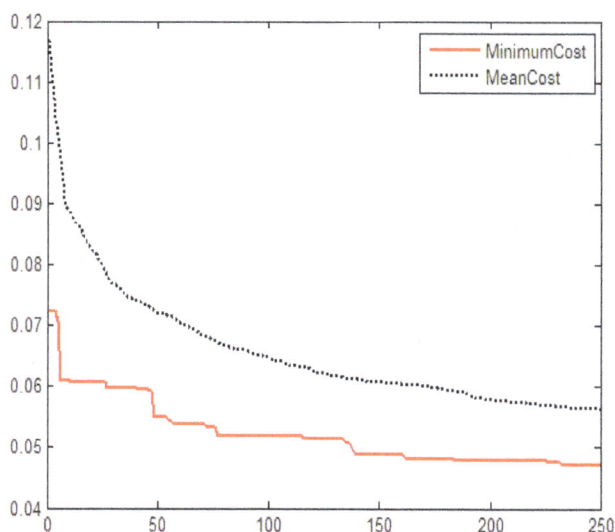

Figure 5. Graph cost for 250 iterations in the 250GEN_5IN model.

4. Comparison with the Genetic Algorithm (GA) and Particle Swarm Optimization (PSO)

In order to evaluate the accuracy of the ICA-ANN model, which has been optimized by the ICA, it was compared with both the GA and PSO, the features of which are provided in Table 6. The ICA-ANN model with a 1-5-6-5 structure and the TANSIG stimulation function were used. According to a study conducted on three algorithms for training data and testing, R^2 coefficients and the slope of the straight line are presented in Table 7. Figure 5 shows graph cost for 250 iterations in the 250GEN_5IN model. Figure 6 shows the comparison of the f_b values for the computational and observational data using the three algorithms in the training stage. According to the equations of lines fitted to the computational and observational values in each model, and also the coefficient of determination that corresponds to them (Table 7), it can be seen that the ANN optimized by the ICA determines the amount of f_b more accurately than the GA and PSO algorithms. Moreover, the determination of f_b using the three models shown in Figure 6 indicates that the ANN optimized by the ICA has higher accuracy and flexibility.

Table 6. Applied GA, ICA, and PSO parameters.

GA		ICA		PSO	
Population	150	Countries	500	Swarm Size	200
Mutation rate	15	Revolution Rate	0.3		
Crossover rate	50	Empires	50	Cognition Coefficient	2
		Uniting threshold	0.02	Social Coefficient	2
Generation	250	Generation	250	Generation	250

Table 7. Results of ANN models optimized by the GA, ICA, and PSO algorithms in training and testing.

Model	Best Fitting Line in Testing		Best Fitting Line in Training		Errors		
	Equation	R^2	Equation	R^2	MSE Test	MSE Train	Best Cost
ICA-ANN	$y = 0.802x + 0.041$	0.858	$y = 0.679x + 0.045$	0.844	0.044	0.047	0.047
GA-ANN	$y = -0.010x + 0.240$	0.000	$y = 0.035x + 0.200$	0.003	0.076	0499	0.049
PSO-ANN	$y = 0.612x + 0.089$	0.530	$y = 0.485x + 0.095$	0.506	0.050	0.045	0.045

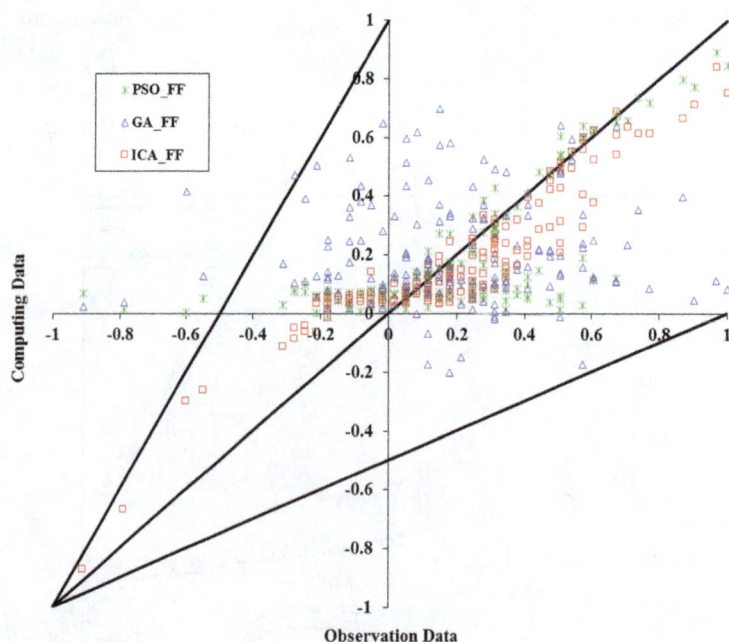

(a)

Figure 6. *Cont.*

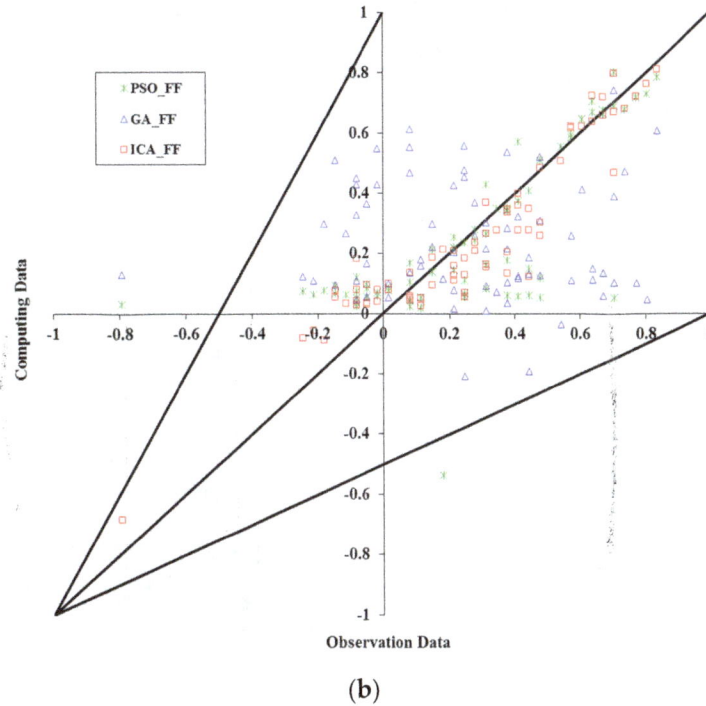

Figure 6. Comparison of the selected ICA-ANN with GA-ANN and PSO-ANN in (**a**) training and (**b**) testing.

5. Results of Validation of the Proposed ICA-ANN Model

In order to conduct the validation of the ICA-ANN model, a set of 36 results was randomly selected. These results were not previously used to train and test the ICA-ANN. As a result, the taught ICA-ANN model generated the value of the pull-off adhesion for each of these areas. Table 8 presents a comparative summary of the values of f_b and $f_{c,b}$, defined, respectively, using the pull-off method and ICA-ANN.

Table 8. Comparative summary of the values of f_b and $f_{c,b}$, defined using the pull-off method and ICA-ANN, respectively.

No.	f_b	$f_{c,b}$	$\lvert \Delta f \rvert$	$\lvert RE \rvert$	No.	f_b	$f_{c,b}$	$\lvert \Delta f \rvert$	$\lvert RE \rvert$	No.	f_b	$f_{c,b}$	$\lvert \Delta f \rvert$	$\lvert RE \rvert$
	MPa	MPa	MPa	%		MPa	MPa	MPa	%		MPa	MPa	MPa	%
1	1.30	1.29	0.01	0.85%	13	1.06	1.06	0.01	0.56%	25	1.55	1.42	0.13	9.09%
2	1.13	1.08	0.04	4.16%	14	1.04	1.06	0.02	1.46%	26	1.19	1.19	0.00	0.25%
3	1.50	1.40	0.10	7.06%	15	1.21	1.19	0.03	2.12%	27	1.23	1.12	0.11	10.01%
4	1.26	1.18	0.08	6.83%	16	1.01	1.05	0.04	4.13%	28	1.07	1.07	0.00	0.28%
5	1.21	1.12	0.09	7.82%	17	0.92	1.06	0.14	13.13%	29	1.19	1.17	0.03	2.44%
6	0.62	0.68	0.06	9.25%	18	1.02	1.06	0.04	3.52%	30	1.06	1.06	0.00	0.20%
7	1.18	1.13	0.05	4.10%	19	0.95	1.06	0.11	10.16%	31	0.95	1.06	0.10	9.60%
8	1.30	1.16	0.14	12.10%	20	1.02	1.06	0.04	3.86%	32	1.11	1.10	0.01	0.79%
9	1.16	1.07	0.09	8.07%	21	1.30	1.29	0.00	0.32%	33	1.21	1.16	0.05	4.68%
10	0.56	0.58	0.02	3.71%	22	1.13	1.16	0.03	2.90%	34	1.11	1.05	0.06	5.63%
11	1.09	1.08	0.01	1.23%	23	1.33	1.34	0.01	0.98%	35	0.94	1.05	0.11	10.89%
12	1.06	1.06	0.01	0.52%	24	0.97	1.06	0.09	8.31%	36	1.30	1.29	0.01	0.85%
–	–	–	–	–	–	–	–	–	–	Mean	1.11	1.11	0.05	4.50%

The results of validation presented in Table 8 indicate the correct identification of validation data. It is evidenced by the obtained low mean value of the absolute error amounting to 0.05 MPa and the satisfactory mean value of the relative error $\lvert RE \rvert$ amounting to 4.89%. Interestingly, the values of $f_{c,b}$ identified by the ICA-ANN are equal to 1.11 MPa and are, on average, the same as the f_b obtained experimentally using the pull-off method. The results of the validation were used to plot the adhesion maps shown in Figure 7 (results in the other points on the tested surface were obtained by means of linear interpolation). This figure illustrates two adhesion maps. The first one was obtained on the basis of experimental tests conducted using the pull-off method (Figure 7a). The second one was identified by the ICA-ANN.

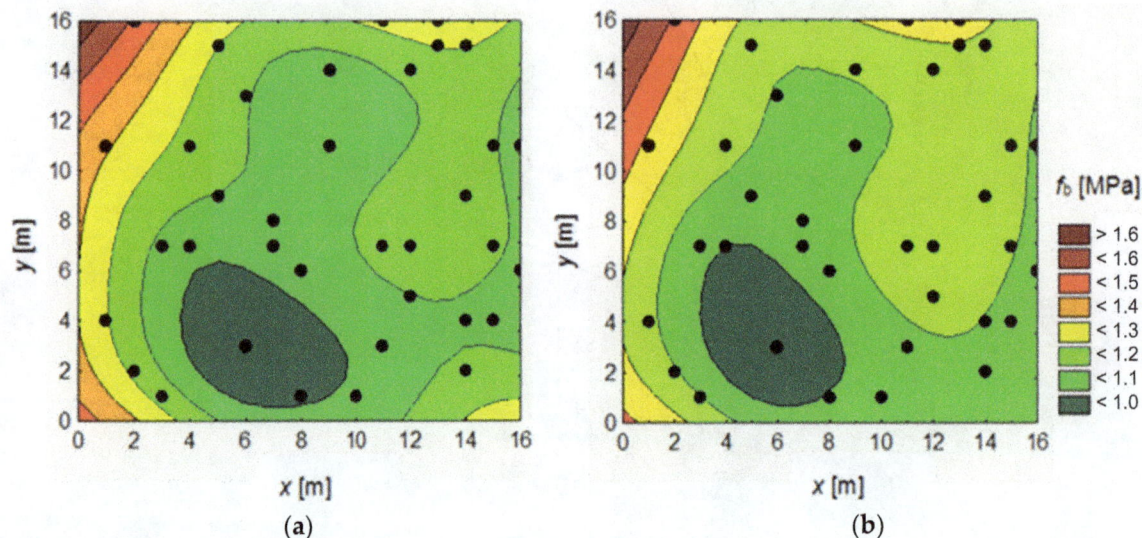

Figure 7. Comparative maps of the value of the f_b, based on validation results yielded by: (**a**) the pull-off method; (**b**) the ICA-ANN model (the symbol "•" represents a randomly-selected test point used for validation).

6. Conclusions

The results of the proposed hybrid method indicate that the metaheuristic ICA is successful in optimizing the ANN. The prediction model requires knowledge of a total of five parameters: K_d, N_{av}, M_p/N, and v, obtained using the IR method, and also parameter f_T, which was obtained using the I-E method. ANN with a 1-5-6-5 structure with the nonlinear tan-sigmoid transfer function and the properties related to the ICA with 500 countries and 50 primary empires and 250 repetitions has a higher ability and accuracy in determining the value of the f_b.

In the best ANN model, optimized by the ICA for determining the value of the f_b, the determination coefficient R^2 at the stages of training and testing is 0.844 and 0.858, respectively. The results indicated that the ICA-ANN model outperforms the other methods, such as the GA and PSO. The R^2 values are also higher than those obtained previously by the simple MLP-ANN [29]. The obtained values of R^2 can be considered as satisfactory, taking into account the fact that it is not necessary to use the 3D roughness parameters of an existing concrete substrate surface.

Various functional issues solved by this algorithm also indicate that the proposed optimization strategy can successfully help to solve practical problems. The pull-off adhesion of the cement mortar in the existing concrete substrate in real cement composites can be obtained with the use of the proposed method. However, this method is applicable when the thicknesses of the layers of the cement mortar and substrate, and also the composition of the cement mortar and existing concrete substrate in the particular cement composite, are similar to the ones that were used to create the database in this article. Otherwise, it is necessary to create a new database for cement composites that are tested in practice.

Acknowledgments: This work was supported by the National Centre of Science, Poland (grant no. 2014/15/D/ST8/00550 "Evaluation of the interlayer bond of variably thick concrete layers based on nondestructive tests using artificial intelligence").

Author Contributions: Łukasz Sadowski conceived and designed the experiments; Mehdi Nikoo and Mohammad Nikoo performed the numerical analysis; Łukasz Sadowski analyzed the data; Łukasz Sadowski, Mehdi Nikoo and Mohammad Nikoo wrote the paper.

Conflicts of Interest: The authors declare no conflict of interest. The founding sponsors had no role in the design of the study; in the collection, analyses, or interpretation of data; in the writing of the manuscript, and in the decision to publish the results.

References

1. Ranjbar, N.; Behnia, A.; Chai, H.K.; Alengaram, J.; Jumaat, M.Z. Fracture evaluation of multi-layered precast reinforced geopolymer-concrete composite beams by incorporating acoustic emission into mechanical analysis. *Constr. Build. Mater.* **2016**, *127*, 274–283. [CrossRef]
2. Szemerey-Kiss, B.; Török, Á. Failure mechanisms of repair mortar stone interface assessed by pull-off strength tests. *Bull. Eng. Geol. Environ.* **2017**, *76*, 159–167. [CrossRef]
3. Niwa, J.; Matsumoto, K.; Sato, Y.; Yamada, M.; Yamauchi, T. Experimental study on shear behavior of the interface between old and new deck slabs. *Eng. Struct.* **2016**, *126*, 278–291. [CrossRef]
4. Wu, J.; Liu, X.; Chew, S.H. Parametric study on cement-based soft-hard-soft (SHS) multi-layer composite pavement against blast load. *Constr. Build. Mater.* **2015**, *98*, 602–619. [CrossRef]
5. EN 12504–3. *Analysis of Concrete in Constructions. Part 3: Determination of the Pull-out Force*; Polish Committee for Standardization: Warsaw, Poland, 2006.
6. Pereira, E.; de Medeiros, M.H.F. Pull-off test to evaluate the compressive strength of concrete: An alternative to Brazilian standard techniques. *Rev. IBRACON Estrut. Mater.* **2012**, *5*, 757–780. [CrossRef]
7. Momayez, A.; Ehsani, M.; Ramezanianpour, A.A.; Rajaie, H. Comparison of methods for evaluating bond strength between concrete substrate and repair materials. *Cem. Concr. Res.* **2005**, *35*, 748–757. [CrossRef]
8. Bungey, J.; Madandoust, R. Factors influencing pull-off tests on concrete. *Mag. Concr. Res.* **1992**, *44*, 21–30. [CrossRef]
9. Hoła, J.; Sadowski, Ł.; Schabowicz, K. Nondestructive identification of delaminations in concrete floor toppings with acoustic methods. *Autom. Constr.* **2011**, *20*, 799–807. [CrossRef]
10. Hoła, J.; Sadowski, Ł.; Schabowicz, K. Nondestructive assessment of the adhesion of concrete screeds in the ventilating ducts of mine shafts. In Proceedings of the 11th European Conference on Non-Destructive Testing (ECNDT 2014), Prague, Czech Republic, 6–10 October 2014.
11. Hoła, J.; Sadowski, Ł. Testing interlayer pull-off adhesion in concrete floors by means of nondestructive acoustic methods. In Proceedings of the 18th World Conference on Non Destructive Testing, Durban, South Africa, 16–20 April 2012.
12. Sadowski, Ł. Nondestructive evaluation of interlayer bond in floors by using artificial neural networks. Ph.D. Thesis, Wrocław University of Science and Technology, Wrocław, Poland, 2012.
13. Sadowski, Ł. Analysys of the effect of concrete base roughness on the pull-off adhesion of the topping layer. *IAPGOŚ* **2013**, *1*, 39–42. (In Polish)
14. Veiskarami, M.; Ghorbani, A.; Alavipour, M. Development of a constitutive model for rockfills and similar granular materials based on the disturbed state concept. *Front. Struct. Civ. Eng.* **2012**, *6*, 365–378.
15. Cheng, M.Y.; Tran, D.H.; Hoang, N.D. Fuzzy clustering chaotic-based differential evolution for resource leveling in construction projects. *J. Civ. Eng. Manag.* **2017**, *23*, 113–124. [CrossRef]
16. Kaloop, M.R.; Hu, J.W. Optimizing the De-Noise neural network model for GPS time-series monitoring of structures. *Sensors* **2015**, *15*, 24428–24444. [CrossRef] [PubMed]
17. Talaei, A.S.; Nasrollahi, A.; Ghayekhloo, M. An automated approach for optimal design of prestressed concrete slabs using PSOHS. *KSCE J. Civ. Eng.* **2017**, *21*, 782–791. [CrossRef]
18. Li, X.; Qiu, J.; Shang, Q.; Li, F. Simulation of Reservoir Sediment Flushing of the Three Gorges Reservoir Using an Artificial Neural Network. *Appl. Sci.* **2016**, *6*, 148. [CrossRef]
19. Kaveh, A.; Talaei, A.S.; Nasrollahi, A. Application of Probabilistic Particle Swarm in Optimal Design of Large-Span Prestressed Concrete Slabs. *Iran. J. Sci. Technol. Trans. Civ. Eng.* **2016**, *40*, 33–40. [CrossRef]
20. Zhang, J.K.; Yan, W.; Cui, D.M. Concrete Condition Assessment Using Impact-Echo Method and Extreme Learning Machines. *Sensors* **2016**, *16*, 447. [CrossRef] [PubMed]
21. Tran, T.H.; Hoang, N.D. Predicting Colonization Growth of Algae on Mortar Surface with Artificial Neural Network. *J. Comput. Civ. Eng.* **2016**, *30*, 04016030. [CrossRef]

22. Khademi, F.; Jamal, S.M.; Deshpande, N.; Londhe, S. Predicting strength of recycled aggregate concrete using Artificial Neural Network, Adaptive Neuro-Fuzzy Inference System and Multiple Linear Regression. *Int. J. Sustain. Built Environ.* **2016**, *5*, 355–369. [CrossRef]

23. Hasanzadehshooiili, H.; Lakirouhani, A.; Medzvieckas, J. Superiority of artificial neural networks over statistical methods in prediction of the optimal length of rock bolts. *J. Civ. Eng. Manag.* **2012**, *18*, 655–661. [CrossRef]

24. Sadowski, Ł. Non-destructive evaluation of the pull-off adhesion of concrete floor layers using RBF neural Network. *J. Civ. Eng. Manag.* **2013**, *19*, 550–560. [CrossRef]

25. Sadowski, Ł.; Hoła, J. New non-destructive way of identifying the values of pull-off adhesion between concrete layers in floors. *J. Civ. Eng. Manag.* **2014**, *20*, 561–569. [CrossRef]

26. Sadowski, Ł. Non-destructive identification of pull-off adhesion between concrete layers. *Autom. Constr.* **2015**, *57*, 146–155. [CrossRef]

27. Sadowski, Ł.; Hoła, J. ANN modeling of pull-off adhesion of concrete layers. *Adv. Eng. Softw.* **2015**, *89*, 17–27. [CrossRef]

28. Sadowski, Ł.; Nikoo, M.; Nikoo, M. Principal component analysis combined with a self organization feature map to determine the pull-off adhesion between concrete layers. *Constr. Build. Mater.* **2015**, *78*, 386–396. [CrossRef]

29. Sadowski, Ł.; Hoła, J.; Czarnecki, S. Non-destructive neural identification of the bond between concrete layers in existing elements. *Constr. Build. Mater.* **2016**, *127*, 49–58. [CrossRef]

30. Archer, N.; Wang, S. Application of the back propagation neural network algorithm with monotonicity constraints for two-group classification problems. *Decis. Sci.* **1993**, *24*, 60–75. [CrossRef]

31. Lenard, M.; Alam, P.; Madey, G. The applications of neural networks and a qualitative response model to the auditor's going concern uncertainty decision. *Decis. Sci.* **1995**, *26*, 209–227. [CrossRef]

32. Veeramachaneni, K.; Peram, T.; Mohan, C.; Osadciw, L. Optimization Using Particle Swarms with Near Neighbor Interactions. In *Genetic and Evolutionary Computation—GECCO 2003*; Cantú-Paz, E., Foster, J.A., Deb, K., Eds.; Lecture Notes in Computer Science, Vol. 2723; Springer: Berlin, Heidelberg, Germay, 2003; pp. 110–121.

33. Khabbazi, A.; Atashpaz-Gargari, E.; Lucas, C. Imperialist competitive algorithm for minimum bit error rate beamforming. *Int. J. Bio-Inspir. Comput.* **2009**, *1*, 125–133. [CrossRef]

34. Kaveh, A.; Talatahari, S. Optimum design of skeletal structures using imperialist competitive algorithm. *Comput. Struct.* **2010**, *88*, 1220–1229. [CrossRef]

35. Lucas, C.; Nasiri-Gheidari, Z.; Tootoonchian, F. Application of an imperialist competitive algorithm to the design of a linear induction motor. *Energy Convers. Manag.* **2010**, *51*, 1407–1411. [CrossRef]

36. Talatahari, S.; Azar, F.B.; Sheikholeslami, R.; Gandomi, A.H. Imperialist competitive algorithm combined with chaos for global optimization. *Commun. Nonlinear Sci. Numer. Simula.* **2012**, *17*, 1312–1319. [CrossRef]

37. Xing, B.; Gao, W. Imperialist Competitive Algorithm. In *Innovative Computational Intelligence: A Rough Guide to 134 Clever Algorithms*; Springer International Publishing: Cham, Switzerland, 2014; pp. 203–209.

38. Tejani, G.G.; Savsani, V.J.; Patel, V.K. Adaptive symbiotic organisms search (SOS) algorithm for structural design optimization. *J. Comput. Des. Eng.* **2016**, *3*, 226–249. [CrossRef]

39. Rafiei, M.H.; Khushefati, W.H.; Demirboga, R.; Adeli, H. Neural Network, Machine Learning, and Evolutionary Approaches for Concrete Material Characterization. *ACI Mater. J.* **2016**, *113*, 781–789. [CrossRef]

40. Kaveh, A.; Nasrollahi, A. A new hybrid meta-heuristic for structural design: Ranked particles optimization. *Struct. Eng. Mech.* **2014**, *52*, 405–426. [CrossRef]

41. Alavi, A.H.; Hasni, H.; Zaabar, I.; Lajnef, N. A new approach for modeling of flow number of asphalt mixtures. *Arch. Civ. Mech. Eng.* **2017**, *17*, 326–335. [CrossRef]

42. Kaveh, A.; Nasrollahi, A. Engineering design optimization using a hybrid PSO and HS algorithm. *Asian J. Civ. Eng.* **2013**, *14*, 201–223.

43. Fedeliński, P.; Górski, R. Optimal arrangement of reinforcement in composites. *Arch. Civ. Mech. Eng.* **2015**, *15*, 525–531. [CrossRef]

44. Kaveh, A.; Nasrollahi, A. A new probabilistic particle swarm optimization algorithm for size optimization of spatial truss structures. *Int. J. Civ. Eng.* **2014**, *12*, 1–13.

45. Atashpaz-Gargari, E. Social optimization algorithm development and performance review. Master's Thesis, School of Electrical and Computer Engineering, Tehran University, Tehran, Iran, 2009.

46. Atashpaz-Gargari, E.; Lucas, C. Imperialist competitive algorithm: An algorithm for optimization inspired by imperialistic competition. *IEEE Congr. Evolut. Comput.* **2007**, *21*, 4661–4667.

47. Hosseini-Moghari, S.M.; Morovati, R.; Moghadas, M.; Araghinejad, S. Optimum operation of reservoir using two evolutionary algorithms: Imperialist competitive algorithm (ICA) and cuckoo optimization algorithm (COA). *Water Resour. Manag.* **2015**, *29*, 3749–3769. [CrossRef]

48. Armaghani, D.J.; Hasanipanah, M.; Mohamad, E.T. A combination of the ICA-ANN model to predict air-overpressure resulting from blasting. *Eng. Comput.* **2016**, *32*, 155–171. [CrossRef]

49. Shirazi, A.Z.; Mohammadi, Z. A hybrid intelligent model combining ANN and imperialist competitive algorithm for prediction of corrosion rate in 3C steel under seawater environment. *Neural Comput. Appl.* **2016**, 1–10.

50. Taghavifar, H.; Mardani, A.; Taghavifar, L. A hybridized artificial neural network and imperialist competitive algorithm optimization approach for prediction of soil compaction in soil bin facility. *Measurement* **2013**, *46*, 2288–2299. [CrossRef]

51. Ahmadi, M.; Ebadi, M.; Shokrollahi, A.; Majidi, S. Evolving artificial neural network and imperialist competitive algorithm for prediction oil flow rate of the reservoir. *Appl. Soft Comput.* **2013**, *13*, 1085–1098. [CrossRef]

52. Pourbaba, M.; Talatahari, S.; Sheikholeslami, R. A chaotic imperialist competitive algorithm for optimum cost design of cantilever retaining walls. *KSCE J. Civ. Eng.* **2013**, *17*, 972–979. [CrossRef]

53. Sadowski, L.; Nikoo, M. Corrosion current density prediction in reinforced concrete by imperialist competitive algorithm. *Neural Comput. Appl.* **2014**, *25*, 1627–1638. [CrossRef] [PubMed]

54. *ASTM C1740–10 Standard Practice for Evaluating the Condition of Concrete Plates Using the Impulse-Response Method*; ASTM: West Conshohocken, PA, USA, 2010.

55. Davis, A.G. The non-destructive impulse response test in North America: 1985–2001. *NDT & E Int.* **2003**, *36*, 185–193.

56. Ottosen, N.S.; Ristinmaa, M.; Davis, A.G. Theoretical interpretation of impulse response tests of embedded concrete structures. *J. Eng. Mech.* **2004**, *130*, 1062–1071. [CrossRef]

57. Sansalone, M.J.; Streett, W.B. *Impact-echo: Nondestructive Evaluation of Concrete and Masonry*; Bullbrier Press: Jersey Shore, PA, USA, 1997.

58. Schubert, F.; Köhler, B. Ten lectures on impact-echo. *J. Nondestr. Eval.* **2008**, *27*, 5–21. [CrossRef]

59. Carino, N.J. The impact-echo method: An overview. In Proceedings of the 2001 Structures Congress & Exposition, Washington, DC, USA, 21–23 May 2001; pp. 21–23.

60. Gavin, J.; Holger, R.; Graeme, C. Input determination for neural network models in water resources applications. Part 2. Case study: Forecasting salinity in a river. *J. Hydrol.* **2005**, *301*, 93–107.

61. Dorofki, M.; Elshafie, A.H.; Jaafar, O.; Karim, O.A.; Mastura, S. Comparison of artificial neural network transfer functions abilities to simulate extreme runoff data. *Int. Proc. Chem. Biol. Environ. Eng.* **2012**, *33*, 39–44.

3

Fabrication of Efficient Cu$_2$ZnSnS$_4$ Solar Cells by Sputtering Single Stoichiometric Target

Hongtao Cui [1],*,†, Xiaolei Liu [2],*,†, Lingling Sun [1], Fangyang Liu [1], Chang Yan [1] and Xiaojing Hao [1],*

[1] School of Photovoltaic and Renewable Energy Engineering, University of New South Wales, Sydney 2052, NSW, Australia; l.sun@unsw.edu.au (L.S.); fangyang.liu@unsw.edu.au (F.L.); c.yan@unsw.edu.au (C.Y.)
[2] UCL Institute for Materials Discovery, University College London, London WC1E 7JE, UK
* Correspondence: h.cui@unsw.edu.au (H.C.); frank_lin_liu@hotmail.com (X.L.); xj.hao@unsw.edu.au (X.H.)

† These authors contributed equally to this work.

Academic Editor: I. M. Dharmadasa

Abstract: Low cost single stoichiometric target sputtering of Cu$_2$ZnSnS$_4$ (CZTS) precursor has been adopted to fabricate CZTS solar cells. The effect of a series of deposition pressures and deposition durations on the device performance has been investigated. A 3.74% efficient solar cell has been achieved at a base pressure of 1×10^{-4} Torr with a stoichiometric target, which to the authors' knowledge, is the record efficiency for such a stoichiometric target.

Keywords: single target; stoichiometry; sputtering; CZTS

---

1. Introduction

Cu$_2$ZnSnS$_4$ (CZTS) is one of the most promising thin film solar cell absorber candidates owing to its natural earth abundance, direct band gap with the optimal value of 1.45 eV, and its environmentally compatible nature [1,2]. Sputtering is a low cost option for the production of CZTS solar cells due to the large material usage, uniform large area deposition, and the ease of scale-up [3,4]. Single target sputtering is even cost effective due to the lower power supply required and easier control of the precursor deposition in comparison with multi-target sputtering; additionally, it requires less diffusion of elements. Therefore, single target sputtering leads to less voids, better film quality, and higher efficiency compared with the co-sputtering technique [5]. There have already been a few attempts in the literature that generally use Cu poor Zn rich targets [4–6]. The highest efficiency of a single target sputtering CZTS solar cell is 6.48%, with a target composition of Cu:Zn:Sn = 1.68:1.1:1 and sulfurization in a H$_2$S atmosphere which has proven to be superior to S powder sulfurization [5,6]. However, for a stoichiometric CZTS target, no efficiency has been reported yet [5,6]. This paper uses a stoichiometric CZTS target to reduce the secondary phases and maintain the single phase in the target, because CZTS has a very narrow single phase composition region [7]. S powder was adopted for sulfurization because it is a relatively low cost option.

2. Materials and Methods

The Mo coated soda lime glass substrate has a sheet resistance of 0.15 $\Omega/\square$. The CZTS precursors were sputtered onto the substrates using a stoichiometric compound target in a single target sputter chamber. The target was directly facing down the substrate with a shutter in the middle and the distance between the substrate and target was 10 cm. The sputter target cleaning was conducted for 5 min prior to precursor deposition, after which the pressure reached a base pressure of 1×10^{-4} Torr with the shutter in the closed position. The shutter was then switched to the open position to allow precursor deposition onto the substrate. A series of deposition pressures and deposition durations were investigated: 5 mTorr for 30 min, 7.5 mTorr for 60 min, 10 mTorr for 30 min and 60 min, and 12.5 mTorr for 60 min. The deposition pressure was adjusted by a throttle valve. The Ar flow rates were kept at 10 standard cubic centimeter per minute (sccm) for target cleaning and precursor deposition. All the precursors were then sulfurized in a dual zone tube furnace OTF-1200 MTI (MTI Corporation, Richmond, VA, USA) at 570 °C in a sulphur (S) atmosphere for 30 min with the S zone temperature held at 250 °C, which was vacuumed to 400 Torr with the two ends of the tube furnace sealed according to the procedure in [8]. The annealed samples were then subject to chemical bath deposition of CdS, and then sputtering deposition of intrinsic ZnO (i-ZO) and Al doped ZnO (AZO) sequentially, the details of which are in [9]. A conductive Ag glue was then pasted on the window layer as the top electrode and the cell area was defined by mechanical scribing to be 0.5 cm^2. Details can be found in [10].

The SEM images were captured using a FEI Nova NanoSEM230 system (FEI Corporation, Hillsboro, MA, USA) equipped with an energy dispersive spectroscopy (EDS) detector for chemical composition measurements as well. PANalytical's X'Pert Pro materials research diffraction system (Panalytical, Almelo, The Netherlands) was used for XRD measurements of crystal structural quality of the CZTS films. A Renishaw inVia spectrometer (Renishaw plc, Gloucestershire, UK) was used to conduct Raman measurements with a 514 nm laser excitation. The QEX10 system (PV MEASUREMENTS Inc, Boulder, CO, USA) was utilised for external quantum efficiency (EQE) measurements. A Perkin Elmer Lambda 1050 spectrophotometer with an integrating sphere (IS) (Perkin Elmer, Singapore, Singapore) was used to measure the hemispherical reflectance (R) of the CZTS films. Light I–V measurement was performed at 25 °C under AM 1.5 G illumination with the JSC calibrated by EQE measurement.

3. Results and Discussion

Figure 1 shows the chemical composition of CZTS absorbers with precursors deposited at different pressures and durations. It indicates that the Zn/Cu ratio decreases and the Cu/(Zn + Sn) ratio increases with increasing deposition pressure. This suggests that the Cu sputtering rate increased, yet the Zn sputtering rate decreased with increasing pressure, assuming a constant Sn deposition rate. The composition of the sulfurised CZTS films is Cu/(Zn + Sn) 0.9, and Zn/Sn 1.55. The Cu/(Zn + Sn) is large whereas Zn/Sn is not in the desired range (Cu/(Zn + Sn) = 0.8–0.9, Zn/Sn = 1.2–1.3) reported for high efficiency CZTS solar cells [11]. The assumption for putting 5 mT-30 min the comparison group is that the film quality would not change over a deposition duration of 60 min, which should be largely valid.

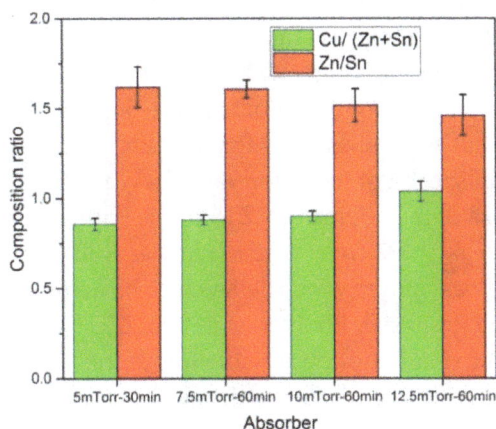

Figure 1. Chemical composition (atomic ratios of Cu/(Zn + Sn) and Zn/Sn) of sulfurized Cu$_2$ZnSnS$_4$ (CZTS) absorbers with precursors deposited at different pressiures and durations: 5 mTorr-30 min is the deposition condition: 50 mTorr pressure and 30 min duration.

Figure 2 shows SEM cross sectional images of the CZTS absorbers fabricated from different precursors. Figure 2a indicates that the 5 mTorr-30 min absorber is mainly composed of large grains spanning the whole thickness growing from the surface to the back contact, yet it has a large amount of voids at the back contact region as well as the junction region, and a film thickness of only 370 nm, almost half of the 7.5 mTorr-60 min absorber. The thickness is as expected for a 30 min deposition duration in comparison with 60 min for 7.5 mTorr-60 min. The voids at the back contact region may be due to the reaction of CZTS with Mo and a small absorber thickness, which led to a very limited amount of time for the formation and growth of the CZTS absorber and a long time for the decomposition reaction [12,13]. The voids at the junction region may reduce the short wavelength absorption and therefore carrier generation for near blue light. Figure 2b–d shows a mixture of large grains and small grains with a few voids for the 7.5 mTorr-60 min absorber, largely small grains for the 10 mTorr-60 min absorber, and mainly small grains with voids at the back contact region for the 12.5 mTorr-60 min absorber, respectively. The voids in the 12.5 mTorr-60 min absorber may be caused by the Cu rich composition, because the excess Cu may either diffuse into the Mo contact [14] or get re-absorbed into the CZTS absorber owing to a high composition tolerance for single phase CZTS at high sulfurization temperature, and leave voids behind [7,12]. The small grain size may result in high recombination at the grain boundary if not properly passivated and could therefore degrade the open circuit voltage (V_{OC}). Voids may not only do similar damage to V_{OC} due to dangling bonds on the surface but could also short circuit the current density (J_{SC}) owing to the voids that block current collection or shunt path formation. In contrast, the large grains from 5 mTorr-30 min and 7.5 mTorr-60 min may have low recombination and high efficiency. Also very interestingly, the absorber film surface is generally smooth (10 mTorr-60 min gives the roughest surface) and the thickness decreased with deposition pressure: 740 nm for 7.5 mTorr-60 min, 590 nm for 7.5 mTorr-60 min, and 555 nm for 12.5 mTorr-60 min, respectively. This would result in different light reflection, absorption, and carrier generation depth profiles. A slightly rough surface may produce an anti-reflection effect due to the double bounce of light. On the other hand, a smooth surface may reduce the series resistance, the formation of possible shunt circuits, and enhance carrier collection.

To check on the structural properties of the films in more detail, the films were characterized by XRD and Raman spectroscopy.

Figure 3 presents the XRD patterns of the CZTS absorbers with varying precursor deposition parameters. It demonstrates that all absorbers share a very similar pattern which agrees well with that of tetragonal kesterite CZTS (JCPDS No. 026-0575) with the preferred orientation of CZTS at (112) consistent with the literature [15,16]; the full width at half maximum (FWHM) of the main peak (112) of CZTS has a small dip down from 0.165° to 0.158°, then increases by 0.17° and ends at 0.228° with increasing the precursor deposition pressure from 5 to 12.5 mTorr. This FWHM trend is consistent with the SEM results as small FWHM reflects large grain size and a low defect level inside the film. It implies improved crystallinity in the following order: 12.5 mTorr-60 min, 10 mTorr-60 min, 5 mTorr-30 min, 7.5 mTorr-60 min. There are a large amount of voids in the 5 mTorr-30 m absorber and therefore its crystallinity degrades, and is not as good as 7.5 mTorr-60 min.

Figure 4 shows the Raman spectra of the absorbers. It reveals that all absorbers share very similar Raman spectra indicating peaks of the dominant CZTS phase with no other apparent phases. It shows an asymmetric Raman main peak at 337 cm^{-1} which matches well with other researchers' findings in the literature [17,18]. The peak asymmetry was believed to be accounted for by phone confinement and strain, especially for small grain sized polycrystalline CZTS film [18]. The $Cu_{2-x}S$ peak marked by dashed lines is equivalent to being not observable for all the absorbers, with only a tiny hillock for 12.5 mTorr-60 min. The FWHM of CZTS's main peak at 337 cm^{-1} also has a small dip down from 7.82 to 7.48 cm^{-1}, and then increases by 8.13 cm^{-1} and ends at 9.75 cm^{-1} with the increasing precursor deposition pressure. This trend matches with the FWHM of the (112) peak in Figure 3. Raman FWHM of the main peak also reflects the crystal structural quality of the absorber and therefore discloses the same crystallinity trend of the absorbers as with XRD FWHM.

The influence of the chemical and structural properties on the optical and electrical properties of completed devices is discussed in the following section.

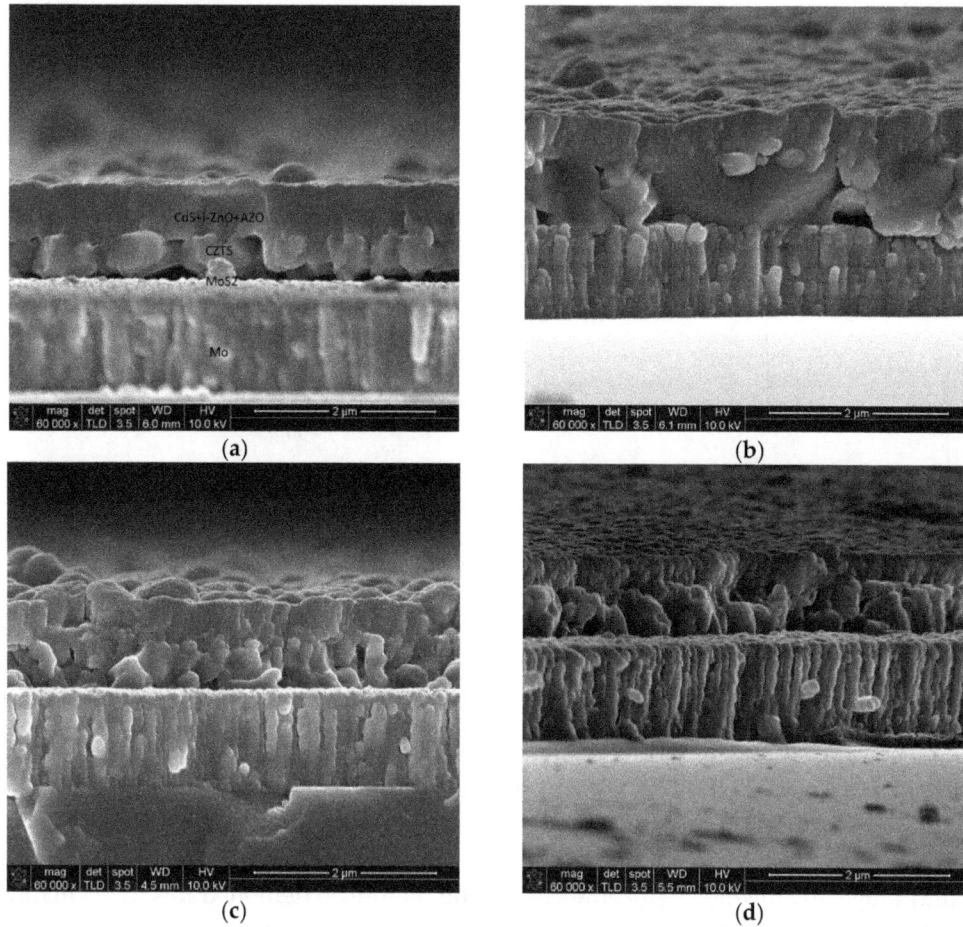

Figure 2. SEM cross section images of CZTS absorbers prepared from different precursors: (**a**) 5 mTorr-30 min, (**b**) 7.5 mTorr-60 min, (**c**) 10 mTorr-60 min, and (**d**) 12.5 mTorr-60 min.

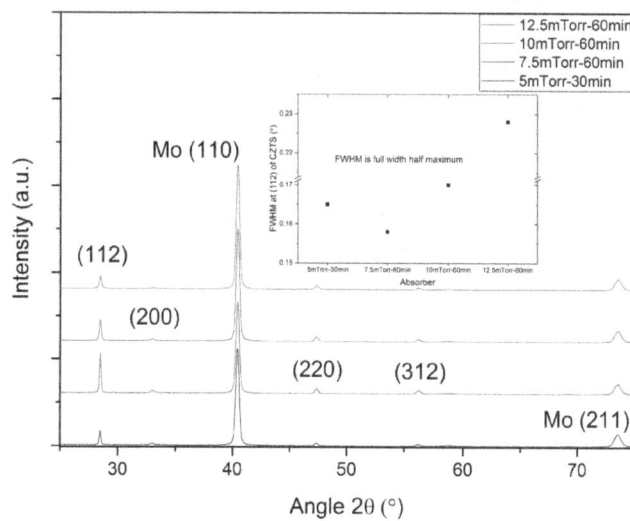

Figure 3. XRD patterns of CZTS absorbers fabricated from various precursors: 5 mTorr-30 min, 7.5 mTorr-60 min, 10 mTorr-60 min, and 12.5 mTorr-60 min.

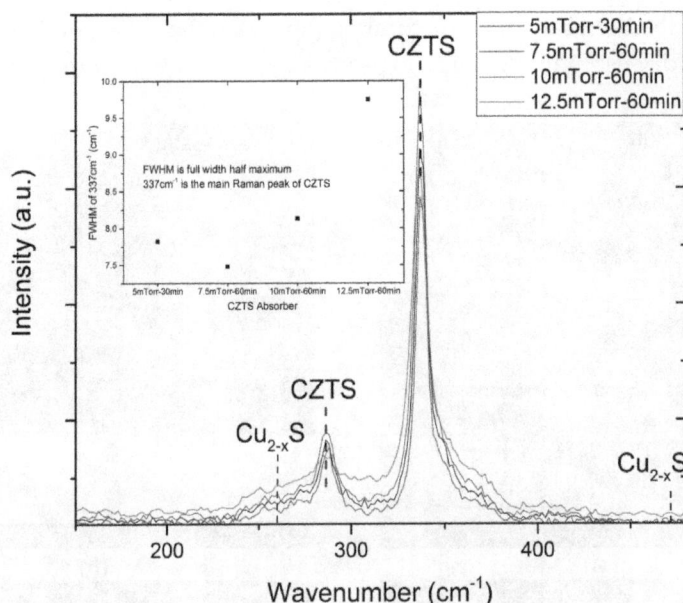

Figure 4. Raman spectra of CZTS absorbers fabricated from varying precursors: 5 mTorr-30 m, 7.5 mTorr-60 min, 10 mTorr-60 min, and 12.5 mTorr-60 min. The inlet picture shows the full width half maximum (FWHM) of the Raman peak at 337 cm^{-1} which is obtained by fitting the peaks with mixed Gaussian–Lorentzian using GRAMS/AI spectroscopic software (version 9.0R2, Thirmo Fisher Scientific Inc, Waltham, MA, USA).

Figure 5 shows external quantum efficiency (EQE), reflection (R), internal quantum efficiency (IQE = EQE/(1 − R)), and (EQE × E)^2 versus photon energy E of all completed devices with varying precursors. Figure 5a shows that low pressure deposited samples give high QE almost in the whole wavelength range of interest, except that 5 mTorr-30 min has a rather low QE for the near blue light (300–400 nm) range due to voids in the junction region. Figure 5b demonstrates that 5 mTorr-30 min has the highest reflection owing to its lowest thickness, and 10 mTorr-60 min features the lowest reflection thanks to its second highest thickness as well as the roughest surface in the group in the wavelength range from 300 to 800 nm. For wavelength ranges above this range, a thick film results in low reflection and therefore high absorption. Figure 5c reflects direct carrier collection efficiency over the whole spectrum of the solar cells excluding optical effects, yet IQE reflects the same with EQE. It means that voids in the junction region of 5 mTorr-30 min also affect the properties of the localized part of the junction and therefore the collection efficiency of the carriers. The minority carrier collection efficiency or the carrier collection depth is generally the depletion region width plus the minority carrier diffusion length assuming zero recombination in the above region. In reality, the heterojunction interface is not perfect and a cliff-like band alignment [19] between the conduction bands of CZTS and CdS generally leads to high recombination which would be deteriorated by lattice mismatch between CZTS and CdS; meanwhile, a back contact region is generally a high recombination region because the detrimental reaction between CZTS and Mo lead to secondary phases and voids. In the literature, a 350 nm diffusion length gives an efficiency up to 8.4% [20]; and the depletion region width is generally only within 100 nm for a 6.2% solar device [21]. This suggests that only within 450 nm thickness of absorber is contributing to carrier generation and collection. Additionally, Electron Beam Induced Current (EBIC) images indicate that even the reduced-void CZTS solar cell shows almost negligible carrier collection at the back contact region [22]. The major carrier generation as well as collection was mainly in the top junction region plus a diffusion length, and the back contact region acted as a recombination sink, not only trapping but also promoting recombination. Therefore, the thick absorber resulted in less effective light absorption, yet high recombination, over the large in-effective regions. This chiefly explains why 5 mTorr-30 min was the best QE performer. On top of the appropriate absorber thickness close to the carrier collection depth, large grain size is also an indispensable factor for device performance.

Figure 5d is based on a linear fitting of $(EQE \times E)^2$ vs. E to obtain the intercept on the E axis, which is the bandgap of the absorber [23]. It shows that the bandgaps of 5 mTorr-30 min, 7.5 mTorr-60 min, 10 mTorr-60 min, and 12.5 mTorr-60 min are 1.505, 1.535, 1.57, and 1.535 eV, respectively. A large band gap ordinarily has high open circuit voltage.

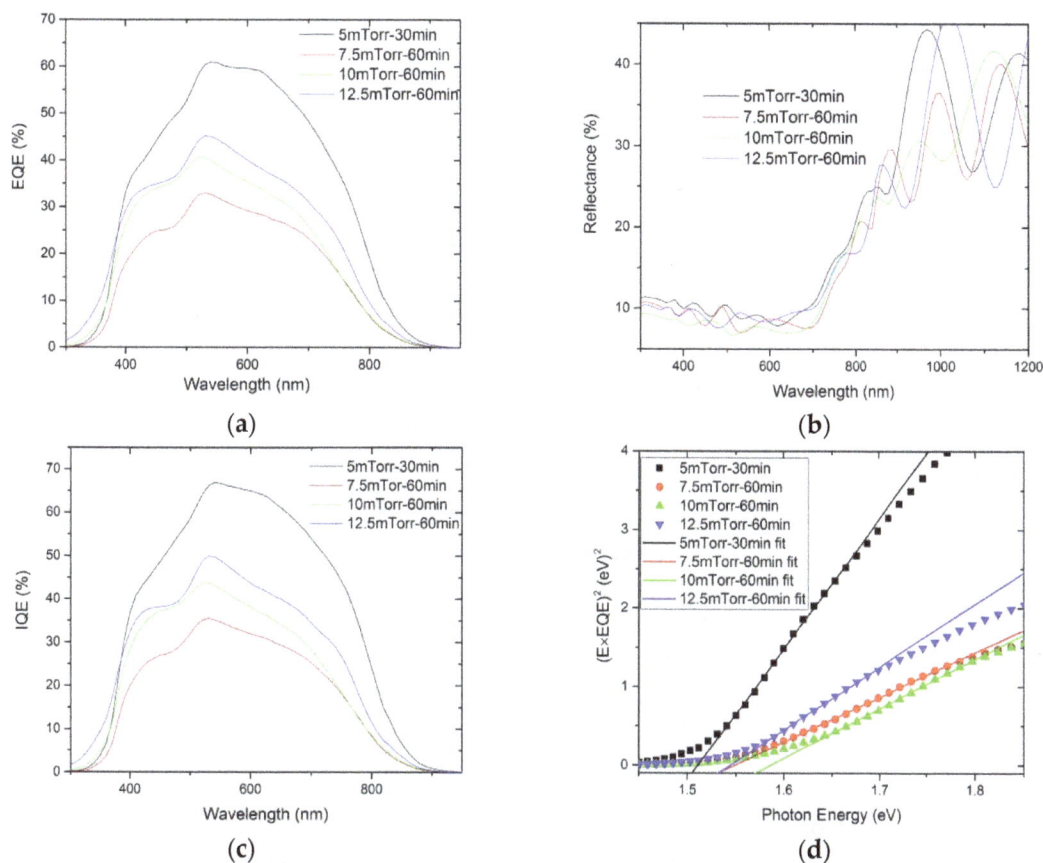

Figure 5. (a) External quantum efficiency (EQE) of CZTS devices with different precursor deposition parameters: 5 mTorr-30 min, 7.5 mTorr-60 min, 10 mTorr-60 min, and 12.5 mTorr-60 min; (b) Reflectance of samples in Figure 5a; (c) Internal quantum efficiency (IQE) of samples in Figure 5a; (d) Band gap estimation plot $(E \times EQE)^2$ versus E, where E is the photon energy.

Figure 6 is the light I-V results of the solar cells. The 5 mTorr-30 min absorber has the highest J_{SC} of 13.4 mA/cm^2, fill factor of 57.1%, and shunt resistance of 1115 $\Omega \cdot$cm^2, yet has the lowest series resistance of 5.9 $\Omega \cdot$cm^2 and therefore the highest efficiency of 3.74% thanks to the essentially similar absorber thickness with the effective carrier collection depth plus large grain size, which to the authors' knowledge is also the record efficiency for stoichiometric single target sputtering CZTS devices; however, it has the second lowest V_{OC} mainly due to the voids and low bandgap as discussed above. 7.5 mTorr-60 min has the largest V_{OC} thanks to the high crystallinity of the material, while it has the lowest J_{SC} caused predominantly by a much thicker high recombinative bottom region together with voids relevant blocking effect or shunting effect. The thick recombinative bottom region not only increased the series resistance dramatically but also reduced the effective light absorption and photogeneration. The lowest V_{OC} of 12.5 mTorr-60 min can be roughly ascribed to the lowest crystallinity of the absorber material and the highest Cu content (a relatively high copper concentration ratio was reported to cause short lifetimes and therefore low V_{OC} [24]) with voids relevant recombination a minor factor.

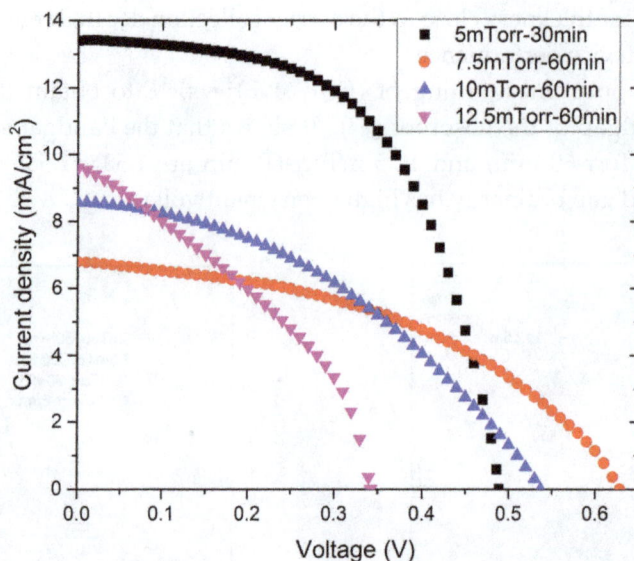

	Voc (mV)	J_{SC} (mA/cm²)	FF (%)	eff (%)	Series Resistance	Shunt Resistance
5mTorr-30min	489	13.4	57.1	3.74	5.9	1115
7.5mTorr-60min	628	6.8	45.4	1.94	23	626
10mTorr-60min	537	8.57	40.3	1.85	25.5	779
12.5mTorr-60min	343	9.6	37	1.22	8.5	99

Figure 6. Light $J-V$ curve of CZTS devices fabricated from different precursors: 5 mTorr-30 min, 7.5 mTorr-60 min, 10 mTorr-60 min and 12.5 mTorr-60 min. The efficiency, V_{OC}, J_{SC}, fill factor (FF), shunt resitance (in $\Omega \cdot cm^2$), and series resistance (in $\Omega \cdot cm^2$) are given in the inset table.

4. Conclusions

A record efficiency of 3.74% was achieved for single stoichiometric target sputtering CZTS solar cells at a low processing pressure of 5 mTorr and a short duration of 30 min. Such parameters yield the highest EQE, J_{SC}, FF, shunt resistance, and efficiency by virtue of mostly the appropriate absorber thickness which is close to the minority carrier collection depth. The relatively low V_{OC} may be due to a large amount of voids within the absorber and the low bandgap. Interestingly, high pressure resulted in thin absorber thickness, leading to high EQE and J_{SC}, yet much lower V_{OC} and FF, which may be chiefly caused by low crystallinity and high Cu concentration. Low pressure leads to large grain size and narrow FWHM of major XRD and Raman peaks. These suggest that thin absorbers of 450 nm thickness and low pressure deposition of the precursor would be necessary for future high efficiency single target CZTS solar cells.

Acknowledgments: This Program has been supported by the Australian Government through the Australian Renewable Energy Agency and the Australian Research Council, and by the China Guodian Corporation. Responsibility for the views, information, or advice expressed herein is not accepted by the Australian Government. The authors acknowledge the facilities of the Electron Microscope Unit of The University of New South Wales. CUI acknowledges the Discovery Early Career Researcher Award, grant # DE160101252.

Author Contributions: Hongtao Cui and Xiaolei Liu conceived and designed the experiments; Xiaolei Liu, Fangyang Liu and Chang Yan performed the experiments; Hongtao Cui and Lingling Sun analyzed the data; Xiaojing Hao contributed reagents/materials/analysis tools; Hongtao Cui wrote the paper.

Conflicts of Interest: The authors declare no conflict of interest. The founding sponsors had no role in the design of the study; in the collection, analyses, or interpretation of data; in the writing of the manuscript, and in the decision to publish the results.

References

1. Peter, L.M. Towards sustainable photovoltaics: The search for new materials. *Phil. Trans. R. Soc. A* **2011**, *369*, 1840–1856. [CrossRef] [PubMed]

2. Mitzi, D.B.; Gunawan, O.; Todorov, T.K.; Barkhouse, D.A.R. Prospects and performance limitations for Cu–Zn–Sn–S–Se photovoltaic technology. *Phil. Trans. R. Soc. A* **2013**, *371*. [CrossRef] [PubMed]

3. Song, X.; Ji, X.; Li, M.; Lin, W.; Luo, X.; Zhang, H. A review on development prospect of CZTS based thin film solar cells. *Int. J. Photoenergy* **2014**. [CrossRef]

4. Xie, M.; Zhuang, D.; Zhao, M.; Zhuang, Z.; Ouyang, L.; Li, X.; Song, J. Preparation and characterization of Cu_2ZnSnS_4 thin films and solar cells fabricated from quaternary Cu-Zn-Sn-S target. *Int. J. Photoenergy* **2013**, *2013*. [CrossRef]

5. Nakamura, R.; Tanaka, K.; Uchiki, H.; Jimbo, K.; Washio, T.; Katagiri, H. Cu_2ZnSnS_4 thin film deposited by sputtering with Cu_2ZnSnS_4 compound target. *J. Jpn. Appl. Phys.* **2014**, *53*. [CrossRef]

6. Katagiri, H.; Jimbo, K. Development of rare metal-free CZTS-based thin film solar cells. In Proceedings of the 37th IEEE Photovoltaic Specialists Conference, Seattle, WA, USA, 19–24 June 2011; pp. 003516–003521.

7. Chagarov, E.; Sardashti, K.; Haight, R.; Mitzi, D.B.; Kummel, A.C. Density-functional theory computer simulations of $CZTS_{0.25}Se_{0.75}$ alloy phase diagrams. *J. Chem. Phys.* **2016**, *145*. [CrossRef]

8. Zhang, K.; Su, Z.; Zhao, L.; Yan, C.; Liu, F.; Cui, H.; Hao, X.; Liu, Y. Improving the conversion efficiency of Cu_2ZnSnS_4 solar cell by low pressure sulfurization. *Appl. Phys. Lett.* **2014**, *104*, 141101. [CrossRef]

9. Cui, H.; Liu, X.; Liu, F.; Hao, X.; Song, N.; Yan, C. Boosting Cu_2ZnSnS_4 solar cells efficiency by a thin Ag intermediate layer between absorber and back contact. *Appl. Phys. Lett.* **2014**, *104*, 041115. [CrossRef]

10. Cui, H.; Li, W.; Liu, X.; Song, N.; Lee, C.; Liu, F.; Hao, X. Optimization of precursor deposition for evaporated Cu_2ZnSnS_4 solar cells. *Appl. Phys. A* **2015**, *118*, 893–899. [CrossRef]

11. Mitzi, D.B.; Gunawan, O.; Todorov, T.K.; Wang, K.; Guha, S. The path towards a high-performance solution-processed kesterite solar cell. *Solar Energy Mater. Solar Cells* **2011**, *95*, 1421–1436. [CrossRef]

12. Scragg, J.J.; Watjen, J.T.; Edoff, M.; Ericson, T.; Kubart, T.; Platzer-Bjorkman, C. A detrimental reaction at the molybdenum back contact in $Cu_2ZnSn(S,Se)_4$ thin-film solar cells. *J. Am. Chem. Soc.* **2012**, *134*, 19330–19333. [CrossRef] [PubMed]

13. Scragg, J.J. *Copper Zinc Tin Sulfide Thin Films For Photovoltaics: Synthesis and Characterisation by Electrochemical Methods*; Springer: Berlin, Germany, 2011.

14. Liu, F.; Sun, K.; Li, W.; Yan, C.; Cui, H.; Jiang, L.; Hao, X.; Green, M.A. Enhancing the Cu_2ZnSnS_4 solar cell efficiency by back contact modification: Inserting a thin TiB_2 intermediate layer at Cu_2ZnSnS_4/Mo interface. *Appl. Phys. Lett.* **2014**, *104*, 051105. [CrossRef]

15. Thota, N.; Gurubhaskar, M.; Sunil, M.A.; Prathap, P.; Subbaiah, Y.P.V.; Tiwari, A. Effect of metal layer stacking order on the growth of Cu_2ZnSnS_4 thin films. *Appl. Surf. Sci.* **2017**, *396*, 644–651. [CrossRef]

16. Sarswat, P.K.; Snure, M.; Free, M.L.; Tiwari, A. CZTS thin films on transparent conducting electrodes by electrochemical technique. *Thin Solid Films* **2012**, *520*, 1694–1697. [CrossRef]

17. Sarswat, P.K.; Free, M.L.; Tiwari, A. Temperature-dependent study of the raman a mode of Cu_2ZnSnS_4 thin films. *Phys. Status Solidi B* **2011**, *248*, 2170–2174. [CrossRef]

18. Sarswat, P.K.; Free, M.L. The effects of dopant impurities on Cu_2ZnSnS_4 system Raman properties. *J. Mater. Sci.* **2015**, *50*, 1613–1623. [CrossRef]

19. Yan, C.; Liu, F.; Song, N.; Ng, B.K.; Stride, J.A.; Tadich, A.; Hao, X. Band alignments of different buffer layers (CdS, Zn(O,S), and In_2S_3) on Cu_2ZnSnS_4. *Appl. Phys. Lett.* **2014**, *104*, 173901. [CrossRef]

20. Shin, B.; Gunawan, O.; Zhu, Y.; Bojarczuk, N.A.; Chey, S.J.; Guha, S. Thin film solar cell with 8.4% power conversion efficiency using an earth-abundant Cu_2ZnSnS_4 absorber. *Prog. Photovolt. Res. Appl.* **2013**, *21*, 72–76. [CrossRef]

21. Dhakal, T.P.; Peng, C.-Y.; Tobias, R.R.; Dasharathy, R.; Westgate, C.R. Characterization of a CZTS thin film solar cell grown by sputtering method. *Solar Energy* **2014**, *100*, 23–30. [CrossRef]

22. Sugimoto, H.; Hiroi, H.; Sakai, N.; Muraoka, S.; Katou, T. Over 8% efficiency Cu_2ZnSnS_4 submodules with ultra-thin absorber. In Proceedings of the 38th IEEE Photovoltaic Specialists Conference, Austin, TX, USA, 3–8 June 2012; pp. 002997–003000.

23. Ahmed, S.; Reuter, K.B.; Gunawan, O.; Guo, L.; Romankiw, L.T.; Deligianni, H. A high efficiency electrodeposited Cu_2ZnSnS_4 solar cell. *Adv. Energy Mater.* **2012**, *2*, 253. [CrossRef]

24. Sugimoto, H.; Liao, C.; Hiroi, H.; Sakai, N.; Kato, T. Lifetime Improvement for High Efficiency Cu_2ZnSnS_4 Submodules. In Proceedings of the 39th IEEE Photovoltaic Specialists Conference, Tampa, FL, USA, 16–21 June 2013; pp. 3208–3211.

4

Damping Optimization of Hard-Coating Thin Plate by the Modified Modal Strain Energy Method

Wei Sun * and Rong Liu

School of Mechanical Engineering & Automation, Northeastern University, Shenyang 110819, China; liurong_neu@126.com
* Correspondence: weisun@mail.neu.edu.cn

Academic Editors: Quanshun Luo and Yongzhen Zhang

Abstract: Due to the medium and small damping characteristics of the hard coating compared with viscoelastic materials, the classical modal strain energy (CMSE) method cannot be applied to the prediction of damping characteristics of hard-coating composite structure directly. In this study, the CMSE method was modified in order to be suitable for this calculation, and then the damping optimization of the hard-coating thin plate was carried out. First, the solution formula of modified modal strain energy (MMSE) method was derived and the relevant calculation procedure was proposed. Then, based on the principle that depositing the hard coating on the locations where modal strain energy is higher, the damping optimization method and procedure were presented. Next, a cantilever thin plate coated with Mg-Al hard coating was taken as an example to demonstrate the solution of the modal damping parameters for the composite plate. Finally, the optimization of coating location was studied according to the proposed method for the cantilever thin plate, and the effect of the coating area on the damping characteristics of hard-coating plate was also discussed.

Keywords: hard coating; modified modal strain energy method; thin plate; damping; optimization

---

1. Introduction

Because hard coating has the advantages of high hardness, high-temperature resistance, and anti-corrosion, it has been widely used in aircraft, aerospace, vehicles, machine tools, and other fields [1–3]. The hard coating is usually prepared by air plasma spraying (APS) and electron beam-physical vapor deposition (EB-PVD) [4] and Stony's approach is commonly used to check the characteristics of coating [5]. Recent studies have shown that the hard coating will contribute to the increase of structural damping and the decrease of resonant stress [6,7]. In other words, hard coating has the effect of vibration reduction.

The study of hard-coating damping design involves the preparation of hard-coating material, the testing of characteristic parameters, the dynamic modeling of hard-coating composite structure, and so on. Among them, the prediction of damping characteristics of hard-coating composite structure is a key content and the obtained results can provide references for the vibration control of thin shell structure. For the viscoelastic composite structure, the CMSE method [8,9] can accurately obtain the modal loss factors of composite structure with no need for calculating complex eigenvalues. The CMSE method has also been used to evaluate the functional dependence of damping on temperature and frequency for a coating beam structure [10]. In terms of calculating principles of the CMSE method, the internal damping of metal substrate is usually neglected, thus different complex stiffness matrices will be produced compared with the practical composite structure. This assumption is reasonable for the viscoelastic composite structure, because the loss factor of viscoelastic material is much larger than the internal damping of metal material, the change of the complex stiffness matrix is small and acceptable. However, the loss factor of hard coating is larger than that of metal but far less than that of viscoelastic material [11]. For this case, the CMSE method is no longer applicable for the prediction of the damping properties of hard-coating composite structure directly. In addition, the CMSE method generally only uses the real part of the complex stiffness matrix to calculate the eigenvectors of the composite structure, which is used to calculate the modal strain energy further. Because the imaginary part of complex stiffness matrix has certain contribution to the solution of these eigenvectors, this computation pattern can also introduce errors for the damping prediction of the composite structure [12].

Furthermore, during the hard-coating damping design, damping optimization also needs to be involved, that is, the hard coating should be coated on the proper locations of the structure to achieve the optimum damping effect. The optimization of coating location is extremely important. On one hand, compared with full coating, the strategy of partial coating can reduce treatment cost; on the other hand, it can minimize the change of the original structure and the increase of structural mass. Some literature [13–16] has proven that the resonant response or resonant stress of the structure can be effectively suppressed if the damping materials are coated on the locations where modal strain energy is higher. Therefore, damping optimization can be easily implemented with the reference of the distribution of modal strain energy. Then, to obtain the most favorable effect of vibration reduction, the key points of this optimal design lie in the accurate prediction of the damping parameters of hard-coating composite structure and the reasonable design of optimization procedure.

Considering the research status above, an MMSE method is developed in this study. It is relatively more suitable for predicting the damping characteristics of hard-coating composite structure, because the material damping of both metal substrate and the hard coating are contained and the imaginary part of the complex stiffness matrix is also considered. On the basis of this, the damping optimization of the composite structure is presented. The article is organized as follows. In Section 2, the solution formula of the MMSE method was derived and the relevant calculation procedure was proposed. In Section 3, based on the principle that depositing the hard coating on the locations where modal strain energy is higher, the damping optimization method and procedure were presented. In Section 4, a cantilever thin plate coated with Mg-Al hard coating was taken as an example to demonstrate solution of the modal damping parameters for the composite plate. In Section 5, the optimization of coating location was studied according to the proposed method for the cantilever thin plate, and the effect of the coating area on the damping characteristics of the hard-coating plate was also discussed. Some important conclusions were acquired from the study and they are listed in Conclusions.

2. MMSE Method for the Hard-Coating Composite Structure

2.1. Calculation Principle

The schematic of a hard-coating composite structure is shown in Figure 1 and it consists of a hard-coating layer and a metal substrate. Symbols H_c, H_s, and E_c^*, E_s^* are the thickness and the complex modulus of the hard coating and the metal substrate, respectively. E_c^* and E_s^* can be expressed as

$$E_c^* = E_c (1 + i\eta_c), E_s^* = E_s (1 + i\eta_s) \tag{1}$$

where, * represents a complex number, $i = \sqrt{-1}$, and E_c, E_s, η_c, η_s are the Young's modulus and loss factor of the hard coating and the metal substrate, respectively.

Hard coating ——————————— E_c^*, H_c

Metal substrate ——— [   ] ——— E_s^*, H_s

Figure 1. Hard-coating damping composite structure.

Since the elastic modulus of the hard coating and the metal substrate are expressed by complex modulus, the finite element dynamic equation of the system in frequency domain becomes

$$(K_R + iK_I - \omega^2 M)X = F \tag{2}$$

where, M is the mass matrix of the composite structure, K_R is the real part of the complex stiffness matrix, K_I is the imaginary part of the complex stiffness matrix, F is the exciting force vector, and X is the displacement vector.

The real and imaginary parts of the complex stiffness matrix can be further expressed as

$$K_R = K_{Rs} + K_{Rc} \tag{3}$$

$$K_I = K_{Is} + K_{Ic} \tag{4}$$

where, K_{RS}, K_{RC} and K_{IS}, K_{IC} represent the contributions of the metal substrate and the hard coating to the real and imaginary parts of the complex stiffness matrix, respectively.

For the CMSE method, usually, only the real part K_R of the complex stiffness matrix is adopted to construct characteristic equation, which is used to solve the real mode and can be expressed as

$$(K_R - \omega_r^2 M)\varphi_r = 0 \quad (r = 1, 2 \cdots n) \tag{5}$$

where, ω_r, φ_r are the r-th order natural frequency and modal shape, respectively. Further, the loss factor of composite structure can be solved using the obtained real modal shape and the solving equation is

$$\eta_r = \frac{\varphi_r^T K_{Ic} \varphi_r}{\varphi_r^T K_R \varphi_r} \quad (r = 1, 2, \cdots n) \tag{6}$$

where, ηr is the r-th order modal loss factor of composite structure. In addition, Equation (6) only considers the contribution of hard coating to the loss factor of the whole system, so the CMSE method cannot be used to predict the damping characteristics directly.

Then, the CMSE method is modified in this study. Two modified methods are proposed here and the main difference between the two methods is the pattern of constructing the characteristic equation. For the method of MMSE 1, the characteristic equation of solving mode shape can be written as

$$(K_R + \beta K_I - \hat{\omega}_r^2 M)\hat{\varphi}_r = 0 \quad (r = 1, 2 \cdots n) \tag{7}$$

where, $\hat{\omega}_r$, $\hat{\varphi}_r$ are the r-th order natural frequency and modal shape respectively obtained by the method of MMSE, and $\hat{a}$ is a modified coefficient which describes the contribution of the imaginary part of the complex stiffness matrix to the solution of mode shapes. The value of $\hat{a}$ can be calculated by [17]

$$\beta = \frac{\mathrm{trace}\ (K_I)}{\mathrm{trace}\ (K_R)} \tag{8}$$

where "trace ()" is a function to gain the trace of a matrix, namely, the sum of matrix diagonal elements.

Compared with the method of MMSE 1, the characteristic equation of the method of MMSE 2 is expressed as

$$(K_R + \beta_r K_I - \hat{\omega}_r^2 M)\hat{\varphi}_r = 0 \quad (r = 1, 2 \cdots n) \tag{9}$$

It can be seen from Equation (9) that the modified coefficient, $\hat{a}$ is replaced by $\hat{a}_r$ in the method of MMSE 2, which means that different order will have different modified coefficient. The solving formula for $\hat{a}_r$ is shown as follows

$$\beta_r = \frac{\varphi_r^T K_I \varphi_r}{\varphi_r^T K_R \varphi_r} \quad (r = 1, 2 \cdots n) \tag{10}$$

It should be noted from the Equation (10) that the modal shape vector corresponding to the CMSE method (shown in Equation (5)) needs to be used during the solution of modified coefficient $\hat{a}_r$ for the method of MMSE 2.

According to the MMSE method, the prediction formula of damping characteristics for the hard-coating composite structure can be finally determined as

$$\hat{\eta}_r = \frac{\hat{\varphi}_r^T K_I \hat{\varphi}}{\hat{\varphi}_r^T K_R \hat{\varphi}} \quad (r = 1, 2, \cdots n) \tag{11}$$

It can be seen from the prediction formula that the loss factor of both the metal substrate and the hard coating are considered and the contribution of the imaginary part of the complex stiffness matrix to the modal shapes are also introduced. So the calculation accuracy of damping characteristics should be higher than the CMSE method.

2.2. Calculation Procedure of MMSE Method

According to the aforementioned calculation principles, the calculation procedure of the two MMSE methods can be described according to Figure 2. It mainly includes five key steps: Solving the mass matrix and complex stiffness matrix, decomposing complex stiffness matrix, computing the modified coefficient, calculating the modified modal shape, and predicting the damping characteristics.

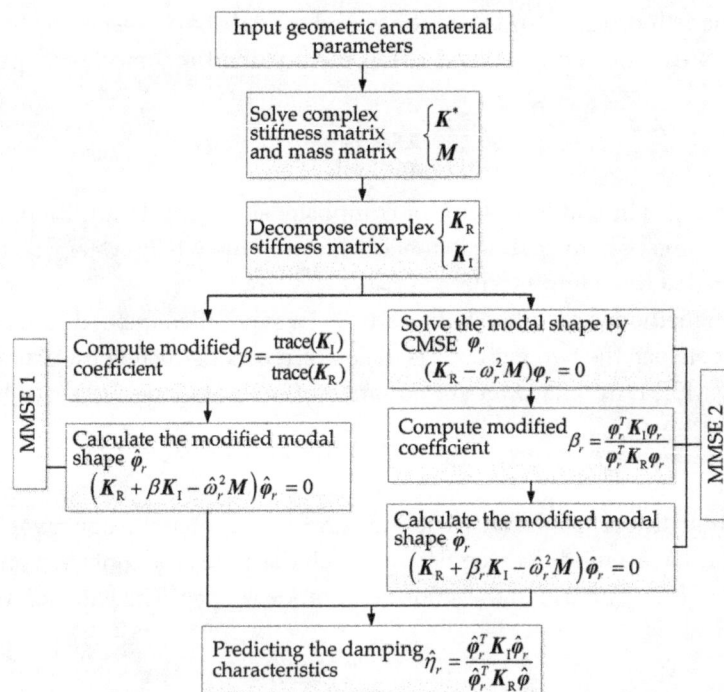

Figure 2. The calculation procedure of MMSE method.

It should be noted that, for the method of MMSE 1, the modified coefficient â can be obtained by the direct utilization of real and imaginary parts of the complex stiffness matrix without using the calculation steps of CMSE method. However, for the method of MMSE 2, the CMSE method should be adopted to obtain the modal shape φ_r, and then $\hat{a}_r$—corresponding to different modal order—can be calculated. Obviously, the computational efficiency of MMSE 1 should be higher than that of MMSE 2. However, due to the fact that the influence of different modal order has been taken into account in MMSE 2, as a general rule, the computational accuracy of MMSE 2 should be higher than that of MMSE 1.

3. Damping Optimization of Hard-Coating Composite Structure by the MMSE Method

3.1. Damping Optimization Principle

The optimization for vibration reduction using hard coating is a comprehensive problem and the following constrained conditions should be considered, such as coating process, damping capacity, the influence of hard coating on the original performance, and the possibility of the peeling off of hard coating, etc. The objectives of optimization can be set as obtaining the optimal material parameters, coating thickness, coating location, and so on. In this study, only the optimization of coating location was considered, and the optimization study of material parameters and thickness of hard coating can be found in the author's previous study [18].

As mentioned in the introduction section, to obtain a better damping effect, the hard coating should be coated on the locations where modal strain energy is higher. This conclusion has been verified by the study of viscoelastic structure. For example, Kumar et al. [13], placed the viscoelastic damping material on the locations with higher modal strain energy for a curved plate and conducted the experiment to prove that this damping strategy can achieve excellent vibration reduction effect. Moreira et al. [14] did a similar study and they took the thin rectangular plate as research object, and also proved that placing the viscoelastic damping material on the locations with higher modal strain energy is most efficient for vibration reduction by experimentation. Masti [15] and Sainsbury [16] also proved the conclusion using numerical studies. Because both hard coating and the viscoelastic damping material can be characterized by complex modulus model, and also because both of them suppress the excessive vibration of the structure and dissipate energy by the material internal damping, the optimal principle of this study can be thought of as similar to the above description. A simple optimization schematic diagram for thin plate structure is shown in Figure 3. It is the distribution of the first order modal strain energy of a cantilever thin plate. In addition, it should be noted that the higher modal strain energy is a relative concept. Here, according to the distribution of modal strain energy, the area with high modal strain energy is defined as the area accounting for 20%–50% of the full structure, the modal strain energy of elements in this area is significantly higher than the elements in the remaining area.

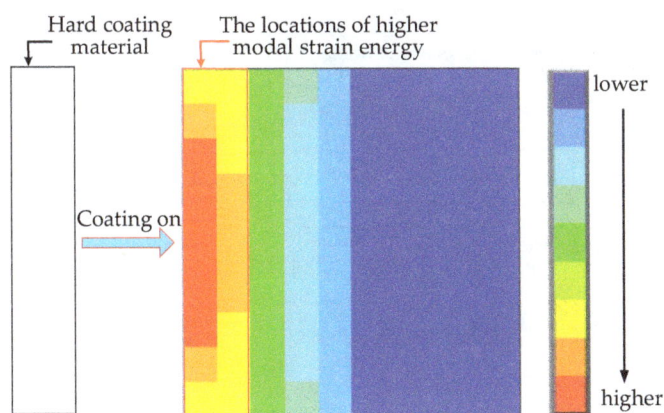

Figure 3. A simple optimization schematic diagram for hard-coating thin plate structure.

Acquiring the modal strain energy distribution of the structure is the basis of implementing the optimization of coating location. To get the distribution, the modal strain energy of each element should be calculated firstly, the solving formula is

$$MSE_{rj} = \varphi_r^{j^{\mathrm{T}}} K_r^j \varphi_r^j \tag{12}$$

where, MSE_{rj} is the modal strain energy corresponding to the r-th order and the j-th element, φ_r^j, K_r^j are the modal shape vector and stiffness matrix of the r-th order and the j-th element, respectively. This calculation shown in Equation (12) is for the metal substrate, thus the complex stiffness matrix was not used here. One can believe that it cannot produce big errors because the internal damping of metal substrate is small.

Then, the damping optimization method of coating locations can be established effectively and the optimal objective can be set for the single mode or multi modes. The proposed MMSE method can be used to evaluate the optimization effect. The detailed optimal procedure can be seen in the following discussion.

3.2. Optimization Procedure

In view of the fact that it is difficult to implement the MMSE method developed in this study by commercial FEM software, such as ANSYS, the optimization programs used here are in-house code developed by MATLAB and the version of MATLAB is MATLAB R2012a. The presented optimization procedure is shown in Figure 4. Here, the optimization of coating thickness and the material parameters are not contained and these parameters are set as constant.

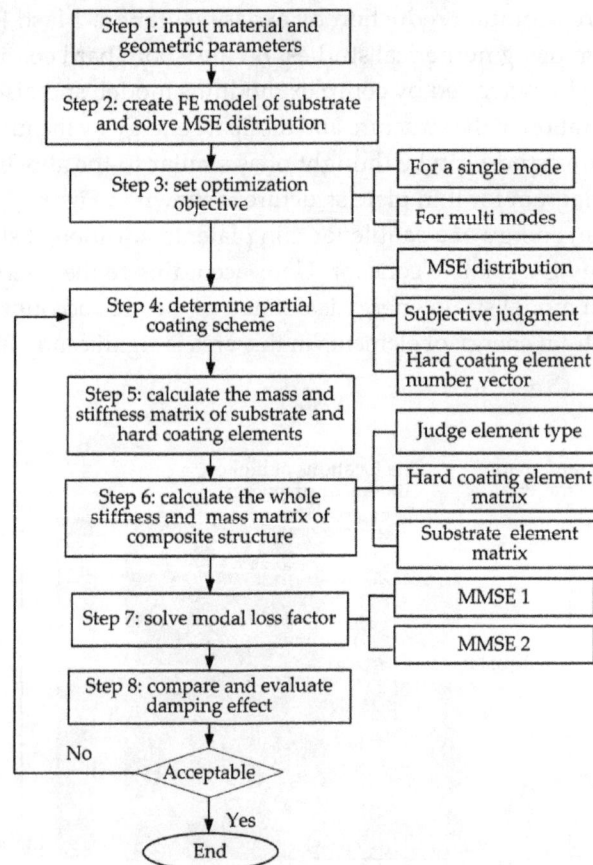

Figure 4. The procedure of damping optimization based on MMSE.

Some key steps need to be explained and show as follows:

In Step 2, when meshing the structure and creating the finite element model, it is preferable to make all the elements have the same size. In this way, the mass and stiffness matrix of the elements need to be calculated only once and the computational efficiency will be dramatically improved.

In Step 3, the optimization objective can be set only for a single mode of the structure. This means that the hard coating has better vibration reduction effect only for the considered order. Also, the optimization objective can be set for multi modes, then, all the considered orders will have higher modal loss factors by implementing partial coating.

In Step 4, according to the distribution of modal strain energy and the optimization objective, the simulation of partial coating needs to be carried out, that is, placing the coating elements on the locations where the modal strain energy is higher. This step is done by subjective judgment and a coating element number vector will come into being after finishing this step.

In Step 6, before assembling the elements, it is necessary to judge the type of element by a condition discriminant function (ismember (..., ...)) firstly. If it is a coating element, the mass and stiffness matrices of hard-coating composite element should be introduced, otherwise, only considering the relevant matrices of substrate element. Then, in the light of the element number, all the elements are combined into the whole stiffness matrix and the whole mass matrix of the composite structure.

In Step 8, the damping effect of partial coating needs to be evaluated based on the optimization objective. The damping effect is usually compared with the full coating structure. If it is near or achieves the damping ability of full coating structure, the optimization can be considered acceptable. If not, return to Step 4 and reselect the coating locations.

4. Prediction of Damping Characteristics for a Cantilever Thin Plate Coated with Mg-Al Hard Coating

To verify the effectiveness of MMSE method, the damping characteristics of a cantilever titanium plate coated with Mg-Al hard coating are predicted by the proposed MMSE method in this section. It can be seen from the surface morphology of Mg-Al hard coating (shown in Figure 5), that there are many cracks and voids in the coating. The internal friction among coating particles will produce damping effects, thus Mg-Al coating can be used as a damping coating.

Figure 5. Surface morphology of Mg-Al hard coating.

The geometry and material parameters of substrate and hard coating are listed in Table 1. One side of the cantilever plate is fully coated with Mg-Al hard coating. Among them, the Young's modulus and loss factor of Mg-Al hard coating are obtained by experiment using vibration beam method [19] and the material parameters of the titanium plate are gotten from the handbook of metallic materials.

Table 1. The geometry and material parameters of substrate and hard coating.

Items	Length (mm)	Width (mm)	Thickness (mm)	Young's Modulus (GPa)	Loss Factor	Poisson Ratio	Density (kg/m^3)
Titanium Plate	110	110	3	110.32	0.0007	0.31	4420
Mg-Al Hard Coating	110	110	0.06	52.9	0.0045	0.3	1829.6

Here, to calculate the damping characteristics of the composite plate, the programs of CMSE, MMSE 1, and MMSE 2 are developed. The correctness of the developed programs is to be verified by ANSYS software, but only verification of CMSE method can be achieved by ANSYS. The finite element model is shown in Figure 6. The four-node plate element is adopted to simulate the hard-coating element. In the model, there are 100 elements and 121 nodes, respectively. The corresponding results are listed in Table 2.

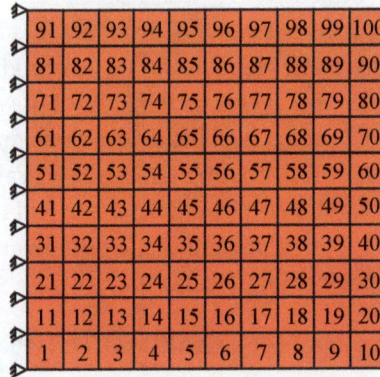

Figure 6. The finite element model of hard-coating thin plate.

Table 2. The obtained modal loss factor of hard-coating thin plate/10^{-3}.

Orders	ANSYS CMSE	MATLAB CMSE	MATLAB MMSE1	MATLAB MMSE2
1	0.12685	0.12776	0.80789	0.80789
2	0.12723	0.12884	0.80880	0.80880
3	0.12623	0.12817	0.80823	0.80823
4	0.12643	0.12788	0.80799	0.80799
5	0.12597	0.12847	0.80848	0.80848

As seen in Table 2, for the calculation of modal loss factors by CMSE method, the results obtained from in-house code and ANSYS are almost consistent. However, the modal loss factor obtained by MMSE method is much different from the results obtained by CMSE method and the results of MMSE method are approximately six times that of CMSE method. Considering the uncertainties of hard-coating loss factors obtained by vibration beam method, the loss factor of hard coating listed in Table 1 is increased and decreased by 5%, respectively. Then, the damping characteristics of hard-coating plate are recalculated using the developed program and the results are listed in Table 3. It can be found from Table 3 that the values of modal loss factors of composite plate increase with the increase of the loss factor of hard coating and the results obtained by CMSE and MMSE method are obviously different.

Table 3. Modal loss factors of composite plate corresponding to different loss factor of hard coating/10^{-3}.

Orders	Loss Factor Increased by 5%			Loss Factor Decreased by 5%		
	CMSE	MMSE1	MMSE2	CMSE	MMSE1	MMSE2
1	0.13416	0.81428	0.81428	0.12138	0.80150	0.80150
2	0.13529	0.81525	0.81525	0.12241	0.80236	0.80236
3	0.13459	0.81465	0.81465	0.12177	0.80183	0.80183
4	0.13428	0.81439	0.81439	0.12149	0.80160	0.80160
5	0.13490	0.81491	0.81491	0.12205	0.80206	0.80206

It can be known from the theoretical derivation in Section 2, for the MMSE method, the loss factor of both the metal substrate and hard coating are considered and the contribution of the imaginary part of the complex stiffness matrix to the modal shape is also introduced, so the MMSE method should be more accurate than the CMSE method. In addition, also it can be noted from Tables 2 and 3 that the results obtained by the two kinds of MMSE methods are almost completely the same for the considered calculation accuracy, so either of the two MMSE methods can be chosen to optimize coating locations of the cantilever thin plate.

5. Damping Optimization of the Cantilever Thin Plate Partially Coated with Mg-Al Hard Coating

According to the optimization procedure described in Figure 4, the damping optimization is performed for the cantilever thin plate partially coated with Mg-Al hard coating in this section. The optimization objective is set for a single mode and multi modes, respectively. For the optimization calculation, only the method of MMSE 1 is adopted to predict the damping characteristics. Before implementing the damping optimization, the modal strain energy distribution of the cantilever thin plate should be determined first, and thus the modal strain energy distribution of the considered orders are also listed here.

5.1. Obtaining the Modal Strain Energy Distribution

For the considered cantilever thin plate, the values of the first five natural frequencies are gotten by the modal analysis program existed in the in-house code and the relevant results are listed in Table 4. The corresponding modal strain energy distribution for each order is shown in Figure 7 and the 100 elements are disposed in a 10×10 matrix in the figure. It should be noted that the value of natural frequency of the structure is not required during the prediction of damping by modal strain energy method and the values listed here can be used as the reference for the damping optimization.

Table 4. The first five natural frequencies of thin plate structure.

Orders	Nature Frequency/Hz
1	207.57
2	506.63
3	1273.6
4	1623.7
5	1846.4

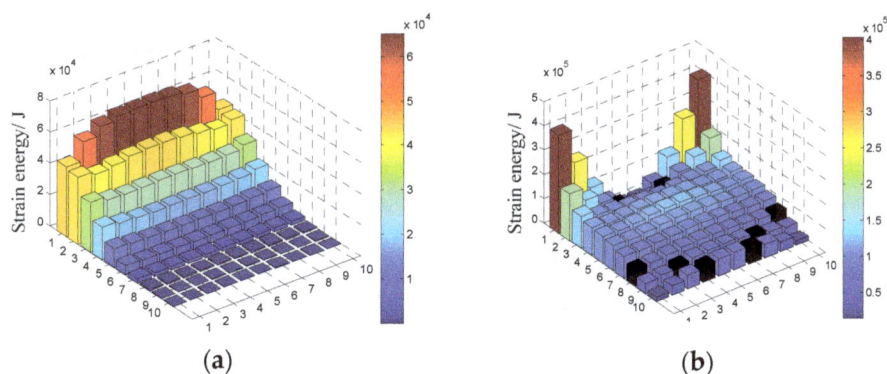

(a)   (b)

Figure 7. *Cont.*

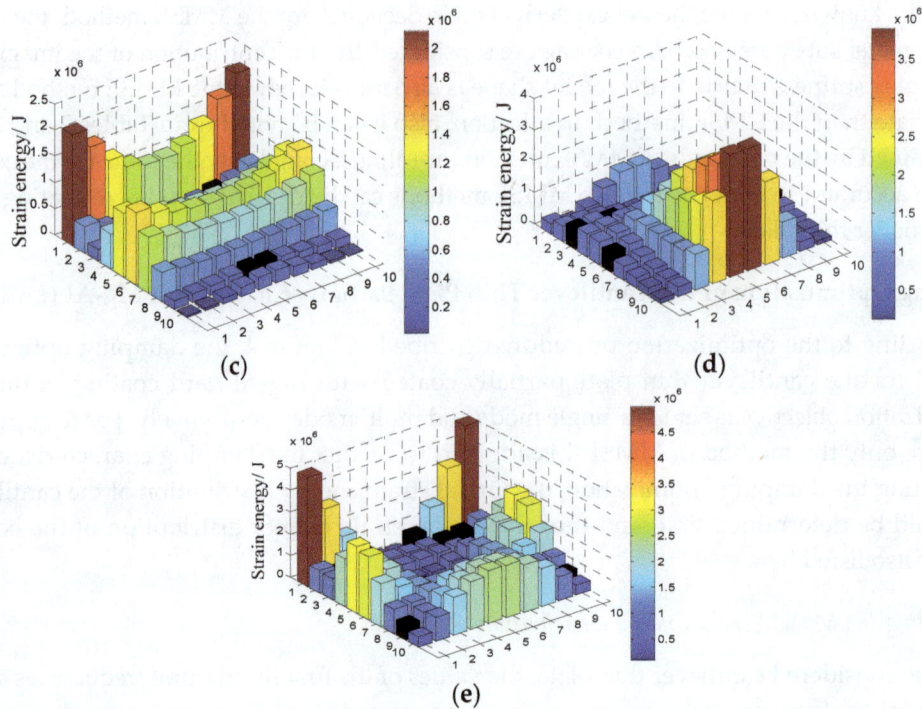

Figure 7. The modal strain energy distribution of the first five orders of cantilever thin plate: (**a**) The first order; (**b**) the second order; (**c**) the third order; (**d**) the forth order; (**e**) the fifth order.

5.2. Damping Optimization for a Single Mode

Here, the first order and fourth order are taken as examples to describe the optimization of coating location for a single mode and only the optimization schemes are shown for the other orders. An optimization criterion can be used as the reference for conducting the single-mode optimization. It is that coating on elements for which accumulated modal strain energy is greater than 40% of the total modal strain energy of the structure.

5.2.1. Damping Optimization for the First Order

It can be seen in Figure 7a, for the first order, the higher modal strain energy of the cantilever thin plate appears on the edge of the fixed end. As a result, a total of 20 elements along the constraint end of the cantilever thin plate are chosen as the coating locations, which are obviously marked by the red shadow area in Figure 8. The calculation shows that cumulated modal strain energy of these coating elements is 59.54% of the total modal strain energy. For this partially coated thin plate, the first five modal loss factors are obtained by MMSE method and listed in Table 5. For comparison, the results corresponding to the full coating structure are also listed in Table 5.

Figure 8. The damping optimization scheme for the first order.

Table 5. Comparison between the damping optimization results for the first order and the results corresponding to the full coating structure.

Orders	Partial Coating/10^{-3}	Full Coating/10^{-3}
1	0.76342	0.80789
2	0.73051	0.80881
3	0.73041	0.80824
4	0.70967	0.80799
5	0.72358	0.80849

As shown in Table 5, when the damping optimization for the first order shown in Figure 8 is performed, the first order modal loss factor of cantilever thin plate is greater than those of the other orders, which means that this scheme can effectively suppress the first order resonance. In addition, compared with the results of full coating, the first order modal loss factor of the partial coating structure decreases only by 5.5%, but the coating area under this scheme decreases by 80%. These results indicate that the vibration reduction scheme shown in Figure 8 is feasible, even if only to suppress the first order resonance of cantilever thin plate.

5.2.2. Damping Optimization for the Fourth Order

It can be seen in Figure 7d, for the fourth order, the higher modal strain energy of cantilever thin plate appears on the middle locations close to the free end. Then the damping optimization scheme for the fourth order can be achieved and shown in Figure 9. A total of 20 elements in the middle locations of the free end are chosen as coating elements which are described by red shadow shown in Figure 9. The calculation shows that cumulated modal strain energy of these coating elements is 49.92% of the total modal strain energy. Similarly, the first five modal loss factors corresponding to this damping optimization scheme are obtained by MMSE method and are listed in Table 6.

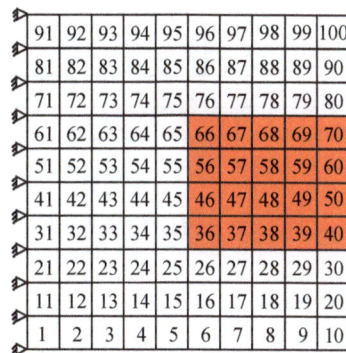

Figure 9. The damping optimization scheme for the fourth order.

Table 6. Comparison between the damping optimization results for the fourth order and the results corresponding to the full coating structure.

Orders	Local Coating/10^{-3}	Full Coating/10^{-3}
1	0.70359	0.80789
2	0.72194	0.80881
3	0.72253	0.80824
4	0.75706	0.80799
5	0.72615	0.80849

It can be noted from Table 6, when the damping optimization for the fourth order shown in Figure 9 is performed, that the fourth order modal loss factor of cantilever thin plate is greater than those of the other orders, which means that this scheme can effectively suppress the fourth order resonance. In addition, compared with the results of full coating, the fourth order modal loss factor of the partial coating structure decreases only by 6.3%, but the coating area under this scheme decreases by 80%. These results indicate that the vibration reduction scheme shown in Figure 9 is acceptable, if only to suppress the fourth order resonance of cantilever thin plate.

5.2.3. Damping Optimization for the Other Orders

The optimizations for the second, third, and fifth orders have been performed, but the detailed contents were omitted here because the study process was almost consistent with the aforementioned orders. Only the damping optimization schemes are presented and shown in Figure 10.

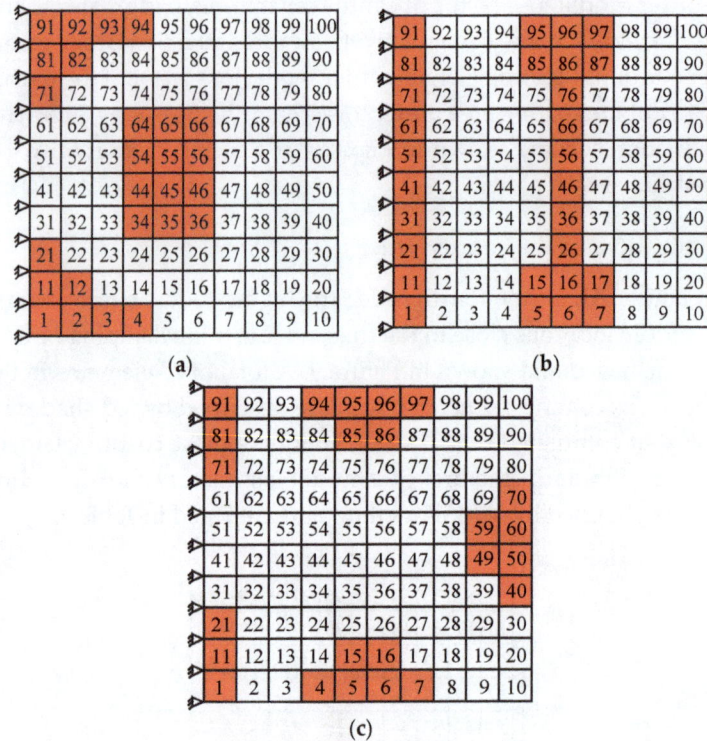

Figure 10. The damping optimization schemes for (a) the second order, (b) the third, and (c) fifth order.

5.3. Damping Optimization for Multi Modes

In the following, the damping optimization simultaneously for the third, fourth, and fifth order are taken as examples to demonstrate the damping optimization method for multi modes of the cantilever thin plate. Referring to the Table 4, it can be known that the third, fourth, and fifth order resonance frequencies are all in the frequency range of 1200–1800 Hz, which belongs to the higher frequency region, so the damping optimization simultaneously for the third, fourth, and fifth order has a certain engineering significance. When performing the damping optimization for multi modes, the location of higher modal strain energy corresponding to each order should be chosen as coating elements. It has been shown in Figure 7 that the locations of higher modal strain energy of the third and fifth orders appear on the two sides of the restrained end and locations of the fourth order are in the middle of the free end. As a result, the final optimization scheme is determined referring to the aforementioned higher modal strain energy distribution and shown in Figure 11. For this optimization scheme, a total of 46 elements are selected as coating elements marked by red shadow. Then, the first five modal loss factors are calculated by MMSE method and the results are listed in Table 7. For comparison, the single-mode optimization results for the relevant orders and the results corresponding to the full coating structure are all listed in Table 7.

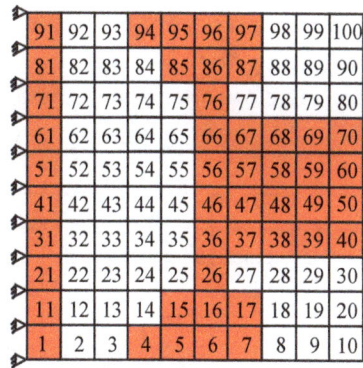

Figure 11. The damping optimization scheme simultaneously for the third, fourth, and fifth order.

Table 7. Comparison among the damping optimization results for a certain order, for multi orders, and the results corresponding to the full coating structure.

Orders	For multi Orders/10^{-3}	For the Third Order/10^{-3}	For the Fourth Order/10^{-3}	For the Fifth Order/10^{-3}	Full Coating/10^{-3}
1	0.74736	0.74341	0.70359	0.72828	0.80789
2	0.75566	0.73774	0.72194	0.73230	0.80881
3	0.77388	0.75742	0.72253	0.74136	0.80824
4	0.77133	0.72271	0.75706	0.71849	0.80799
5	0.77106	0.74311	0.72615	0.74308	0.80849

It can be found from Table 7 that the damping optimization scheme simultaneously for the third, fourth, and fifth order can make the damping values of the three orders closer to that of full coating structure. In the meantime, it can also make the first and second order have higher damping values. Although this kind of scheme uses more hard-coating material than the scheme for a certain order, the coating area still decreases by 54% compared with the full coating. Thus, during the damping design practice, if there are no special requirements (for example, just considering the suppression of the resonance of a certain order), the damping optimization for multi modes should be chosen preferentially, because this optimization objective can achieve a better vibration reduction effect.

5.4. The Influence of Coating Area on Damping Performance of Thin Plate

To better implement the damping optimization, the influence of hard coating area on damping performance of the whole composite plate is further discussed as follows. The hard coating is still coated on the location where the modal strain energy is higher, and then the influence of the changes of coating element number, which is corresponding to the coating area, on the damping performance of the composite plate is discussed. Here, the damping optimization scheme for the fourth order is chosen as an example to analyze the influence of coating area. In Section 5.2, a total of 20 elements in the middle of the free end are chosen to suppress the fourth order resonance of cantilever thin plate. Now the number of coating elements is increased to 36 and the relevant damping scheme is shown in Figure 12. Compared with the scheme shown in Figure 9, the coating area of this scheme increases by 80%. The first five modal loss factors are calculated by MMSE method and listed in Table 8. For comparison, the optimization results of the original scheme and the full coating structure are also listed in Table 8.

It can be noted from Table 8 that when the coating area increases by 80%, compared with the original scheme, the modal loss factors of partial coating plate increase, but the rate of increase is not very evident. Taking the fourth order modal loss factor as an example, the modal loss factor has increased only by 2.1%. Analogous results have been obtained, also, for the other orders. Therefore, an important conclusion can be drawn, and that is, the damping effect of the composite plate is not sensitive to the coating area. This conclusion is very favorable for vibration reduction using hard coating, which indicates that only choosing the zone with higher modal strain energy to deposit hard coating material, rather than paying close attention to the coating area, can bring a satisfactory damping effect.

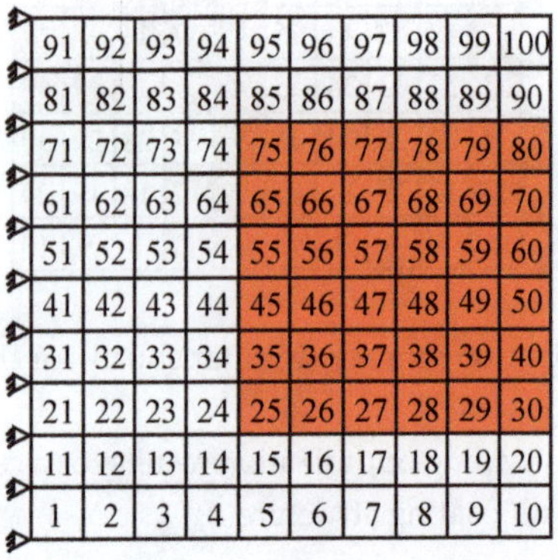

Figure 12. The new damping optimization scheme for the fourth order (bigger coating area).

Table 8. Comparison of modal loss factors for the new, original scheme and the full coating structure.

Orders	New Scheme/10^{-3}	Original Scheme/10^{-3}	Full Coating/10^{-3}
1	0.70740	0.70359	0.80789
2	0.73552	0.72194	0.80881
3	0.73854	0.72253	0.80824
4	0.77316	0.75706	0.80799
5	0.74076	0.72615	0.80849

6. Conclusions

This paper proposes the MMSE method to predict the damping characteristics of hard-coating composite structure. Then, the damping optimization method about coating location is studied. Some important conclusions are listed as follows:

- The thin plate coated with Mg-Al hard coating was chosen to demonstrate the developed MMSE method and the uncertainties of the loss factors of hard coating obtained by vibration beam method were also considered. All the calculation results show that the results obtained by CMSE and MMSE method are obviously different. Because more influencing factors are contained, the calculation accuracy of the MMSE method should be higher than that of the CMSE method.

- Compared with the full coating, a similar damping effect can be obtained by partial coating using the proposed optimization method. For example, in the study of the damping optimization for the first order and for the fourth order, under the premise of decreasing the coating area by 80%, the loss factor of the objective order decreased by only 5.5% and 6.3%, respectively.

- Compared with the damping optimization for a certain order, it can be found that if there are no special requirements (for example, just considering the suppression of the resonance of a certain order), the damping optimization for multi modes should be chosen preferentially in the actual damping design, because this optimization objective can achieve a better vibration reduction effect.

- The damping effect of composite structure is not sensitive to the coating area. This conclusion is very favorable for vibration reduction using hard coating, which indicates that only choosing the zone with higher modal strain energy to deposit hard coating material, rather than paying close attention to the coating area, can bring a satisfactory damping effect.

Acknowledgments: This project was supported by National Natural Science Foundation of China (Grant No. 51375079) and the Fundamental Research Funds for the Central Universities of China (Grant No. N140301001).

Author Contributions: Wei Sun and Rou Liu carried out the modeling and simulation of the work, Wei Sun analyzed the calculation results and wrote the paper.

Conflicts of Interest: The authors declare no conflict of interest.

References

1. Limarga, A.M.; Duong, T.L.; Gregori, G.; Clarke, D.R. High-temperature vibration damping of thermal barrier coating materials. *Surf. Coat. Technol.* **2007**, *202*, 693–697. [CrossRef]
2. Grzesik, W.; Zalisz, Z.; Nieslony, P. Friction and wear testing of multilayer coatings on carbidesubstrates for dry machining applications. *Surf. Coat. Technol.* **2002**, *155*, 37–45. [CrossRef]
3. Fernández-Abia, A.I.; Barreiro, J.; de Lacalle, L.N.L.; Martínez-Pellitero, S. Behavior of austenitic stainless steels at high speed turning using specific force coefficients. *Int. J. Adv. Manuf. Technol.* **2012**, *62*, 505–515. [CrossRef]
4. Patsias, S.; Tassini, N.; Lambrinou, K. Ceramic coatings: Effect of deposition method on damping and modulus of elasticity for yttria-stabilized zirconia. *Mater. Sci. Eng. A* **2006**, *442*, 504–508. [CrossRef]
5. Rodríguez-Barrero, S.; Fernández-Larrinoa, J.; Azkona, I.; Lacalle, L.N.L.D.; Polvorosa, R. Enhanced Performance of Nanostructured Coatings for Drilling by Droplet Elimination. *Mater. Manuf. Process.* **2016**, *31*, 593–602. [CrossRef]
6. Blackwell, C.; Palazotto, A.; George, T.J.; Cross, C.J. The evaluation of the damping characteristics of a hard coating on titanium. *Shock Vib.* **2007**, *14*, 37–51. [CrossRef]
7. Ivancic, F.; Palazotto, A. Experimental considerations for determining the damping coefficients of hard coatings. *J. Aerosp. Eng.* **2005**, *18*, 8–17. [CrossRef]
8. Johnson, C.D.; Kienholz, D.A. Finite element prediction of damping in structures with constrained viscoelastic layers. *AIAA J.* **1982**, *20*, 1284–1290.
9. Hwang, S.J.; Gibson, R.F. The use of strain energy-based finite element techniques in the analysis of various aspects of damping of composite materials and structures. *J. Compos. Mater.* **1992**, *26*, 2585–2605. [CrossRef]
10. Casadei, F.; Bertoldi, K.; Clarke, D.R. Finite element study of multi-modal vibration damping for thermal barrier coating applications. *Comput. Mater. Sci.* **2013**, *79*, 908–917. [CrossRef]
11. Patsias, S.; Saxton, C.; Shipton, M. Hard damping coatings: An experimental procedure for extraction of damping characteristics and modulus of elasticity. *Mater. Sci. Eng.* **2004**, *370*, 412–416. [CrossRef]
12. Hu, B.H.; Dokainish, M.A.; Mansour, W.M. A modified MSE method for viscoelastic systems: A weighted stiffness matrix approach. *J. Vib. Acoust.* **1995**, *117*, 226–2231. [CrossRef]
13. Kumar, N.; Singh, S.P. Experimental study on vibration and damping of curved panel treated with constrained viscoelastic layer. *Compos. Struct.* **2010**, *9*, 233–243. [CrossRef]
14. Moreira, R.A.S.; Rodrigues, J.D. Partial constrained viscoelastic damping treatment of structures: A modal strain energy approach. *Int. J. Struct. Stab. Dyn.* **2006**, *6*, 397–411. [CrossRef]
15. Masti, R.S.; Sainsbury, M.G. Vibration damping of cylindrical shells partially coated with a constrained viscoelastic treatment having a standoff layer. *Thin-Walled Struct.* **2005**, *43*, 1355–1379. [CrossRef]
16. Sainsbury, M.G.; Masti, R.S. Vibration damping of cylindrical shells using strain-energy-based distribution of an add-on viscoelastic treatment. *Finite Elements Anal. Des.* **2007**, *43*, 175–192. [CrossRef]
17. Kung, S.W.; Singh, R. Complex eigensolutions of rectangular plates with damping patche. *J. Sound Vib.* **1998**, *216*, 1–28. [CrossRef]
18. Sun, W.; Han, Q.; Qi, F. Optimal design of damping capacity for hard-coating thin plate. *Adv. Vib. Eng.* **2013**, *12*, 179–192.
19. *E756-04 Standard Test Method for Measuring Vibration-Damping Properties of Materials*; American Society for Testing and Materials: New York, NY, USA, 2004.

5

Ozone Resistance, Water Permeability, and Concrete Adhesion of Metallic Films Sprayed on a Concrete Structure for Advanced Water Purification

Jin-Ho Park, Jitendra Kumar Singh and Han-Seung Lee *

Department of Architecture Engineering, Hanyang University, 55, Hanyangdaehak-ro, Sangrok-gu, Ansan-si, Gyeonggi-do 15588, Korea; jinho9422@naver.com (J.-H.P.); jk200386@hanyang.ac.kr (J.K.S.)
* Correspondence: ercleehs@hanyang.ac.kr

Academic Editor: Paul Lambert

Abstract: We evaluated the applicability of metal spray coating as a waterproofing/corrosion protection method for a concrete structure used for water purification. We carried out an ozone resistance test on four metal sprays and evaluated the water permeability and bond strength of the metals with superior ozone resistance, depending on the surface treatment method. In the ozone resistance test, four metal sprays and an existing ozone-proof paint were considered. In the experiment on the water permeability and bond strength depending on the surface treatment method, the methods of no treatment, surface polishing, and two types of pore sealing agents were considered. The results showed that the sprayed titanium had the best ozone resistance. Applying a pore sealing agent provided the best adhesion performance, of about 3.2 MPa. Applying a pore sealing agent also provided the best waterproofing performance. Scanning electron microscope analysis showed that applying a pore sealing agent resulted in an excellent waterproofing performance because a coating film formed on top of the metal spray coating. Thus, when using a metal spray as waterproofing/corrosion protection for a water treatment concrete structure, applying a pore sealing agent on top of a film formed by spraying titanium was concluded to be the most appropriate method.

Keywords: metal spray system; advanced water treatment; ozone resistance; water permeability; bond strength

1. Introduction

At present, water is purified by using chlorine. Chlorine input processes in the mixing tank, coagulation tank, sedimentation tank, and filtering tank remove harmful materials (e.g., Pb, As, Sn, and Cd) from the water supply and acidic materials from the wastewater, which are otherwise harmful to living things [1,2]. However, some harmful materials cannot be treated by the existing purification method. Thus, the use of an advanced water purification method that uses ozone to effectively remove such harmful materials has recently been increasing. The ozone used in advanced water purification facilities forms a stable oxygen molecule in water and generates one oxygen radical. Then, the oxygen radical generates two hydroxyl radicals (OH$^-$) with a strong oxidizing power by decomposing a polar molecule of water. Organic materials react with ozone directly or with the hydroxyl radical (OH$^-$) generated by ozone decomposition, which changes their properties. This is called oxidation [3–9]. Advanced water purification uses the properties of ozone to purify wastewater, producing clean water [10]. The oxidation of ozone and strong oxidizing power of harmful acids in wastewater affect the quality of the concrete of the water treatment structure. This can become a cause of deterioration, resulting in the waterproofing/anti-corrosion material of the concrete structure breaking away and causing the concrete to crack [1,2,11–14]. In 2010, Korea established recommended standards for ozone-proof paint to resolve this problem, and the development of organic/inorganic paint materials has been studied [15]. However, existing waterproofing/anti-corrosion materials (i.e., organic materials based on epoxy resins) cause corrosion, breakaway, exfoliation, and cracks in the finishing material, in an environment with high concentrations of ozone and chemicals; they can even affect the concrete itself and degrade the long-term durability [16,17]. To address this problem of organic/inorganic paint materials, the stainless steel (SUS) panel method has recently been applied to some facilities; this method uses SUS panels with excellent waterproofing/corrosion protection performance, ozone resistance, and chemical resistance [18,19]. However, the SUS panel method has problems with adhesion to concrete structures and corrosion caused by the welding of the joint sections, and it requires very challenging construction technology. In addition, there are enormous costs for the initial construction and maintenance, and there are further costs for waste disposal and construction during repairs because of the complete removal and reconstruction of the paint film [1].

In this study, we evaluated the ozone resistance of metal spray coatings and examined the bond strength, water impermeability, and impact resistance depending on the surface treatment method, in order to develop a finishing method that can fundamentally prevent the deterioration of concrete structures in water treatment facilities. Our hope was to demonstrate the feasibility of metal sprays for constructing metal panels with excellent ozone resistance and chemical resistance more easily than the existing construction method. The basic data from this study should help with the development of a finishing method for water treatment concrete structures.

2. Experimental Plan

2.1. Outline

Table 1 outlines the experiment and experimental factors. A basic procedure was conducted to apply the metal spray to a concrete structure for water treatment. Figure 1 shows that the arc thermal spraying process using the coating was applied on the concrete surface with a circular slit of hot and compressed air [20,21]. Four kinds of metal sprays were selected, and their ozone resistance was evaluated. A zinc–aluminum alloy, SUS304 (commonly used structural steel), SUS316L (commonly used with the SUS panel method), and titanium (high corrosion resistance) were considered, along with the existing ozone-proof paint. Then, the adhesion performance of the metal with the best ozone resistance was evaluated. Because water treatment concrete structures are in contact with water at all times, ozonated water or moisture infiltrates the concrete and finishing material, which degrades the adhesion performance and durability. Thus, the water impermeability of the metal spray coating is an important factor. However, because a metal spray forms a film from stacking solid metal particles on the concrete in powder form, many pores exist between particles and on the surface [22–26]. Thus, different surface treatments were performed on the metal spray coating, and their effects on the water impermeability were evaluated. Concrete with a compressive strength of 24 MPa was used as the substrate, and the metal spray coating was 200 μm thick, which is the average thickness used in construction.

Table 1. Experimental factors and measurement catalogs. Substrate: concrete (24 MPa); Coating thickness: 200 μm.

No.	Experimental Factor	Specimen	Measurement Catalog	Size
1		Zn/Al		
2		SUS304	Ozone resistance	100 mm × 100 mm × 30 mm;
3	Metal sprays	SUS316L	(change in appearance,	150 mm × 70 mm × 2 mm
4		Ti	weight reduction)	
5		Ozone-proof paint		

Table 1. *Cont.*

No.	Experimental Factor	Specimen	Measurement Catalog	Size
6		Sprayed		
7	Surface treatment method	Abraded	Bond strength,	300 mm × 300 mm × 50 mm;
8		Sealed (A) [a]	water permeability	Ø100 mm × 30 mm
9		Sealed (B) [b]		

[a] Epoxy-based sealing; [b] Teflon-based sealing.

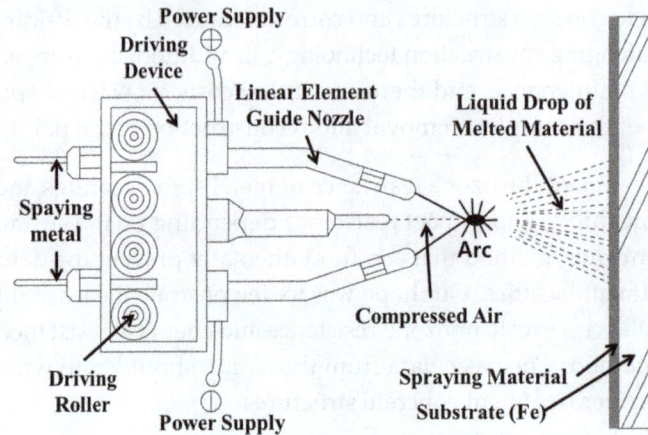

Figure 1. Schematic of the arc thermal spraying process.

2.2. Ozone Treatment

Ozone treatment was carried out in accordance with the SPS KWWA M211 method provided by the Korea Water and Wastewater Works Association [27]. The dissolved ozone concentration of the ozonated water was maintained at 10 ± 1 ppm by using an ozone tester that can be applied to both underwater and airborne settings. The temperature of the chamber was regulated according to an 8-h cycle, comprising 3 h at 20 ± 2 °C, 3 h at 40 ± 2 °C, and 1 h of transition time between each temperature setting. The cycle was repeated 84 times. The ozonated water level was regulated according to an 8-h cycle, comprising a full water level for 1 h and a low water level for 1 h, as shown in Figure 2. This cycle was repeated 84 times and the experiment was then conducted, after the specimens had been washed with clean water.

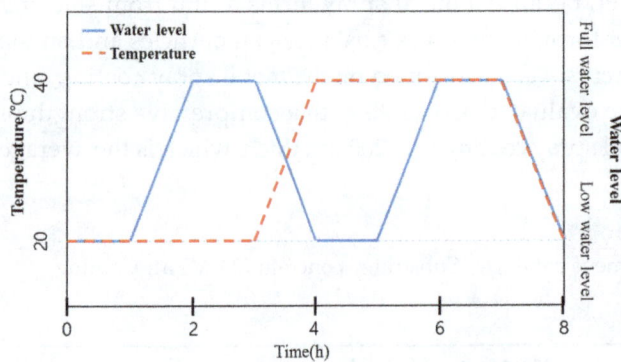

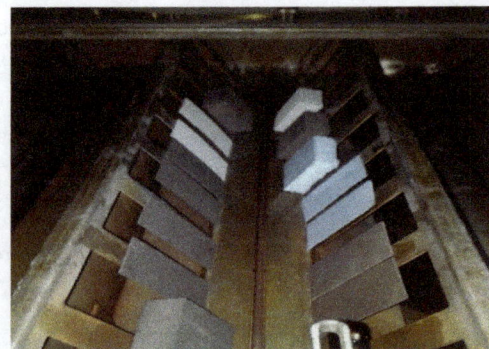

(a)   (b)

Figure 2. (a) Cyclic condition of ozone treatment (Ozone concentration: 10 ± 1 ppm) and (b) view of the ozone resistance test.

2.3. Specimen Production

Test pieces were produced with different sizes using 24 MPa concrete, as shown in Figure 3. To secure the bond strength between the metal spray coating and the concrete, the surfaces of the prepared concrete were sand-blasted. To increase the tensile strength of the concrete surfaces after the sand-blasting treatment, a pervious surface hardener was applied. Then, a surface roughness agent was applied, to increase the roughness. The different surface treatments were then applied to the test pieces.

| Sand blast | pervious surface hardener | Surface Roughness Agent | Metal Spraying | Specimen |

Figure 3. Specimen production process.

2.4. Method

2.4.1. Weight Reduction after Ozone Treatment

The weight of each specimen was measured before ozone treatment in accordance with SPS KWWA M211 [27], by using a scale with a precision of one decimal place. After ozone treatment, each specimen was cleaned using running distilled water and dried in a 60 °C chamber for 24 h or longer. Then, the specimen was stored under standard conditions for 30 min, before the weight was measured with the same method used before ozone treatment. The weight reduction rate was calculated as follows:

$$g = \frac{g_1 - g_2}{A} \tag{1}$$

where g is the weight reduction (g/m^2), g_1 is the weight of the specimen before ozone treatment (g), g_2 is the weight of the specimen after ozone treatment (g), and A is the area of the specimen (m^2).

2.4.2. Appearance after Ozone Treatment

The appearance was evaluated by visual observation in accordance with SPS KWWA M211 [27]. The test pieces produced by spraying metal on all six faces were put through 84 cycles of preprocessing in accordance with the ozone treatment method and were then visually observed for phenomena such as cracks, swelling, pinholes, rust, or discoloration on the surface.

2.4.3. Bond Strength Depending on the Surface Treatment Method

The bond strength depending on the surface treatment method was evaluated in accordance with KS F 4716 [28]. For the tensile bond strength test, a square attachment measuring 40 mm × 40 mm was glued onto each surface-treated specimen by using an epoxy adhesive, as shown in Figure 4. After 24 h, the perimeter of the attachment was cut to the concrete surface, and the maximum load was measured. Then, the bond strength was calculated with Equation (2). An average of nine measurements was regarded as sufficient for calculating the bond strength, and the adhesion failure behavior was visually checked.

$$\text{Bond Strength} \left(\text{N/mm}^2 \right) = \frac{T}{A} \tag{2}$$

Here, T is the maximum tensile load (N) and A is the attachment area (1600 mm^2).

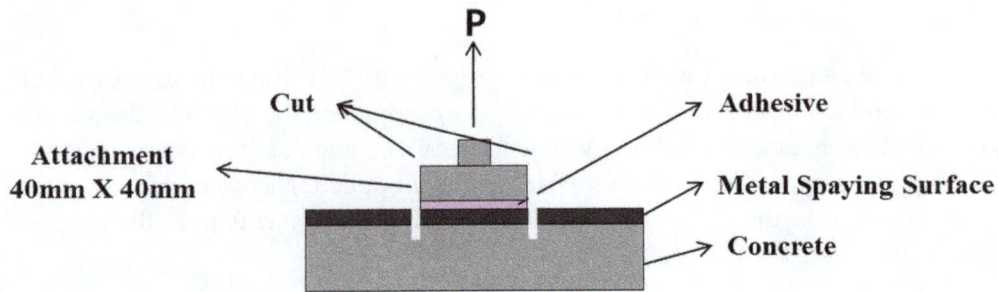

Figure 4. Outline of the bond strength test.

2.4.4. Impermeability Depending on the Surface Treatment Method

The water impermeability was evaluated in accordance with KS F 4919 [29]. The lateral faces, excluding the surface on which the metal was sprayed and the opposite face, were completely sealed with an epoxy resin. A water pressure of 0.3 N/mm^2 was applied to the specimen for 3 h by using the equipment shown in Figure 5. Then, moisture was lightly removed for about 10 s by using a piece of filter paper [30], as per KS M 7602. The mortar bottom plate beneath the sprayed metal surface at the center of the specimen was visually checked to see if it was wet with water. The difference between the weight before the water pressure was applied (W_1) and the weight after the water pressure was applied (W_2) was set as the amount of water permeation based on KS F 4930 [31], and the permeability ratio was calculated as follows:

$$\text{Permeability ratio} = \frac{\text{Amount of water permeation into the specimen onto which metal was sprayed (g)}}{\text{Amount of water permeation into the specimen on which no metal was sprayed (g)}} \quad (3)$$

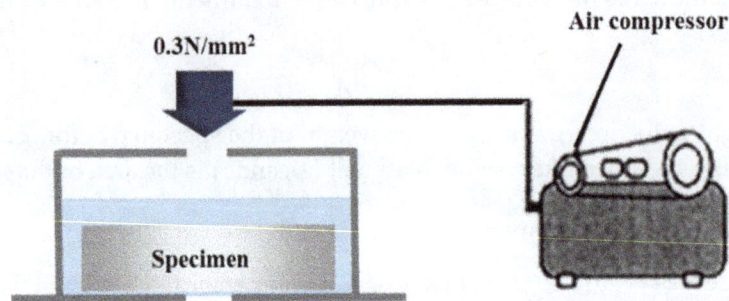

Figure 5. Outline of the impermeability test.

3. Results and Discussion

3.1. Weight Reduction after Ozone Treatment

To determine the metal spray with the highest ozone resistance, the weight reduction after ozone treatment was evaluated. Figure 6 shows the results. The Zn/Al alloy showed the largest weight reduction of 340.4 g/m^2, which confirmed that it is most vulnerable to ozone. SUS304 and SUS316L, which are known to have a strong chemical resistance, showed reductions of 14.2 and 17.8 g/m^2, respectively. These materials are vulnerable to ozone because the metal is arc-discharged at a high temperature when sprayed, and the Fe–Ni–Cr structure of stainless steel is destroyed when heat-treated at a high temperature. This causes the Fe component to move to the surface of the sprayed metal in large quantities [32]. The ozone-proof paint showed a weight reduction of 8.3 g/m^2. Ti showed almost no weight reduction, with a value of 1.3 g/m^2. Thus, only Ti and the ozone-proof paint satisfied the performance criterion of SPS KWWA M211 (i.e., 10 g/m^2).

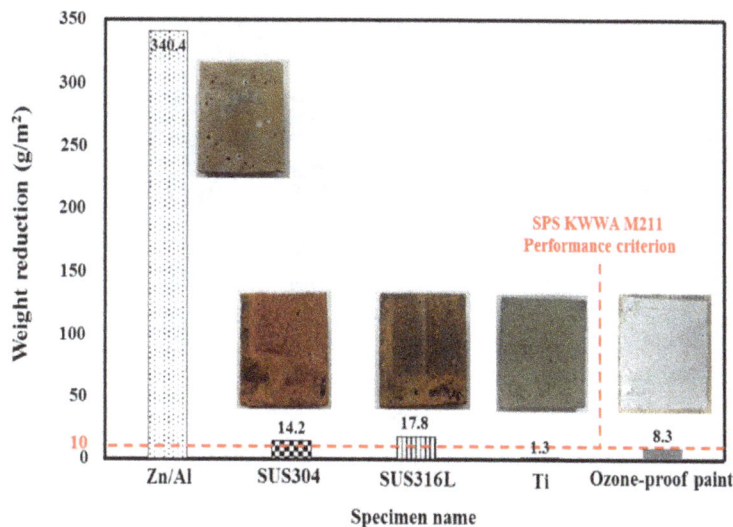

Figure 6. Weight reduction after ozone treatment depending on the sprayed metal.

3.2. Change in Appearance after Ozone Treatment

Table 2 presents the changes in appearance of the specimens before and after ozone treatment. The Zn/Al specimen showed the most serious deterioration. The SUS304 specimen was partially exfoliated and heavily rusted. The SUS316L specimen was also visually confirmed to have rusted. The specimen may have failed to withstand the ozone because the Fe–Ni–Cr structure was destroyed by heat treatment, similar to the case of the weight reduction, which increased the intergranular corrosion susceptibility [32]. In the case of the ozone-proof paint, although there was no exfoliation, cracks, or rust, the specimen was discolored. Thus, the ozone-proof paint may be unable to withstand high concentrations of ozone for a long time. The Ti specimen showed no exfoliation, cracks, rust, or discoloration. Thus, it had the highest ozone resistance.

Table 2. Change in appearance after ozone treatment depending on the metal spray.

Ozone Treatment	Zn/Al	SUS304	SUS316L	Ti	Ozone-Proof Paint
Before					
After					
Phenomenon	Exfoliation, discoloration	Rust, discoloration	Rust, discoloration	No change	Discoloration

3.3. Adhesion Performance of the Metal Spray Depending on the Surface Treatment Method

After the ozone resistance of different metal sprays had been evaluated, the bond strength depending on the surface treatment method was evaluated. Ti was used to test the bond strength because it had the best ozone resistance. Figure 7 presents the results depending on the surface treatment method.

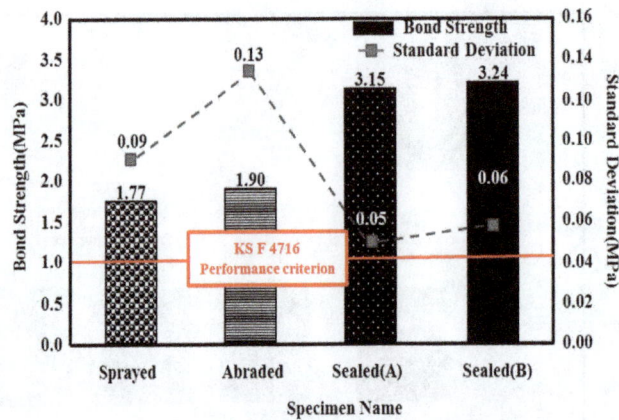

Figure 7. Bond strength depending on the surface treatment method

The sealed (A) and sealed (B) specimens showed a superior bond performance to the sprayed specimen, by about 1.8 times (3.15 and 3.24 MPa, respectively), regardless of the type of pore sealing agent. This may be because many pores are generated on the film by the metal spray method [25]; when a pore sealing agent is applied to such a film, it permeates and plays the role of an adhesive between the metal spray coating and concrete, as shown in Figure 8. This enhances the adhesion performance. The sprayed specimen showed the lowest bond strength of 1.77 MPa. However, it satisfied the bond strength performance criterion of 1.1 MPa, as stated in in KS F 4716. Although the abraded specimen showed a bond strength of 1.90 MPa, which was higher than that of the sprayed specimen, the standard deviation was relatively high, at about 0.13. This may be because the error range of the experimental values increased due to the impact applied to the film during the surface polishing process.

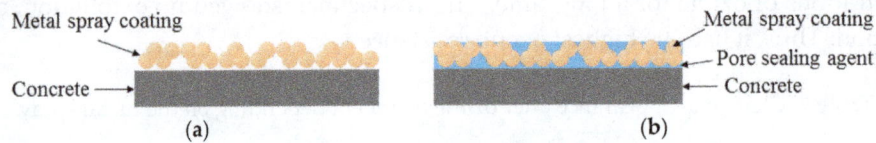

Figure 8. Mimetic diagram of the pore sealing agent. (**a**) Sprayed mimetic diagram; (**b**) Sealed mimetic diagram.

Figure 9 shows the failure modes of the specimens after the bond strength test. The sprayed specimen with a bond strength of 1.77 MPa showed failure between the metal spray coating and concrete. The abraded specimen showed the same failure mode. However, both the sealed (A) and sealed (B) specimens showed non-interfacial failure, where the failure occurred in the concrete. Thus, within the scope of this study, applying a pore sealing agent was found to be very effective at securing the bond strength, regardless of the agent type.

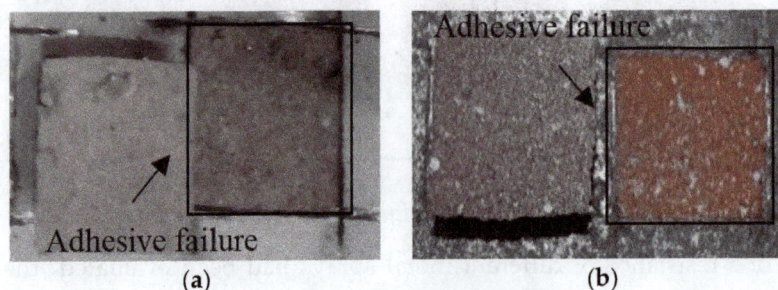

Figure 9. *Cont.*

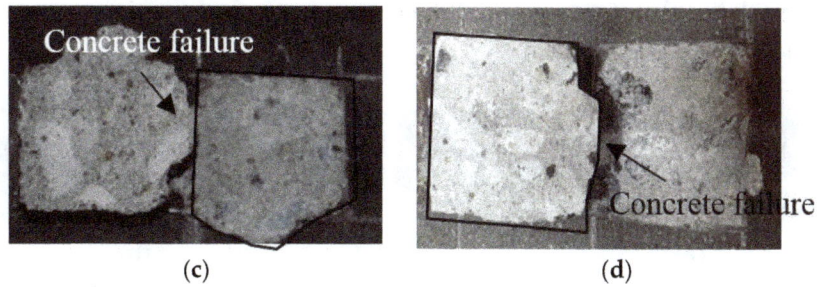

(c)   (d)

Figure 9. Failure mode depending on the surface treatment method: (**a**) sprayed, (**b**) abraded (**c**) sealed (A), and (**d**) sealed (B).

3.4. Impermeability of the Metal Spray Depending on the Surface Treatment Method

Because a water treatment concrete structure is in contact with water at all times, the water impermeability of the metal spray coating is an important factor. Figure 10 shows the related ratio of water permeability. The water permeation of the mortar specimen was about 12.1 g, and the sprayed specimen showed the highest water permeation, at about 11.0 g. This may be because water permeated through the pores in the film formed by the metal spray method, as shown in Figure 7a [25]. The water permeation of the sealed (B) specimen was about 0.8 g, which showed a 92.6% improvement in the water impermeability compared to the sprayed specimen. The water permeation of the sealed (B) specimen was 1.0 g, which showed a 91.1% improvement. The water permeation of the abraded specimen was about 6.4 g, which revealed a 42.9% improvement.

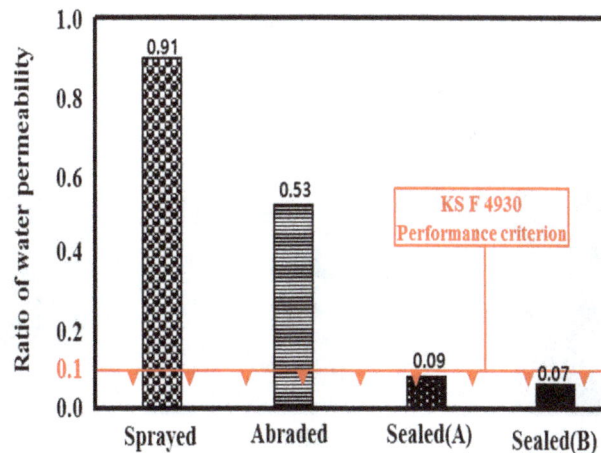

Figure 10. Permeability ratio depending on the surface treatment method.

Based on the experimental results, the permeability ratios were in the order of sprayed > abraded > sealed (A) > sealed (B). Both of the sealed specimens satisfied the KS F 4930 performance criterion, exhibiting a ratio of less than 0.1. A visual check showed that water permeated the whole area of the mortar specimen, as shown in Figure 11. In the case of the sprayed specimen, water permeated to a depth of about 22.5 mm. Thus, its permeability was 25% of that of the mortar specimen. In the case of the abraded specimen, water permeated to a depth of about 7.5 mm, so the water permeability was about 66.6% of that of the sprayed specimen and about 75% of that of the mortar specimen. In contrast, although the sealed (A) and sealed (B) specimens showed changes in weight after the experiment, the visual check showed that no water permeated the specimens. Thus, the impermeability depending on the surface treatment method of the sprayed metal was in the order of sealed (A) = sealed (B) > abraded > sprayed. No difference was found regarding the type of pore sealing agent. This may be because the pore sealing agents formed a thin film on the metal spray coating to prevent water from permeating the specimens.

Figure 11. Results of the impermeability test: (**a**) mortar; (**b**) sprayed; (**c**) abraded; (**d**) sealed (A); and (**e**) sealed (B).

3.5. Analysis of the Surface and Cross-Section of the Sprayed Metal Depending on the Surface Treatment Method

Based on the results of the impermeability test, the specimens to which a pore sealing agent was applied were found to be superior. The sprayed metal surfaces of each specimen were compared by using a scanning electron microscope (SEM) and digital optical microscope. Figure 12 shows the surface images taken with the digital optical microscope, and Figures 13 and 14 show the surface and cross-sectional images taken with the SEM.

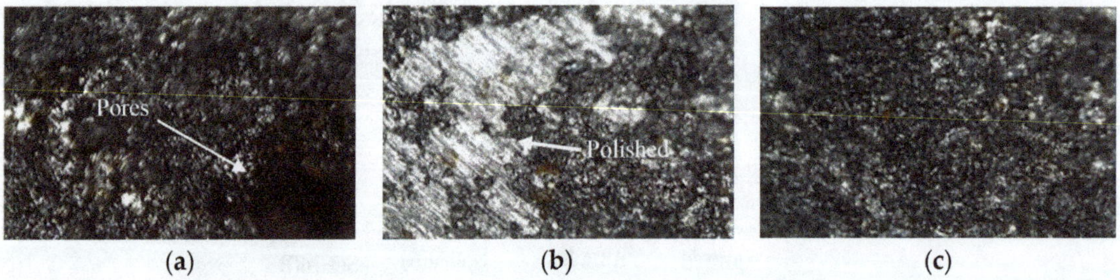

Figure 12. Surface analysis using a digital optical microscope (X160): (**a**) sprayed; (**b**) abraded; and (**c**) sealed.

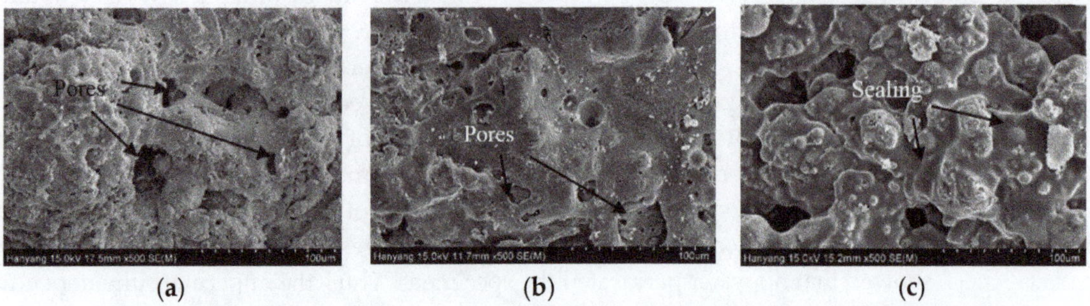

Figure 13. Surface analysis using an SEM: (**a**) sprayed; (**b**) abraded; and (**c**) sealed.

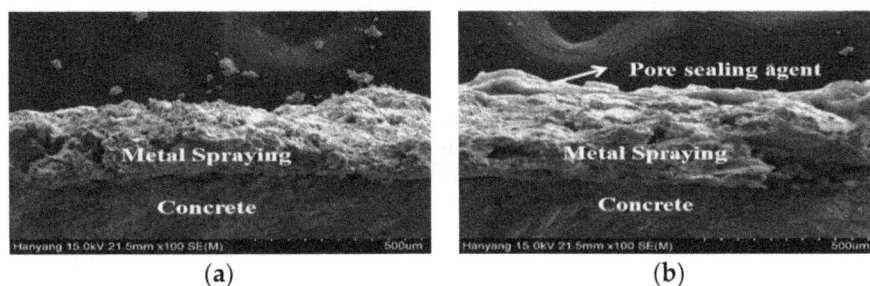

Figure 14. Cross-sectional analysis using an SEM: (**a**) sprayed and (**b**) sealed.

4. Conclusions

The following conclusions were derived after evaluating the ozone resistance, bond strength, and impermeability of metal sprays for their applicability to finishing water treatment concrete structures:

- With regard to the ozone resistance, Ti showed the smallest weight reduction of 1.3 g/m^2 among the metal sprays and no observable deterioration.

- The stainless steel family (SUS304 and SUS316L) showed large weight reductions and deterioration phenomena, such as rust and exfoliation, in contrast to Ti. These materials may have failed against ozone because the Fe–Ni–Cr structure was destroyed when they were heat-treated, due to the characteristics of the metal spray method. This caused the Fe component to move to the surface. Thus, Ti is the most suitable metal spray for finishing a water treatment facility.

- With regard to the bond strength depending on the surface treatment method, all of the specimens satisfied the KS standard. Using a pore sealing agent (i.e., sealed (A) and sealed (B)) produced the best adhesion performance. This may be because the pore sealing agent permeates the pores generated by the metal spray method and acts as an adhesive to enhance the adhesion performance.

- When the failure modes were compared, while the sprayed and abraded specimens had a relatively low bond strength and showed interfacial failure between the metal spray coating and concrete, the sealed (A) and sealed (B) specimens exhibited a high bond strength and showed non-interfacial failure that occurred in the concrete.

- With regard to the impermeability depending on the surface treatment method, the sprayed specimen showed the lowest impermeability, similar to the case of the bond strength. This may be because water permeates through the pores generated by the metal spray method.

- The sealed (A) and sealed (B) specimens showed that no water permeated the structure. This may be because the pore sealing agents formed a thin film on the metal spray coating that prevented water permeation.

- Surface analysis confirmed many pores on the surface of the sprayed specimen, while the pores were filled up to some extent, in the case of the abraded specimen, from polishing of the metal spray coating. However, polishing could not fill up the pores across the whole area and reduced the bond strength. Thus, applying a pore sealing agent is the most efficient method.

- Ti should be used with the proposed metal spray method to finish water treatment concrete structures, and a pore sealing agent is the most suitable surface treatment method to prevent water from permeating the concrete. However, because the main ingredient of the pore sealing agent of the sealed (A) specimen is epoxy, similar to paint, deterioration by ozonation is expected [17]. Accordingly, using a Teflon-based pore sealing agent is the most appropriate and efficient surface treatment method.

Acknowledgments: This research was supported by Korea Ministry of Environment (MOE) as Public Technology Program based on Environmental Policy (No. 2015000700002) and Basic Science Research Program through the National Research Foundation of Korea (NRF) funded by the Ministry of Science, ICT & Future Planning (No. 2015R1A5A1037548).

Author Contributions: Jin-Ho Park is the main author, who planned and conducted the experiments and created this paper. Han-Seung Lee is a corresponding author, who managed the general control of the entire paper. Jitendra Kumar Singh performed the experiments of this paper.

Conflicts of Interest: The authors declare no conflict of interest.

References

1. Lee, H.S.; Park, J.H.; Singh, J.K.; Ismail, M.A. Protection of reinforced concrete structures of waste water treatment reservoirs with stainless steel coating using arc thermal spraying technique in acidified water. *Materials* **2016**, *9*, 753. [CrossRef]

2. Giergiczny, Z.; Krol, A. Immobilization of heavy metals (Pb, Cu, Cr, Zn, Cd, Mn) in the mineral additions containing concrete composites. *J. Hazard. Mater.* **2008**, *160*, 47–255. [CrossRef] [PubMed]

3. Dodo, M.C.; Buffle, M.O.; Ginten, U.V. Oxidation of antibacterial molecules by aqueous ozone: Moiety-specific reaction kinetics and application to ozone-based wastewater treatment. *Environ. Sci. Technol.* **2006**, *40*, 1969–1977. [CrossRef]

4. Hirsch, R.; Ternes, T.A.; Haberer, K.; Kratz, K.L. Occurrence of antibiotics in the environment. *Sci. Total Environ.* **1999**, *225*, 109–118. [CrossRef]

5. Golet, E.M.; Xifra, I.; Siegrist, H.; Alder, A.C.; Giger, W. Environmental exposure assessment of fluoroquinolone antibacterial agents from sewage to soil. *Environ. Sci. Technol.* **2003**, *37*, 3243–3249. [CrossRef] [PubMed]

6. Göbel, A.; Thomsen, A.; Mcardell, C.S.; Joss, A.; Giger, W. Occurrence and sorption behavior of sulfonamides, macrolides, and trimethoprim in activated sludge treatment. *Environ. Sci. Technol.* **2005**, *39*, 3981–3989. [CrossRef] [PubMed]

7. Kim, S.; Eichorn, P.; Jensen, J.N.; Weber, A.S.; Aga, D.S. Removal of antibiotics in wastewater: Effect of hydraulic and solid retention times on the fate of tetracycline in the activated sludge process. *Environ. Sci. Technol.* **2005**, *39*, 5816–5823. [CrossRef] [PubMed]

8. Brain, R.A.; Johnson, D.J.; Richards, S.M.; Hanson, M.L.; Sanderson, H.; Lam, M.W.; Young, C.; Mabury, S.A.; Sibley, P.K.; Solomon, K.R. Microcosm evaluation of the effects of an eight pharmaceutical mixture to the aquatic macrophytes *Lemna gibba* and *Myriophyllum sibiricum*. *Aquat. Toxicol.* **2004**, *70*, 23–40. [CrossRef] [PubMed]

9. Wilson, C.J.; Brain, R.A.; Sanderson, H.; Johnson, D.J.; Bestari, K.T.; Sibley, P.K.; Solomon, K.R. Structural and functional responses of plankton to a mixture of four tetracyclines in aquatic microcosms. *Environ. Sci. Technol.* **2004**, *38*, 6430–6439. [CrossRef] [PubMed]

10. Park, J.H.; Lee, H.S.; Shin, J.H. An experimental study on evaluation of bond strength of arc thermal metal spraying according to treatment method of water facilities concrete surface. *J. Korea Inst. Build. Construct.* **2016**, *16*, 107–115. [CrossRef]

11. Lupsea, M.; Tiruta-Barna, L.; Schiopu, N. Leaching of hazardous substances from a composite construction product—An experimental and modelling approach for fibre-cement sheets. *J. Hazard. Mater.* **2014**, *264*, 236–245. [CrossRef] [PubMed]

12. Guo, Q. Increases of lead and chromium in drinking water from using cement-mortar-lined pipes: Initial modeling and assessment. *J. Hazard. Mater.* **1997**, *56*, 181–213. [CrossRef]

13. Jensen, H.S.; Lens, P.N.L.; Nielsen, J.L.; Bester, K.; Nielsen, A.; Haaning, H.-J.; Thorkild, V.J. Growth kinetics of hydrogen sulfide oxidizing bacteria in corroded concrete from sewers. *J. Hazard. Mater.* **2011**, *189*, 685–691. [CrossRef] [PubMed]

14. Owaki, E.; Okamoto, R.; Nagashio, D. Deterioration of concrete in an advanced water treatment plant. In *Concrete Under Severe Condition: Environment and Loading*; Gjorv, E., Odd, E., Sakai, K., Banthia, N., Eds.; E & FN Spon: London, UK, 1998; p. 438.

15. Jang, B.H. A study on the ozone resistance test of concrete. *Korea Soc. Civ. Eng.* **2014**, *10*, 327–328.

16. Swamy, R.N.; Tanikawa, S. An external surface coating to protect the concrete and steel from aggressive environments. *Mater. Struct.* **1993**, *26*, 465–478. [CrossRef]

17. Vera, R.; Apablaza, J.; Carvajal, A.M.; Vera, E. Effect of surface coatings in the corrosion of reinforced concrete in acid environments. *Int. J. Electrochem. Sci.* **2013**, *8*, 11832–11846.

18. Crowe, D.; Nixon, R. Corrosion of stainless steels in waste water applications. Available online: http://www.hwea.org/wp-content/uploads/2015/07/150204_Corrosion_of_Stainless_Steels_in_Wastewater_Applications.pdf (accessed on 19 January 2016).

19. Bhalerao, B.B.; Arceivala, S.J. Application of corrosion control techniques in municipal water and waste water engineering. Available online: http://eprints.nmlindia.org/5825/1/129--139.PDF (accessed on 9 January 2016).

20. Cinca, N.; Lima, C.R.C.; Guilemany, J.M. An overview of intermetallics research and application: Status of thermal spray coatings. *J. Mater. Res. Technol.* **2013**, *2*, 75–86. [CrossRef]

21. Bettridge, D.F.; Ubank, R.G. Quality control of high-temperature protective coatings. *Mater. Sci. Technol.* **1986**, *2*, 232–242. [CrossRef]

22. Jandin, G.; Liao, H.; Feng, Z.Q.; Coddet, C. Correlations between operating conditions, microstructure and mechanical properties of twin wire arc sprayed steel coatings. *Mater. Sci. Eng. A* **2003**, *349*, 298–305. [CrossRef]

23. Chaliampalias, D.; Vourlias, G.; Pavlidou, E.; Stergioudis, G.; Skolianos, S.; Chrissafis, K. High temperature oxidation and corrosion in marine environments of thermal spray deposited coatings. *Appl. Surf. Sci.* **2008**, *255*, 3104–3111. [CrossRef]

24. Choi, H.-B.; Lee, H.-S.; Shin, J.-H. Experimental study on the electrochemical anti-corrosion properties of steel structures applying the arc thermal metal spraying method. *Materials* **2014**, *7*, 7722–7736. [CrossRef]

25. Paredes, R.S.C.; Amico, S.C.; d'Oliveira, A.S.C.M. The effect of roughness and pre-heating of the substrate on the morphology of aluminium coatings deposited by thermal spraying. *Surf. Coat. Technol.* **2006**, *200*, 3049–3055. [CrossRef]

26. Lee, H.S.; Singh, J.K.; Ismail, M.A.; Bhattacharya, C. Corrosion resistance properties of aluminum coating applied by arc thermal metal spray in SAE J2334 solution with exposure periods. *Metals* **2016**, *6*, 55. [CrossRef]

27. *Ozone Resistance Test Method of Material for Waterproof and Anticorrosion*; SPS KWWA M211; Korea Water and Wastewater Works Association (KWWA): Seoul, Korea, 2015.

28. *Cement Filling Compound for Surface Preparation*; Korean Standard KS K 4716; Korean Agency for Technology and Standards (KATS): Seoul, Korea, 2016.

29. *Cement-Polymer Modified Waterproof Coatings*; Korean Standard KS K 4919; Korean Agency for Technology and Standards (KATS): Seoul, Korea, 2008.

30. *Filter Paper (for Chemical Analysis)*; Korean Standard KS M 7602; Korean Agency for Technology and Standards (KATS): Seoul, Korea, 2007.

31. *Penetrating Water Repellency of Liquid Type for Concrete Surface Application*; Korean Standard KS F 4930; Korean Agency for Technology and Standards (KATS): Seoul, Korea, 2012.

32. Lai, J.K.L. A review of precipitation behaviour in AISI type 316 stainless steel. *Mater. Sci. Eng.* **1983**, *61*, 101–109. [CrossRef]

6

Photocatalytic Properties of Doped TiO$_2$ Coatings Deposited Using Reactive Magnetron Sputtering

Parnia Navabpour *, Kevin Cooke and Hailin Sun

Miba Coating Group, Teer Coatings Ltd., West Stone House, Berry Hill Industrial Estate, Droitwich WR9 9AS, UK; kevin.cooke@miba.com (K.C.); hailin.sun@miba.com (H.S.)
* Correspondence: parnia.navabpour@miba.com

Academic Editor: Joaquim Carneiro

Abstract: Mechanically robust photocatalytic titanium oxide coatings can be deposited using reactive magnetron sputtering. In this article, we investigate the effect of doping on the activity of reactively sputtered TiO$_2$. Silver, copper and stainless steel targets were used to co-deposit the dopants. The films were characterised using XRD, SEM and EDX. Adhesion and mechanical properties were evaluated using scratch testing and nano-indentation, respectively, and confirmed that the coatings had excellent adhesion to the stainless steel substrate. All coatings showed superhydrophilicity under UV irradiation. A methylene blue degradation test was used to assess their photocatalytic activity and showed all coatings to be photoactive to varying degrees, dependent upon the dopant, its concentration and the resulting coating structure. The results demonstrated that copper doping at low concentrations resulted in the coatings with the highest photocatalytic activity under both UV and fluorescent light irradiation.

Keywords: photocatalytic; TiO$_2$; magnetron sputtering

1. Introduction

Titanium dioxide (TiO$_2$) is a widely investigated, semi-conducting metal oxide photocatalyst which is active under ultraviolet irradiation due to its band gap of 3.0–3.2 eV [1]. Several studies have been carried out to investigate the effect of the crystal structure of TiO$_2$ on its photocatalytic performance. Some of these studies have found a higher activity in the anatase form [2,3] whilst others have reported the mixed-phase anatase/rutile to have a better photocatalytic performance [4]. Comparative studies of single-phase anatase and rutile TiO$_2$ have concluded that the photocatalytic activity is dependent on the reaction being studied and different kinetics and intermediaries may be produced in each case [5,6].

Applications of TiO$_2$ as a photocatalyst range from direct water splitting to create hydrogen [7–9] to water purification and remediation [10] and self-cleaning, as well as in anti-microbial surfaces for glass [11,12], textiles [13], the food industry [14,15] and medical devices [16]. Whilst in some applications, the surface of interest is placed outdoors and can be illuminated and activated by the UV radiation from sunlight, surface activation by visible light is preferred both for indoor applications and in order to maximise the effectiveness of sunlight (UV only accounts for 4% of the sunlight spectrum). Reducing the band gap of TiO$_2$ through doping can enhance its activity under both UV and visible light irradiation [17,18]. Several metallic and non-metallic dopants have been used. Low doping with silver [19], copper [20,21] and iron [22] have been reported to enhance the photocatalytic activity of TiO$_2$.

For application in food and beverage processing equipment and medical devices, TiO$_2$ with good mechanical and chemical resistance is required. Magnetron sputtering can be used to deposit coatings with superior mechanical resistance compared with, for example, sol-gel coatings [23,24]. However, there have been reports that the deposition rate of magnetron-sputtered coatings is slow and that the as-deposited coatings are amorphous (or have low crystallinity) and require heat treatment to become photoactive [25].

In our previous studies [24,26], we presented the potential of closed-field, unbalanced magnetron sputtering (CFUBMS) for the production of photocatalytic TiO_2 and Ag-doped TiO_2 coatings at a high deposition rate. In this work, we attempted to improve the photocatalytic activity of the coatings through the use of other dopants and to gain a better understanding of the effect of the coating structure on its activity.

2. Materials and Methods

2.1. Preparation of Coated Surfaces

TiO_2 and doped TiO_2 coatings were deposited using reactive magnetron sputtering in a Teer Coatings UDP 450 coating system (Teer Coatings, Droitwich, UK). A schematic representation of the coating system is shown in Figure 1. Two titanium targets (99.5% purity) were used for the deposition of TiO_2. Argon (99.998% purity) was used as the working gas and oxygen (99.5% purity) as the reactive gas. Ag, Cu and grade 316 stainless steel targets were used as dopants, the latter obviously providing mixed metallic dopants, primarily Fe, Cr and Ni, with other minor constituents, but still maintaining a high sputtering rate as a non-ferromagnetic material. Advanced Energy Pinnacle Plus pulsed DC power supplies were used to power the substrate and the titanium magnetrons and an Advanced Energy DC power supply was applied to the dopant material's target. The substrate materials were grade 304 stainless steel plaques with a 2B finish (20 mm × 10 mm). All substrates were ultrasonically cleaned in acetone prior to loading in the deposition chamber to remove surface contaminants. The substrates were aligned on a flat plate parallel to the surface of the metal targets at a distance of 150 mm from the target plane and rotated at a speed of 10 rpm.

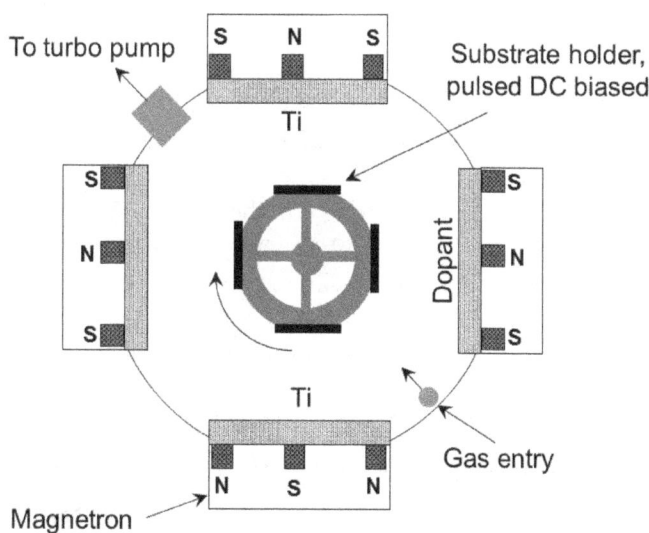

Figure 1. Schematic representation of the coating process.

The substrates were ion-cleaned for a period of 20 min prior to the coating deposition using a pulsed-DC bias voltage of −400 V (77.5% duty cycle) and a low current of 0.2–0.35 A on the targets. The coatings were deposited at a pulsed-DC bias voltage of −40 V. The partial pressure of oxygen was controlled using an optical emission monitor and was set to 25% of the pure metallic emission line's original intensity to obtain stoichiometric TiO_2. The deposition time was 60 min for all coatings. Table 1 shows the magnetron target currents used for the deposition of coatings.

Table 1. Target currents used for the deposition of coatings.

Coating	Ti (A)	Dopant (A)
Ti	2×6	–
Ti-Ag1	2×6	0.5
Ti-Ag2	2×6	0.7
Ti-St1	2×6	0.5
Ti-St2	2×6	0.7
Ti-Cu1	2×6	0.5
Ti-Cu2	2×6	0.7

2.2. Heat Treatment

Heat treatment was carried out in air at 600 °C in a Prometheus Kiln for 30 min after which the coated samples were taken out of the kiln and allowed to cool to room temperature.

2.3. Coating Characterisation

Coating thickness was assessed using the ball crater taper section method (ASTM standard E1182-93 [27] and [28]) on coated substrates. Three measurements were taken for each sample and the average values are reported. Adhesion of the as-deposited and annealed coatings was evaluated using a Teer Coatings ST3001 Scratch Tester (Teer Coatings, Droitwich, UK) [29]. A 200 μm radius conical diamond indenter was used and the coated samples underwent a progressive load of 10–40 N at a rate of 10 N min^{-1} and velocity of 10 mm min^{-1}. The scratch tracks were investigated using SEM in order to detect any flaking.

SEM (Cambridge Stereoscan 200, Cambridge Instruments Ltd., Cambridge, UK) was used to investigate the morphology of the coatings. An acceleration voltage of 15 kV was utilised. The compositional analysis of the doped coatings was performed using EDX at a minimum of five locations and was reported as the mean $\pm 2\sigma$.

Hardness measurements were performed on films deposited onto stainless steel substrate using a Fischerscope™ HM2000 micro-indentation system (Helmut Fischer GmbH, Sindelfingen, Germany). Tests were carried out with a Vickers diamond indenter with loads from 0.4 to 10 mN. During the penetration of the test surface by the indenter under load, hardness can be determined from the resultant load vs indentation depth curve (loading/unloading) which gives the value of composite hardness (comprising effects from both the coating and the substrate). The hardness reported here is the plastic hardness, HUplast, which is based on the lasting indentation after unloading. At least five indentation cycles were performed to create a mean value graph from which the value of hardness was calculated.

Advancing drop contact angle measurements with water were carried out at room temperature using a contact angle measuring instrument. At least six measurements were taken for each surface and the reported values are mean $\pm 2\sigma$.

X-ray diffraction (XRD) (Bruker D8 Advance X-ray Diffractometer, Bruker AXS, Karlsruhe, Germany) was used to evaluate the crystal structure of the coatings. XRD measurements were performed using a Cu Kα radiation ($\lambda = 0.154$ nm) in 2θ steps of 0.014°, and a low scan speed of $0.01°\cdot s^{-1}$.

2.4. Photocatalytic Properties

The photocatalytic activities of the coatings were analysed using the methylene blue (MB) degradation assay under UV and fluorescent light sources [30]. In brief, MB solutions were made up to an initial concentration of 0.0105 mMol L^{-1}. Photocatalytic surfaces were placed in 10 mL of the MB solution stirred using a magnetic stirrer at a rate of 120 rpm and irradiated at an integrated power flux of 40 W/m^2 with two 15 W UV lamps (365 nm wavelength). Tests were also carried out using two 15 W fluorescent tubes in place of the UV tubes to simulate typical lighting environments.

The integrated power flux to the coatings with the fluorescent tubes was 64 W/m^2, of which the UV component (300–400 nm) was 13 W/m^2. A 10 cm distance between the light source and MB solution was used. Absorption spectra were taken from the solution at 10 min intervals using an HR2000+ Ocean Optics spectrophotometer and the absorption peak at 650–668 nm was monitored over a period of 2 h. Figure 2 shows the setup used for the evaluation of methylene blue degradation using the coated surfaces.

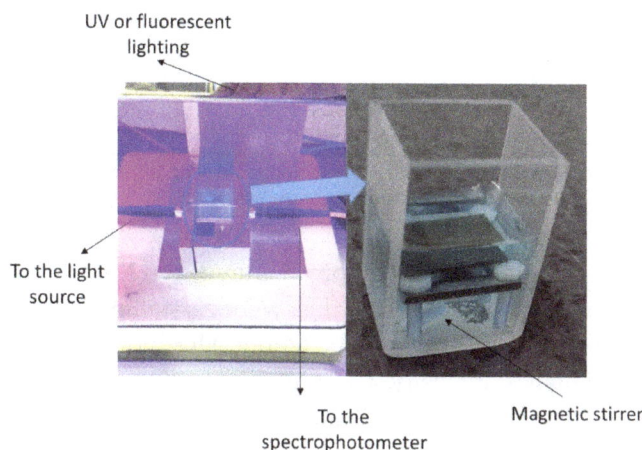

Figure 2. The experimental setup for the evaluation of MB degradation in the presence of coated surfaces.

According to the Lambert-Beer law, the concentration of dye is proportional to the absorbance value:

$$A = \varepsilon c l \tag{1}$$

where A is absorbance, ε is the molar absorbance coefficient; l is the optical length of the cell where the photocatalyst is immersed into MB.

The photocatalytic decomposition of MB was approximated to first-order kinetics, as shown in the equation:

$$\ln(C_0/C) = k_a t \tag{2}$$

where C_0 and C are the concentrations of MB solution at time 0 and time t of the experiment, respectively.

Since the absorbance decay is proportional to the concentration decay, the first-order rate constant, k_a can be found from the slope of the plot $\ln(A_0/A)$ against time. The rate constant should be normalised to take into account both the volume of the MB solution and the surface area of the substrate. The surface normalised rate constant, k'' was calculated by multiplying k_a with the reactor volume to catalyst area ratio [31].

3. Results and Discussion

Silver, copper and iron (as the main constituents of 316 steel) were selected as dopants for TiO$_2$ and were co-deposited with titanium at two levels. As iron is ferromagnetic and hence difficult to sputter using magnetron sputtering and because one of the primary aims of this study was to deposit industrially scalable and economical coatings, a stainless steel target was used as the source of iron. Deposition times were adjusted to obtain coatings with a thickness of around 2 μm. The relative concentrations of titanium and dopant in the coatings were measured using EDX. It was not possible to measure the oxygen content in the coatings accurately using EDX. Table 2 shows the relative concentration of titanium and dopant in the coating, as well as its thickness and plastic hardness. Silver doping of TiO_2 reduced its hardness whilst the addition of copper and low doping levels of stainless steel increased it.

Table 2. Composition and mechanical properties of the coatings.

Coating	Ti (at.%)	Dopant (at.%)	Thickness (μm)	HU (GPa)
TiO_2	100.0	0.0	2.0	7.6 ± 0.1
Ti-Ag1	96.4 ± 1.0	3.6 ± 1.0	2.0	7.0 ± 1.1
Ti-Ag2	94.3 ± 1.8	5.7 ± 1.8	2.1	6.9 ± 0.6
Ti-Cu1	99.3	0.7	1.9	8.7 ± 0.3
Ti-Cu2	88.4	11.6	2.1	9.2 ± 0.5
Ti-St1	N/A *	N/A *	2.3	10.3 ± 0.9
Ti-St2	N/A *	N/A *	2.4	7.2 ± 0.4

* It was not possible to measure the concentration of the dopant in stainless steel–doped coatings as the EDX peak from the dopant could not be separated from that of the substrate.

SEM was used to analyse the surface topography (Figure 3) and XRD to analyse the microstructure (Figure 4) of the as-deposited coatings. All coatings were crystalline but the crystal structure and orientation were affected by the dopant type and content. Un-doped TiO_2 showed several anatase peaks with (101) having the highest intensity. The addition of silver reduced the intensity of the (101) peak as well as the (004), (105) and (211) peaks but the coatings remained in anatase.

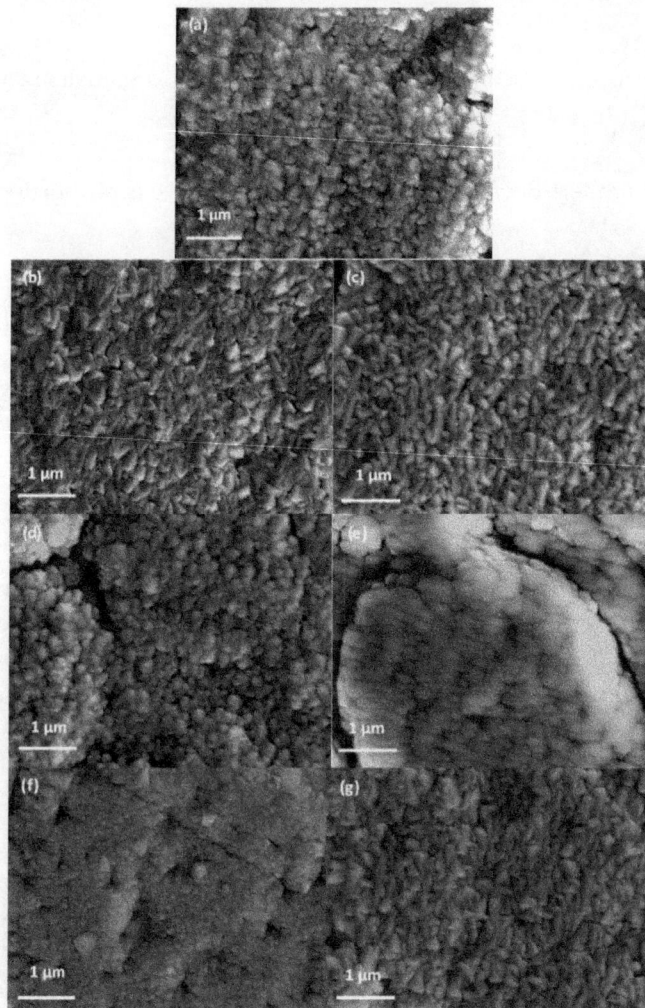

Figure 3. SEM images of the surface of coatings: (**a**) Ti; (**b**) Ti-Ag1; (**c**) Ti-Ag2; (**d**) Ti-Cu1; (**e**) Ti-Cu2; (**f**) Ti-St1; and (**g**) Ti-St2.

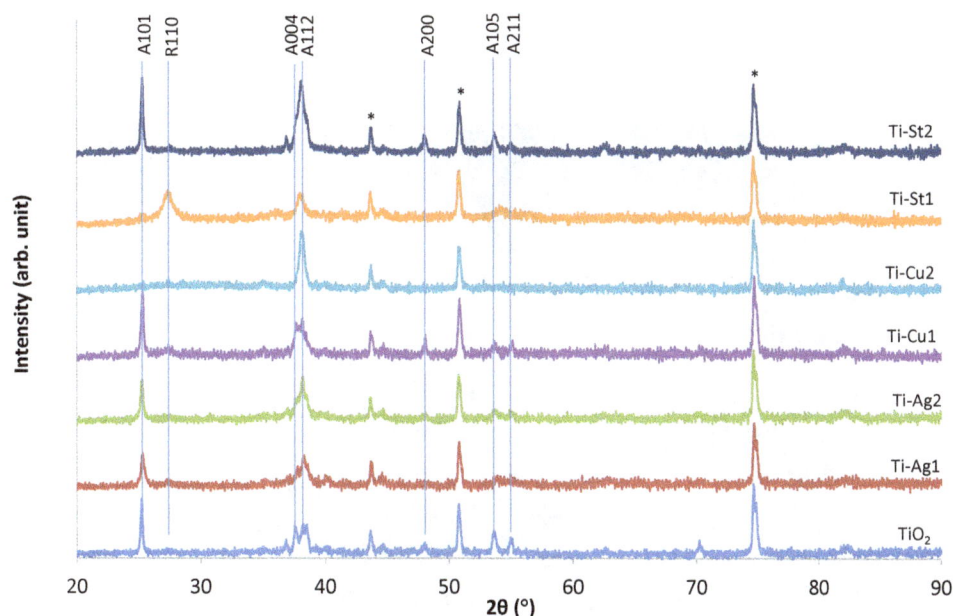

Figure 4. XRD patterns of as-deposited TiO_2 and doped TiO_2 coatings.

The addition of a small amount of copper did not change the structure of TiO_2 significantly but further Cu resulted in the (101) peak disappearing and the occurrence of a strong peak at 38°, which could be a result of the TiO_2 (112) orientation or also the appearance of the CuO (111) orientation. Coating with low concentrations of stainless steel dopants had a mixed anatase (112) and rutile (110) structure. The addition of higher concentrations of dopant resulted in the disappearance of the rutile peak and changed the crystal structure to anatase with (101) and (112) orientation. The magnified (101) anatase peak for TiO_2 and doped TiO_2 coatings is shown in Figure 5. A small broadening and shifting of this peak is seen in the doped coatings, indicating the distortion of the crystal lattice by these dopants.

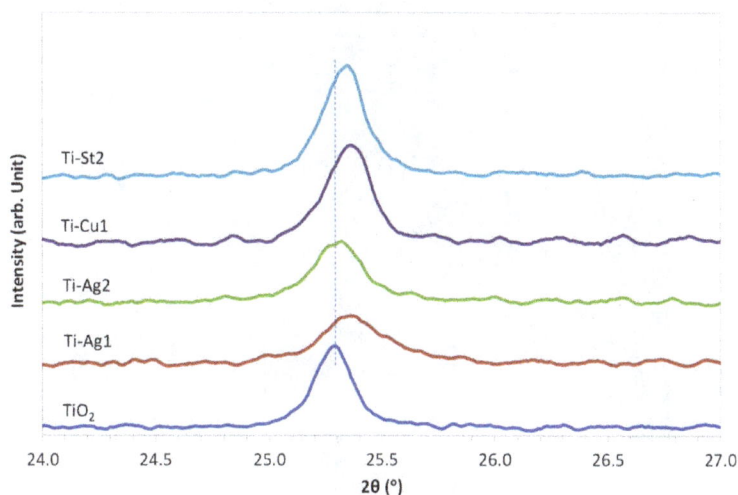

Figure 5. The position of (101) anatase peak in the XRD patterns of TiO_2 and doped TiO_2 coatings.

Heat treatment of TiO_2 has been reported to increase its crystallinity and photocatalytic performance [25,32]. The coatings in this work were therefore annealed at 600 °C in air to investigate the effect on the structure and properties. However, an evaluation of the surface topography revealed that heat treatment resulted in the segregation of significant amounts of the co-sputtered material from the TiO_2 matrix (Figure 6), especially in the coatings with the higher dopant concentrations, suggesting that at least part of the material was not incorporated as a dopant for TiO_2, but rather a separate phase.

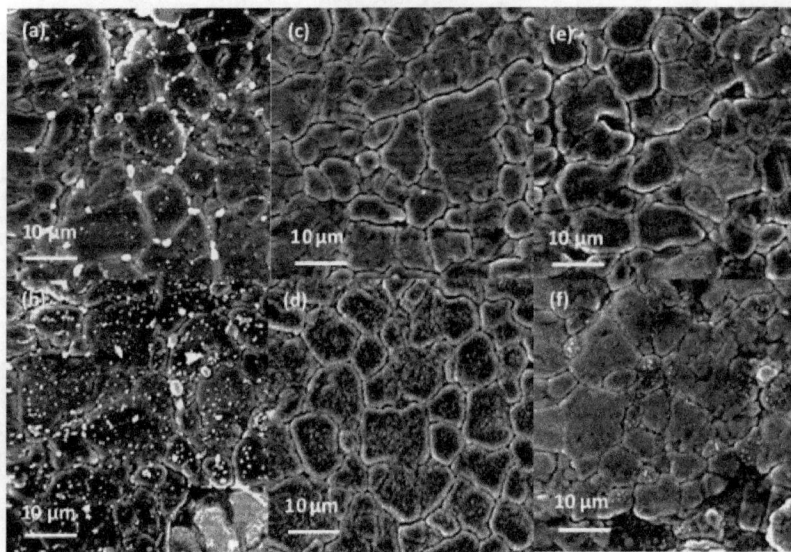

Figure 6. SEM images of coated surfaces after heat treatment: (**a**) Ti-Ag1; (**b**) Ti-Ag2; (**c**) Ti-Cu1; (**d**) Ti-Cu2; (**e**) Ti-St1; and (**f**) Ti-St2.

Scratch testing of the coatings revealed that the as-deposited coatings had excellent adhesion to the substrate and no flaking was seen in the coatings with applied loads of up to 40 N. Annealing at high temperature, however, resulted in stressed coatings which had poor adhesion to the substrate. Figure 7 shows the scratch tracks of Ti and Ti-Cu1 before and after annealing.

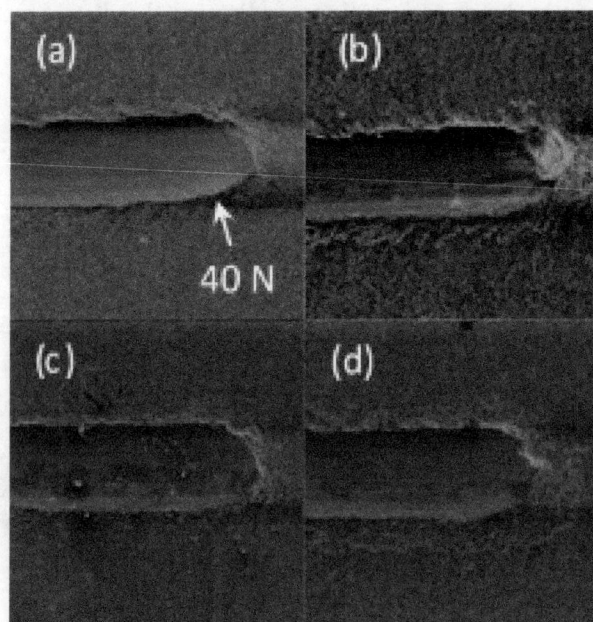

Figure 7. Scratch tracks on (**a**) as-deposited Ti, (**b**) heat-treated Ti, (**c**) as-deposited Ti-Cu1, and (**d**) heat-treated Ti-Cu1.

Due to the fact that the as-deposited coatings were crystalline and had good mechanical properties, and due to the deterioration of the physical and mechanical properties after heat treatment, it was clear that there was no advantage in heat-treating these coatings and therefore further evaluation was limited to the as-deposited coatings.

Table 3 presents the water contact angle on the as-deposited coatings after being stored for a minimum of four weeks in the dark. The addition of silver and copper resulted in an increase in the water contact angle whilst the addition of stainless steel resulted in a decrease in the contact angle. After 30 min of irradiation with UV light, all coatings became superhydrophilic with a water contact angle of <10°.

Table 3. Water contact angles.

Coating	Water Contact Angle (°)
Ti	55 ± 2
Ti-Ag1	66 ± 3
Ti-Ag2	79 ± 3
Ti-Cu1	56 ± 1
Ti-Cu2	81 ± 3
Ti-St1	36 ± 2
Ti-St2	41 ± 1

The photocatalytic efficiency of the surfaces was evaluated from the decay in the absorbance peak using the slope of line $\ln(A_0/A)$ vs. time. Table 4 presents the rate constant, k_a, and the surface-normalised rate constant, k'' [31].

Table 4. Efficiency of coated surfaces under UV and fluorescent irradiation.

Coating	k_a (10^{-5} s^{-1})		k'' (10^{-6} m s^{-1})	
	Ultraviolet	Fluorescent	Ultraviolet	Fluorescent
Ti	2.9	2.5	1.5	1.3
Ti-Ag1	3.6	2.2	1.8	1.1
Ti-Ag2	5.2	2.5	2.6	1.3
Ti-Cu1	7.2	3.3	3.6	1.7
Ti-Cu2	1.8	2.9	0.9	1.5
Ti-St1	2.2	2.0	1.1	1.0
Ti-St2	2.0	3.3	1.0	1.7

Rafieian et al. [31] collated k'' values for MB degradation in the presence of TiO$_2$ prepared using various sputtering methods. The reported k'' values range between 7.99×10^{-8} and 2.87×10^{-6} m s^{-1}. The authors have also demonstrated the use of a microfluidic reactor for the evaluation of the rate constant which allowed them to extract the intrinsic rate constant without limitations due to inefficient mass transport and light distribution. The k'' value obtained in their work was 5×10^{-5} m s^{-1}. As in many of the applications of photocatalytic coatings, it was not possible to eliminate the mass transport and light distribution limitations; the use of reactors, such as that used in ISO Standard 10678:2010 [33] or that used in this work, is appropriate for the comparison of different coatings, especially as we tried to counter some of the effects of mass transport limitations through stirring the MB solution. The k'' value for the TiO$_2$ coating in this work under UV illumination was 1.5×10^{-6} m s^{-1} which was broadly in line with other sputtered TiO$_2$ coatings and was hence used as a control to investigate the effectiveness of the different dopants.

Ti-Cu1 had the highest activity under both UV and fluorescent lights, with Ti-St2 matching its activity under fluorescent light. Other studies in the literature have also found the potential of copper in enhancing the photocatalytic activity of TiO$_2$ [20,21]. The crystal structure of Ti-Cu1 was similar to that of Ti with the exception of a shift in the position of the anatase peak due to the distortion of the crystal lattice by Cu. It is possible that Cu has a role in preventing the recombination of photo-excited electrons and holes. Further work, however, is needed to establish the exact mechanism by which doping improves the photocatalytic activity of TiO_2.

Most coatings were less active under fluorescent irradiation with the exception of Ti-Cu2 and Ti-St2, although Cu doping, even at a low concentration, enhanced the activity of TiO_2 under fluorescent lighting.

Our previous work [24,26] showed the potential of magnetron sputtering for the deposition of photoactive TiO_2 at high deposition rates and with excellent mechanical properties without a need for subsequent heat treatment. It also presented the potential of silver in improving the photocatalytic properties and imparting antimicrobial properties. The current work shows that co-sputtering a small amount of Cu during the reactive sputtering of TiO_2 can enhance its UV activity by about 150% and its activity under fluorescent lighting by 32%. Furthermore, as in the case of Ag, Cu has antimicrobial properties [34] and could be expected to further contribute to the hygiene and self-cleaning properties of the coating, even under dark conditions.

4. Conclusions

The TiO_2 coating was deposited using reactive CFUBMS at high deposition rates. Its structure and photocatalytic activity were compared with doped TiO_2 coatings deposited by co-sputtering from a silver, copper or stainless steel (AISI 316) target. Two levels of dopant content were evaluated for each dopant material. All coatings were photocatalytic and showed superhydrophilicity after irradiation. The coating with low concentrations of Cu doping showed the highest photocatalytic activity compared with unmodified TiO_2.

Acknowledgments: The authors acknowledge the support from InnovateUK (formerly Technology Strategy Board) for the MATERA+ Project "Disconnecting", ref MFM-1855, Project No. 620015 for parts of this work.

Author Contributions: Parnia Navabpour and Kevin Cooke conceived and designed the experiments; Parnia Navabpour performed the experiments and analysed the data; all authors contributed to the writing of the manuscript.

Conflicts of Interest: The authors declare no conflict of interest.

References

1. Dunnill, C.W.; Parkin, I.P. Nitrogen-doped TiO_2 thin films: Photocatalytic applications for healthcare environments. *Dalton Trans.* **2011**, *40*, 1635–1640. [CrossRef] [PubMed]

2. Miao, L.; Tanemura, S.; Kondo, Y.; Iwata, M.; Toh, S.; Kaneko, K. Microstructure and bactericidal ability of photocatalytic TiO_2 thin films prepared by rf helicon magnetron sputtering. *Appl. Surf. Sci.* **2004**, *238*, 125–131. [CrossRef]

3. Tanemura, S.; Miao, L.; Wunderlich, W.; Tanemura, M.; Mori, Y.; Toh, S.; Kaneko, K. Fabrication and characterization of anatase/rutile–TiO_2 thin films by magnetron sputtering: A review. *Sci. Technol. Adv. Mater.* **2005**, *6*, 11–17. [CrossRef]

4. Jiang, D.; Zhang, S.; Zhao, H. Photocatalytic degradation characteristics of different organic compounds at TiO_2 nanoporous film electrodes with mixed anatase/rutile phases. *Environ. Sci. Technol.* **2007**, *41*, 303–308. [CrossRef] [PubMed]

5. Andersson, M.; Österlund, L.; Ljungström, S.; Palmqvist, A. Preparation of nanosize anatase and rutile TiO_2 by hydrothermal treatment of microemulsions and their activity for photocatalytic wet oxidation of phenol. *J. Phys. Chem. B* **2002**, *106*, 10674–10679. [CrossRef]

6. Yin, H.; Wada, Y.; Kitamura, T.; Kambe, S.; Murasawa, S.; Mori, H.; Sakata, T.; Yanagida, S. Hydrothermal synthesis of nanosized anatase and rutile TiO_2 using amorphous phase TiO_2. *J. Mater. Chem.* **2001**, *11*, 1694–1703. [CrossRef]

7. Fakhouri, H.; Pulpytel, J.; Smith, W.; Zolfaghari, A.; Mortaheb, H.R.; Meshkini, F.; Jafari, R.; Sutter, E.; Arefi-Khonsari, F. Control of the visible and UV light water splitting and photocatalysis of nitrogen doped TiO_2 thin films deposited by reactive magnetron sputtering. *Appl. Catal. B: Environ.* **2014**, *144*, 12–21. [CrossRef]

8. Wang, C.; Hu, Q.; Huang, J.; Wu, L.; Deng, Z.; Liu, Z.; Liu, Y.; Cao, Y. Efficient hydrogen production by photocatalytic water splitting using N-doped TiO$_2$ film. *Appl. Surf. Sci.* **2013**, *283*, 188–192. [CrossRef]

9. Wang, C.; Hu, Q.; Huang, J.; Deng, Z.; Shi, H.; Wu, L.; Liu, Z.; Cao, Y. Effective water splitting using N-doped TiO$_2$ films: Role of preferred orientation on hydrogen production. *Int. J. Hydrogen Energ.* **2014**, *39*, 1967–1971. [CrossRef]

10. Černigoj, U.; Štangar, U.L.; Trebše, P.; Ribič, P.R. Comparison of different characteristics of TiO$_2$ films and their photocatalytic properties. *Acta Chim. Slov.* **2006**, *53*, 29–35.

11. Rampaul, A.; Parkin, I.P.; O'Neill, A.O.; DeSouza, J.; Mills, A.; Elliott, N. Titania and tungsten doped titania thin films on glass; active photocatalysts. *Polyhedron* **2003**, *22*, 35–44. [CrossRef]

12. Abdollahi Nejad, B.; Sanjabi, S.; Ahmadi, V. Sputter deposition of high transparent TiO$_{2-x}$N$_x$/TiO$_2$/ZnO layers on glass for development of photocatalytic self-cleaning application. *Appl. Surf. Sci.* **2011**, *257*, 10434–10442.

13. Samal, S.S.; Jeyaraman, P.; Vishwakarma, V. Sonochemical coating of Ag-TiO$_2$ nanoparticles on textile fabrics for stain repellency and self-cleaning-the Indian scenario: A review. *J. Mineral. Mater. Char. Eng.* **2010**, *9*, 519–525.

14. Chawengkijwanich, C.; Hayata, Y. Development of TiO$_2$ powder-coated food packaging film and its ability to inactivate *Escherichia coli* in vitro and in actual tests. *Int. J. Food Microbiolog.* **2008**, *123*, 288–292. [CrossRef] [PubMed]

15. Cushnie, T.P.T.; Robertson, P.K.J.; Officer, S.; Pollard, P.M.; Prabhu, R.; McCullagh, C.; Robertson, J.M.C. Photobactericidal effects of TiO$_2$ thin films at low temperatures—A preliminary study. *J. Photoelec. Photobiolog. A* **2010**, *216*, 290–294. [CrossRef]

16. Maneerat, C.; Hayata, Y. Antifungal activity of TiO$_2$ photocatalysis against *Penicillium expansum* in vitro and in fruit tests. *Int. J. Food. Microbiolog.* **2006**, *107*, 99–103. [CrossRef] [PubMed]

17. Daghrir, R.; Drogui, P.; Robert, D. Modified TiO$_2$ for environmental photocatalytic applications: A review. *Ind. Eng. Chem. Res.* **2013**, *52*, 3581–3599. [CrossRef]

18. Zaleska, A. Doped-TiO$_2$: A review. *Recent Pat. Eng.* **2008**, *2*, 157–164. [CrossRef]

19. Cao, Y.; Tan, H.; Tang, T.; Li, J. Preparation of Ag-doped TiO$_2$ nanoparticles for photocatalytic degradation of acetamiprid in water. *J. Chem. Technol. Biotechnol.* **2008**, *83*, 546–552. [CrossRef]

20. Colon, G.; Maicu, M.; Hidalgo, M.C.; Navıo, J.A. Cu-doped TiO$_2$ systems with improved photocatalytic activity. *Appl. Catal. B Environ.* **2006**, *67*, 41–51. [CrossRef]

21. Park, H.S.; Kim, D.H.; Kim, S.J.; Lee, K.S. The photocatalytic activity of 2.5 wt% Cu-doped TiO$_2$ nano powders synthesised by mechanical alloying. *J. Alloys Compd.* **2006**, *415*, 51–55. [CrossRef]

22. Andriamiadamanana, C.; Laberty-Robert, C.; Sougrati, M.T.; Casale, S.; Davoisne, C.; Patra, S.; Sauvage, F. Room-temperature synthesis of iron-doped anatase TiO$_2$ for lithium-ion batteries and photocatalysis. *Inorg. Chem.* **2014**, *53*, 10129–10139. [CrossRef] [PubMed]

23. Takeda, S.; Suzuki, S.; Odaka, H.; Hosono, H. Photocatalytic TiO$_2$ thin film deposited onto glass by DC magnetron sputtering. *Thin Solid Films* **2001**, *392*, 338–344. [CrossRef]

24. Navabpour, P.; Ostovarpour, S.; Tattershall, C.; Cooke, K.; Kelly, P.; Verran, J.; Whitehead, K.; Hill, C.; Raulio, M.; Priha, O. Photocatalytic TiO$_2$ and Doped TiO$_2$ Coatings to Improve the Hygiene of Surfaces Used in Food and Beverage Processing—A Study of the Physical and Chemical Resistance of the Coatings. *Coatings* **2014**, *4*, 433–449. [CrossRef]

25. Ratova, M.; West, G.T.; Kelly, P. Optimisation of HiPIMS photocatalytic titania coatings for low temperature deposition. *Surf. Coat. Technol.* **2014**, *250*, 7–13. [CrossRef]

26. Navabpour, P.; Ostovarpour, S.; Hampshire, J.; Verran, J.; Cooke, K. The effect of process parameters on the structure, photocatalytic and self-cleaning properties of TiO$_2$ and Ag-TiO$_2$ coatings deposited using reactive magnetron sputtering. *Thin Solid Films* **2014**, *571*, 75–83. [CrossRef]

27. *ASTM E1182-93(1998). Standard Test Method for Measurement of Surface Layer Thickness by Radial Sectioning;* ASTM International: West Conshohocken, PA, USA, 1998.

28. Walls, J.M.; Hall, D.D.; Sykes, D.E. Composition-depth profiling and interface analysis of surface coatings using ball cratering and the scanning auger microprobe. *Surf. Interf. Anal.* **1979**, *1*, 204–210. [CrossRef]

29. Stallard, J.; Poulat, S.; Teer, D.G. The study of the adhesion of a TiN coating on steel and titanium alloy substrates using a multi-mode scratch tester. *Tribol. Int.* **2006**, *39*, 159–166. [CrossRef]

30. Ratova, M.; West, G.T.; Kelly, P. Optimisation studies of photocatalytic tungsten-doped titania coatings deposited by reactive magnetron co-sputtering. *Coatings* **2013**, *3*, 194–207. [CrossRef]

31. Rafieian, D.; Driessen, R.T.; Ogieglo, W.; Lammertink, R.G.H. Intrinsic photocatalytic assessment of reactively sputtered TiO_2 films. *ACS Appl. Mater. Interfaces* **2015**, *7*, 8727–8732. [CrossRef] [PubMed]

32. Eufinger, E.; Poelman, D.; Poelman, H.; De Gryse, R.; Martin, G.B. Effect of microstructure and crystallinity on the photocatalytic activity of TiO_2 thin films deposited by dc magnetron sputtering. *J. Phys. D Appl. Phys.* **2007**, *40*, 5232. [CrossRef]

33. *ISO 10678:2010. Fine Ceramics (Advanced Ceramics, Advanced Technical Ceramics)—Determination of Photocatalytic Activity of Surfaces in an Aqueous Medium by Degradation of Methylene Blue*; International Organisation for Standardization (ISO): Geneva, Switzerland, 2010.

34. Wu, B.; Huang, R.; Sahu, M.; Feng, X.; Biswas, P.; Tang, Y.J. Bacterial responses to Cu-doped TiO_2 nanoparticles. *Sci. Total Environ.* **2010**, *408*, 1755–1758. [CrossRef] [PubMed]

7

Erosion Wear Investigation of HVOF Sprayed WC-$_{10}$Co$_4$Cr Coating on Slurry Pipeline Materials

Kaushal Kumar [1,*], Satish Kumar [2], Gurprit Singh [2], Jatinder Pal Singh [2] and Jashanpreet Singh [2]

[1]   Department of Mechanical Engineering; Guru Jambheshwar University of Science & Technology, Hisar 125001, India

[2]   Department of Mechanical Engineering; Thapar University, Patiala 147004, India; satish.kumar@thapar.edu (S.K.); gurprit.singh@thapar.edu (G.S.); jatinderpal.singh@thapar.edu (J.P.S.); jashanpreet.singh@thapar.edu (J.S.)

*   Correspondence: ghanghaskaushal@gmail.com

Academic Editor: James Kit-hon Tsoi

Abstract: In the present work, erosion wear due to slurry mixture flow has been investigated using a slurry erosion pot tester. Erosion tests are conducted on three different slurry pipe materials, namely, mild steel, SS202, and SS304, to establish the influence of rotational speed, concentration, and time period. In order to increase erosion wear resistance, a high-velocity oxy-fuel (HVOF) coating technique is used to deposit a WC-$_{10}$Co$_4$Cr coating on the surface of all piping materials. Experimental results show that rotational speed is a highly-influencing parameter for the erosion wear rate as compared to solid concentration, time duration, and weighted mean diameter. WC-$_{10}$Co$_4$Cr HVOF coating improved the erosion resistance of piping materials up to 3.5 times. From experimental data, the exponents of solid concentration, velocity, and the size of particles are calculated for the empirical erosion wear equation. A functional equation of the erosion wear rate is developed. The predicted erosion wear is in agreement with the experimental data and found to be within a deviation of $\pm 12\%$.

Keywords: erosion wear; bottom ash; slurry pipe materials; powder coating

---

1. Introduction

Slurry pipelines are used in various industrial applications, such as disposal of waste material, coal ash, and tailing materials [1]. However, knowledge, as well as technology, associated with the design and operation of such pipelines is still developing. Thus, many aspects of the design have to be done with the help of semi-empirical correlations based on experimental data. This is due to large number of variable factors involved in the design [2–4]. Accurate prediction of erosion wear rate is very important for the slurry pipeline [5,6]. Over the years, various bench scale setups have been developed to simulate the erosion wear mechanism at the laboratory scale [3,4]. Various researchers have found that particle size distribution, rotational speed, and solid concentration are the major factors that influence the material's erosion wear rate [7–9]. Some investigators have suggested that the erosion wear can be minimizing by using protective coatings on piping materials. These coatings can provide a hard layer of carbides, oxides, and nitrides of Cr, W, Al, and Ti [10,11]. The high-velocity oxy-fuel (HVOF) technique can be applied to deposit coatings with high cohesive strength and superior mechanical properties [10–14]. In the present work, erosion wear behavior of three piping materials, namely, mild steel, SS202, and SS304, have been analyzed with and without a HVOF-sprayed WC-$_{10}$Co$_4$Cr coating. The present study is conducted with the motivation of measuring the average erosion wear at specimen placed in horizontal direction. The erodent material is taken as bottom ash having a solid concentration that varies from 30 wt % to 60 wt %. Experiments are carried out at four different speeds, namely, 600, 900, 1200, and 1500 rpm with time durations of 90, 120, 150, and 180 min.

2. Material and Methods

2.1. Specimen

For the study of erosion wear, the base material is cut into flat pieces of 75 mm × 25 mm × 5 mm and each drilled with a central hole for holding purposes in the rotating spindle of the tester. Chemical composition of the base material is measured by using an optical emission spectrometer (Foundry Master, Oxford instruments, Uedem, Germany). The optical emission spectrometer was installed to ensure the precise chemical composition of ferrous metals. The surface roughness of specimen is determined by using a roughness tester (Model SJ400, Surftest, Mitutoyo America Corporation, CA, USA) with a lowest count of 0.001 mg. The mean roughness values are determined before and after erosion wear experimentation. The mean hardness of all coated and uncoated specimens are determined by using a micro Vicker hardness tester (Model MVH1, Metatech Industries, Pune, India) with a load of 1000 g. The microhardness was tested at four different spots on the specimen's surface by the Vicker's diamond pyramid at regular separation distances. The purpose of regular indentation was to avoid cracking of the specimen under the applied load (by the indenter).

2.2. Coating Deposition

Commercially available coating powder WC-$_{10}$Co$_4$Cr has been used in the present study to provide resistance against slurry erosion. Energy-dispersive spectroscopy (EDS) analysis of the coating powder confirms the presence of different elements in the powder. A high velocity oxy-fuel technique is used to deposit the coating on piping materials by applying the HIPOJET 2700 (M/S Metalizing Equipment Company Private Limited, Jodhpur, Rajasthan, India) thermal spray process. A compressed air jet is used for cooling the test specimens during and after the coating process. The process variables used for the thermal spray process are listed in Table 1. Prior to powder coating deposition, the surface of steel specimens are grit-blasted with Al$_2$O$_3$ grit, which increases the quality of coating adhesion by enhancing the surface roughness of the specimens.

Table 1. The process variables for the thermal spray process.

Medium	Spray Distance (mm)	Flow Rate (L/min)	Pressure (kPa)	Feed Rate (g/min)	Particle Size (μm)
Air	138	640	10	30	15
Oxygen	138	260	5	30	15
Fuel	138	75	6.2	30	15

2.3. Slurry Preparation

The erodent material is taken as bottom ash, evacuated from the Rajiv Gandhi thermal power plant (Hisar, Haryana, India). The particle size distribution (PSD) of bottom ash sample is evaluated with the help of standard sieves. The solid concentration of the slurry varied from 30 wt % to 60 wt %. The specific gravity of the bottom ash sample is measured by using pycnometer equipment. The static settled concentration value is determined by preparing a solid-liquid suspension of an initial solid concentration, i.e., 30 wt %. The surface morphology and composition of the bottom ash is measured by using scanning electron microscopy (SEM) and energy dispersive X-ray spectroscopy (EDX) (Model JSM-6510LV, JOEL Ltd., Nieuw-Vennep, The Netherlands).

2.4. Slurry Erosion Test Rig

Slurry erosion wear tests are performed with the help of an erosion pot tester with 1.8 L capacity (model: TR-41, Ducom instruments, Bangalore, India) in a similar approach as adopted by [8]. The schematic diagram of the test rig is shown in Figure 1. The tester consists of a rotating spindle, cylindrical pot, propeller, and screw jack. The rotating spindle, itself, holds the specimen and propeller fixed to it. With the rotation of the spindle, the specimen and propeller also rotate. A propeller is fixed at the end of the spindle so that slurry particles do not settle but remain suspended in the mixture throughout the working process. A proximity sensor disc is used to measure the speed of the spindle. Specimens were washed with acetone after each test run. An electronic weight microbalance is used to calculate the weight loss of specimens with 0.001 mg resolution.

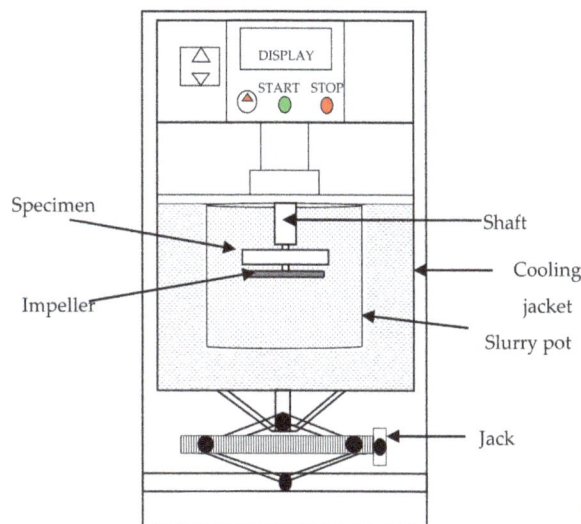

Figure 1. Schematic diagram of the erosion wear pot tester used for the investigation.

3. Results and Discussion

3.1. Characterization of Erodent Material

Bottom ash was used as the erodent material to perform the erosion wear experiments. The surface morphology of the bottom ash sample is shown in Figure 2a. It seems that particles of bottom ash are coarser, asymmetrical, darker grey in color due to the existence of unburned carbon, and have an uneven surface texture. The chemical composition of the bottom ash sample is shown in Figure 2b. The chemical composition of the bottom ash sample is expressed as follows: SiO_2-52.11 wt %, Al_2O_3-36.24 wt %, FeO-2.15 wt %, TiO_2-1.68 wt %, CaO-1.28 wt %, CO_2-3.19 wt %, and LOI-3.35 wt %.

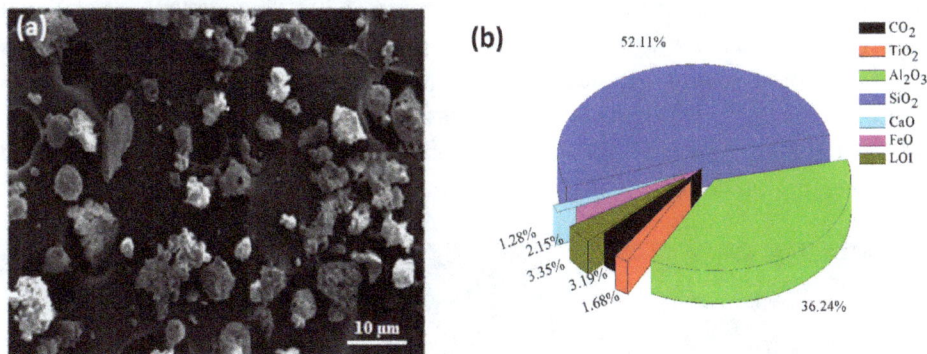

Figure 2. (**a**) SEM morphology and (**b**) chemical composition of bottom ash sample.

Figure 3 represents the particle size distribution (PSD) of the bottom ash sample. It was observed that more than 19.41% particles are coarser than 250 μm, 65.20% particles are in the range of 75–250 μm, and only 15.39% particles are finer than 75 μm. The specific gravity of the bottom ash sample was measured as 1.94. Table 2 represents the final static settled concentration of the slurry suspension, which was recorded as 52.15 wt %.

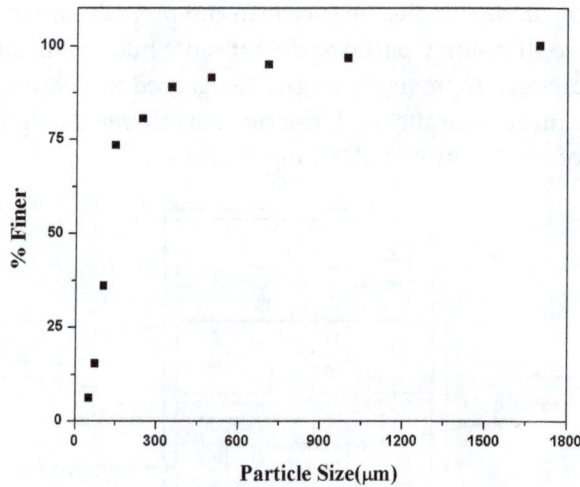

Figure 3. Particle size distribution (PSD) of bottom ash.

Table 2. Static settled concentration of bottom ash (wt %).

Time (min)	Static Settled Concentration (wt %)	Time (min)	Static Settled Concentration (wt %)
0	30	20	37.75
1	30.12	30	41.05
2	30.35	60	45.73
3	30.72	180	49.92
4	31.16	420	51.36
5	31.95	660	52.15
10	34.92	–	–

3.2. Effect of WC-$_{10}$Co$_4$Cr Coating Powder Deposition on Substrate

The morphology of the coating powder is studied by performing scanning electron microscopy, as shown in Figure 4. It is observed that the coating powder has an angular morphology. The coatings, with thicknesses in the range of 166–175 µm, were deposited on the surfaces of the substrates. The chemical compositions of the base materials were measured by using an optical emission spectrometer, summarized in Table 3. The percentage of Ni and Cr was found to be at their maxima in SS304, whereas Mn was found to be at its maximum in SS202. The percentage of Ni, Cr, and Mn were found highest in SS202, which is beneficial in providing better microhardness. The average microhardness and roughness of uncoated and WC-$_{10}$Co$_4$Cr-coated materials is presented in Table 4. From Table 4, it seems that SS202 has better hardness and roughness as compared to mild steel and SS304. However, the roughness of SS202 was increased approximately three times with the deposition of WC-$_{10}$Co$_4$Cr coating powder.

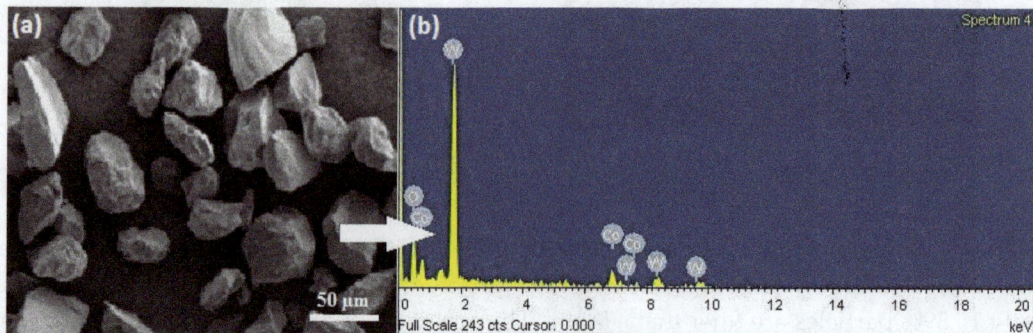

Figure 4. (a) SEM and (b) EDS micrographs of WC-$_{10}$Co$_4$Cr coating powder.

Table 3. Chemical composition of base materials (wt %).

Material	Fe	Cr	Ni	Mn	C	Si	Co	P	Al	S	Cu
Mild Steel	98.90	0.04	0.05	0.45	0.14	0.20	–	0.07	0.02	0.06	0.07
SS202	74.54	13.42	0.18	9.67	0.09	0.42	0.05	0.07	0.04	0.04	1.48
SS304	69.65	18.65	8.94	1.43	0.13	0.55	0.18	0.11	–	0.06	0.30

Table 4. Effect of WC-$_{10}$Co$_4$Cr coating on different properties of piping materials.

Materials	Indentation Depth (μm)		Average Hardness (HV)		Average Roughness (μm)	
	Uncoated	Coated	Uncoated	Coated	Uncoated	Coated
Mild steel	103	37	138	952	2.23	5.48
SS202	82	25	276	1158	1.63	5.81
SS304	96	29	237	1129	1.56	5.78

3.3. Effect of Rotational Speed/Velocity on Average Erosion Wear

Many researchers have found that rotational speed, solid concentration and particle size distribution are some of the major factors that influence the erosion wear rate [4,15,16]. The following empirical correlation has been used to predict the erosion wear by using experimental data, and has also been discussed [15].

$$E_w = kv^a d_e{}^b C_w{}^c \qquad (1)$$

where E_w is the erosion rate, v is the velocity, d_e is the particle size, C_w is the solid concentration of slurry, and four constants k, a, b, and c. The regression analysis is carried out to find the value of exponents "a", "b", and "c" [17]. In the present work, erosion wear on the different piping materials has been investigated with and without coating to study the influence of three different parameters, namely, the speed of rotation, solid concentration, and test duration. The values of exponent "a" have been determined as 2.06, 1.53, and 1.69 for mild steel, SS202, and SS304, respectively, which shows good agreement with researchers [2,15]. Truscott [18] reported that value of the velocity exponent lies between 1 and 3.5. Values of $a < 2$ (1.53 and 1.69) represent that the bottom ash particles rebound from the surface of SS202 and SS304 at high impact velocity. However, $a = 2.06$ represents that erosion wear approaches the ideal value, which is directly proportional to the kinetic energy of the impacting particles. The erosion wear experimentation performed at four different rotational speeds of 600, 900, 1200 and 1500 rpm with a time duration of 180 min, and a solid concentration of 30 wt %. The erosion wear of piping material is evaluated in the terms of the cumulative weight loss per unit surface area of the test specimen (g/m^2). The linear speed on the surface of the specimen increases with the rotation radius, thus, the erosion wear of the specimen is average. The effect of the rotational speed on the average erosion wear of coated and uncoated piping materials is shown in Figure 5a.

It has been observed that uncoated SS202 and SS304 exhibit higher erosive resistances as compared to mild steel at all speeds. The order of decreasing erosion wear was found to be mild steel > SS304 > SS202. Results indicated that when the rotational speed increases from 600 to 1500 rpm, erosion wear increases to about 107%, 87.41%, and 90.41% for mild steel, SS202, and SS304, respectively. Hence, it is clear that erosion wear shows high dependence on rotational speed. The kinetic energy of erodent particles increases with rotational speed. This increase in kinetic energy will lead to more localized attacks at several spots on the target material and can cause severe plastic deformation, which leads to more erosion wear. It has also been realized that the mild steel specimen exhibits higher erosion wear rather than SS202 and SS304 specimens at all rotational speed. This is anticipated due to much harder nature of SS202 and SS304 as compared to mild steel. Experimental data shows good agreement with the findings of [2,11,15].

3.4. Effects of Solid Concentration on Average Erosion Wear

The erosion wear rate of different piping materials at a concentration range of 30%–60%, at a speed of 600 rpm, for 180 min test run duration, is shown in Figure 5b. It has been observed that erosion wear increases nonlinearly with the increase in the solid concentration of bottom ash slurry. From the experimental results it has been concluded that erosion wear increases by 53.63% for mild steel, 41.72% for SS202, and 45.33% for SS304 when the solid concentration increases from 30 wt % to 60 wt %. The amount of solid particles per unit volume increases due to increases in the solid concentration of the slurry. Therefore, higher concentration allows a larger number of solid particles to strike on the surface of the wear specimen, which tends to increase the erosion wear of the material. Similar observations have also been made by other researchers with fly ash and sand slurry [4,6]. The value of exponent "c" for mild steel, SS202, and SS304 was determined as 0.642, 0.515, and 0.537, respectively, for Equation (1). The value of the exponents for all three uncoated piping materials show better agreement with researchers [15].

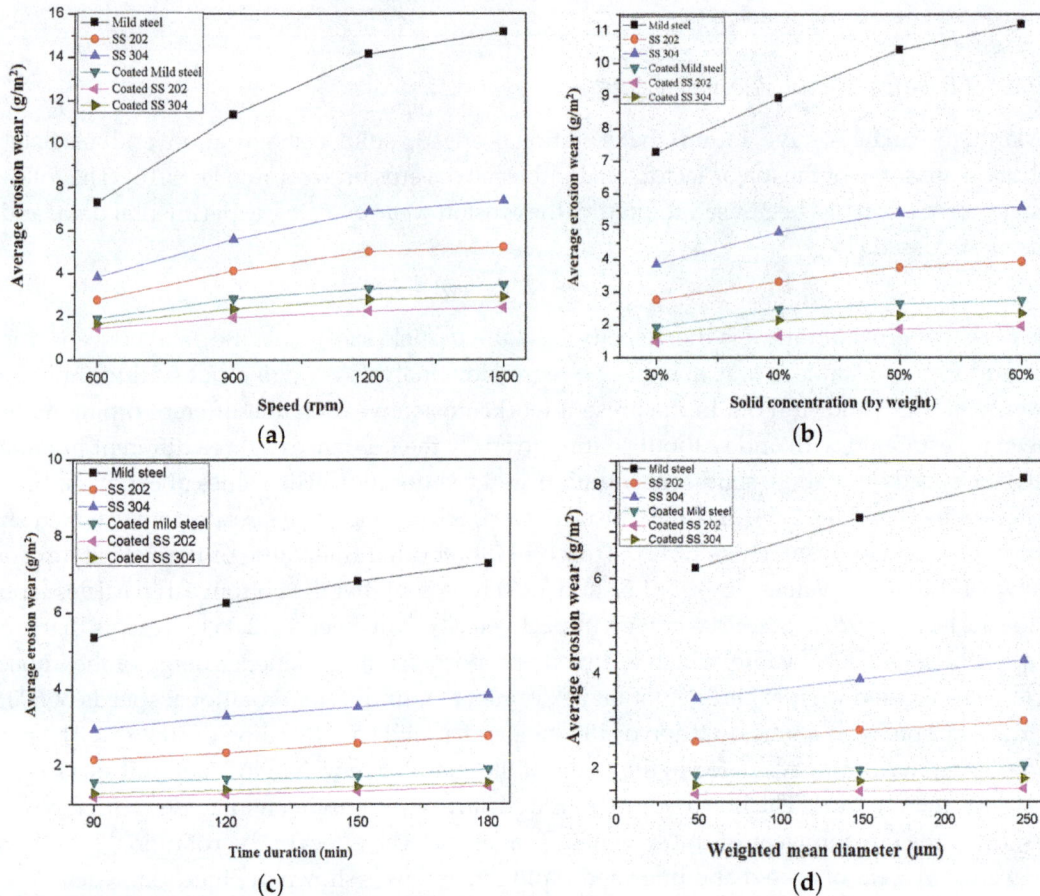

Figure 5. Effect of rotational speed on average erosion wear of uncoated and coated steel (**a**) for $C_w = 30\%$ and time = 180 min; (**b**) speed = 600 rpm and time = 180 min; (**c**) speed = 600 rpm and $C_w = 30\%$ and (**d**) speed = 600 rpm, $C_w = 30\%$ and time = 180 min.

3.5. Effects of Time Duration on Average Erosion Wear

The effect of time duration on the erosion wear performance of piping materials, with and without coating, was also examined. The erosion wear experiments were performed at four different time durations: 90, 120, 150 and 180 min, at a rotational speed of 600 rpm and solid concentration of 30 wt %, which is shown in Figure 5c. The erosion wear of different piping materials, namely, mild steel, SS202, and SS304, increases with time variation for coated and uncoated steels [15]. It has been concluded from the above results that, with the increase in time duration, the erosion wear of mild steel, SS202, and SS304 increases by 36.26%, 28.11% and 30.40%, respectively. This can be due to the fact that the continuous impinging action of the erodent on the target surface results in higher erosion wear. Similar trends were reported by the authors [4,11,15].

3.6. Effect of Weighted Mean Diameter on Average Erosion Wear

The effect of mean particle diameter on erosion wear was studied for both uncoated and coated piping materials, as shown in Figure 5d. The researchers concluded that the weighted mean diameter of multi particulate slurry is the most appropriate parameter as compared to d_{50} and the median diameter (arithmetic mean of the particle diameter) [15]. In the present investigation the weighted mean diameter was taken as 48, 148 and 248 μm.

From Figure 5d, it is seen that erosion wear increases with the increase in particle size for all materials, whether coated or uncoated. In other words, the finer particles show less erosion wear as compared to coarser particles. With an increase in the particle size of the erodent, the kinetic energy of the impact particles increases, which is further responsible for increases in the erosion wear. Additionally, in the present study, specimens are fixed in the horizontal direction and impingement angles are very low, i.e., approaches to 0°. Thus, mild steel shows the rebounding of particles as compared to SS202 and SS304. The values of exponent "b" are determined as 0.414, 0.295 and 0.316, respectively, for mild steel, SS202, and SS304 for Equation (1). The values of exponent "b" show good agreement with researchers [2,15,19]. The value of exponent "a" is much higher as compared to exponent "b". The results show that rotational speed exhibits a highly-influencing parameter as compared to particle size and solid concentration.

3.7. Effect of WC-$_{10}$Co$_4$Cr Coating on Average Erosion Wear

To study the effect of coating WC-$_{10}$Co$_4$Cr on piping materials of mild steel, SS202, and SS304, the erosion wear experiments were performed at different rotational speeds, solid concentrations, and time durations, which are shown in Figure 5a–d. Experimental results indicated that coated mild steel, SS202 and SS304 exhibit high slurry erosive wear resistance as compared to uncoated materials. However, the trend of increasing wear is nonlinear throughout the entire experiment. Nevertheless, a remarkable resistance in erosion wear has been observed after WC-$_{10}$Co$_4$Cr coating for all three piping materials. It is worth noting that coated specimens followed a similar trend as uncoated specimens in the case of increasing wear rate, i.e., mild steel > SS304 > SS202. The coated mild steel reported higher erosion wear than coated SS202 and SS304 for all sets of parameters. The coated SS202 shows the least erosion wear at 30% concentration with 600 rpm for 90 min as compared to mild steel and SS304. After WC-$_{10}$Co$_4$Cr coating, it was clear that the erosion resistance of mild steel improved in the range of 2.5 to 3.5 times, whereas in SS202 and SS304 improved nearly 1.5 to 2.3 times. Furthermore, it can be also concluded that the rate of increase in erosion wear of coated and uncoated materials also resulted from the increase in variable parameters. The value of exponent "a" was determined as 1.21, 1.14, and 1.16 for coated mild steel, SS202, and SS304, respectively, which shows good agreement with researchers [2,15]. A value of $a < 2$ represents that the bottom ash particles rebound from the specimen surface at high impact velocity. Moreover, the value of exponent "a" drops after deposition of WC-$_{10}$Co$_4$Cr powder. This indicates that WC-$_{10}$Co$_4$Cr coating adds to the brittle nature of ductile stainless steels.

3.8. Visual Examination of Eroded Coated Surfaces

Scanning electron microscopy (SEM) of eroded surfaces is used to study the surface morphology of coated specimens. After the erosion experimentation the specimen undergoes roughness and hardness testing again. It was observed that the roughness of the entire specimens decreases, whereas no change in harness is observed. The average roughness of uncoated mild steel, SS202, and SS304 is observed as 2.45, 1.35, and 1.38 μm, whereas the values were 4.89, 4.63, and 4.58 μm after deposit coating. The investigators suggested that cutting, fracture, and abrasion are the major modes of erosion wear mechanisms that are relevant to slurry pipelines [2,3,13,14]. The eroded images of WC-$_{10}$Co$_4$Cr-coated mild steel, SS202, and SS304 with EDS (energy dispersive spectroscopy) are shown in Figures 6–8. It has been observed that the tungsten carbide phase is visualized in more areas on the surface of the SS202 substrate as compared to mild steel and SS304. From Figures 6 and 7, it is noticed that mild steel was eroded more than SS202; this may be due to less affinity of mild steel towards tungsten-based coatings as compared SS202, since less tungsten carbide phases were detected on the surface of coated mild steel substrate. A small amount of erodent was also found in SEM micrographs on the surface of materials, which may be due to embedment of hard erodent particles in the relatively soft Co matrix. A long peak of tungsten confirms its existence in the dark phase of the topography. The presence of wear marks can be observed on the worn surface caused by micro-cutting, carbide-fracture, and lip formation. It has been stated that surface wear occurring through micro-cratering is responsible for material removal.

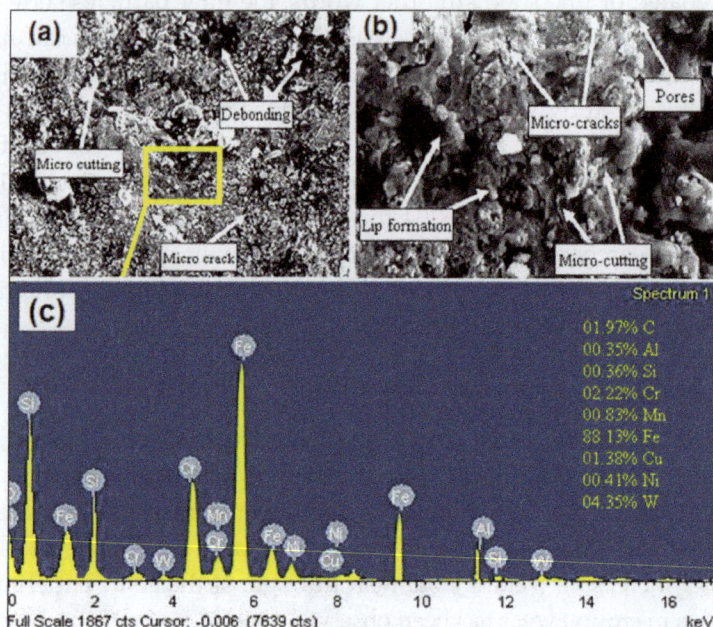

Figure 6. Eroded surface morphology of WC-$_{10}$Co$_4$Cr coated mild steel: SEM micrographs at (**a**) 1000×, (**b**) 100× magnification and (**c**) EDS micrographs.

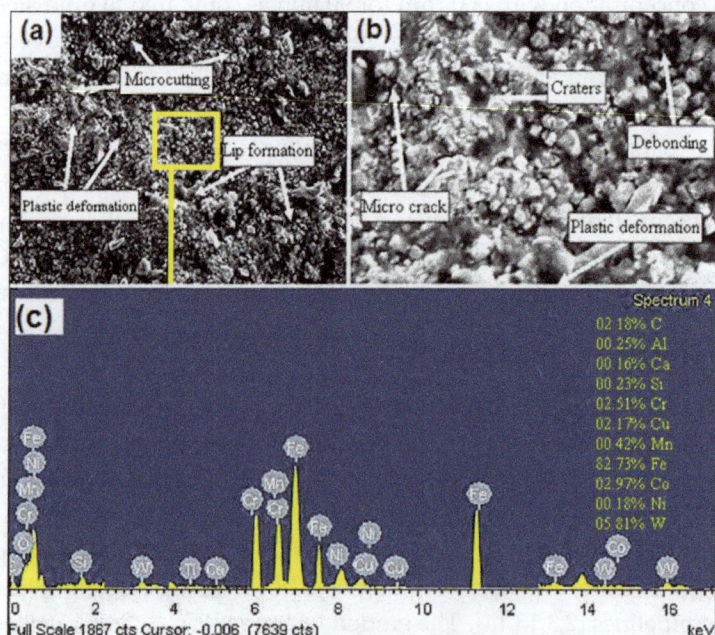

Figure 7. Eroded surface morphology of WC-$_{10}$Co$_4$Cr coated SS202: SEM micrographs at (**a**) 1000×, (**b**) 100× magnification and (**c**) EDS micrographs.

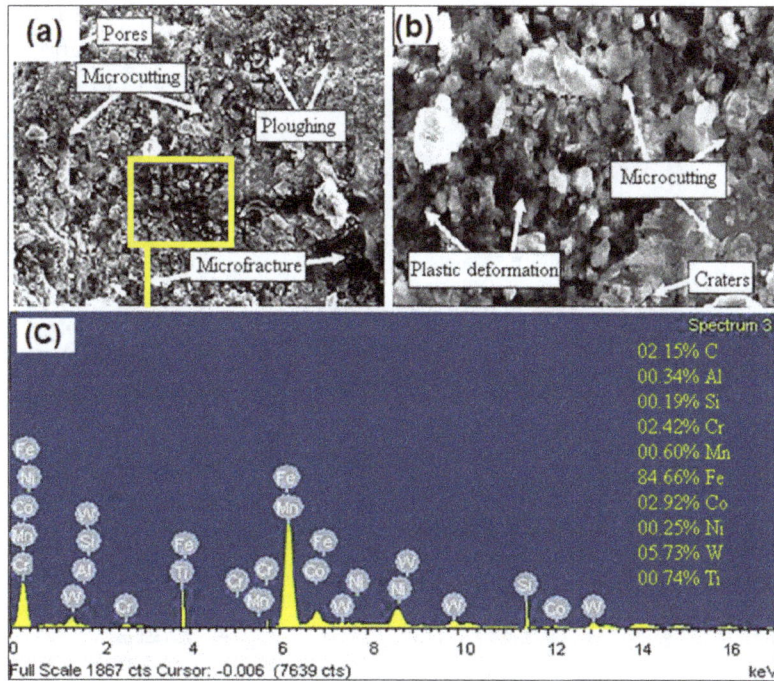

Figure 8. Eroded surface morphology of WC-$_{10}$Co$_4$Cr coated SS304: SEM micrographs at (**a**) 1000×, (**b**) 100× magnification and (**c**) EDS micrographs.

3.9. Correlation for Erosion Wear

The effects of velocity, solid concentration, and particle size on erosion wear of piping materials, namely, mild steel, SS202 and SS304, are analyzed. An attempt has been made to develop a correlation to estimate the erosion wear in terms of weight loss (mm·year^{-1}) as given below:

$$E_Y = \frac{W_L}{S \times A} \times \frac{8760}{T} \times 10^3 \tag{2}$$

where W_L is the measured weight loss (kg), A is the surface area, S is the density of pipe material and T is the duration of test (h).

Extensive experimentation was performed to determine the functional relationship of dependent parameters, like solid concentration, velocity, and particle size, on the erosion wear. The weighted mean diameter of bottom ash is calculated as 48, 148 and 248 µm. A total of 48 data points were generated for each material. Initially, the erosion wear variation of bottom ash slurry with solid concentration, velocity, and particle size for mild steel, SS202 and SS304 are plotted on log-log graphs to obtain the functional relationship between them.

Equation (1) of erosion wear for mild steel, SS202 and SS304 can be written as:

$$E_{Y\ MS} = 0.211 v^{2.06} d_e^{0.414} C_w^{0.642} \tag{3}$$

$$E_{Y\ SS202} = 0.27 v^{1.53} d_e^{0.295} C_w^{0.515} \tag{4}$$

$$E_{Y\ SS304} = 0.223 v^{1.69} d_e^{1.69} C_w^{0.537} \tag{5}$$

The predicted erosion wear is calculated from the abovementioned correlation. To achieve better precision, actual experimental values of erosion wear have been compared to the predicted values obtained from correlations. Figure 9 shows the variation of erosion wear for experimental and theoretical values obtained with 48 data points for mild steel, SS202 and SS304. Predicted erosion wear values ideally match with experimental values at the 45° line. A reasonable agreement has been observed. It is observed that erosion wear was found within the deviation limits from ±12%, ±9% and ±8% for mild steel, SS304 and SS202, respectively. This deviation range was found to be acceptable by researchers [2,3,15].

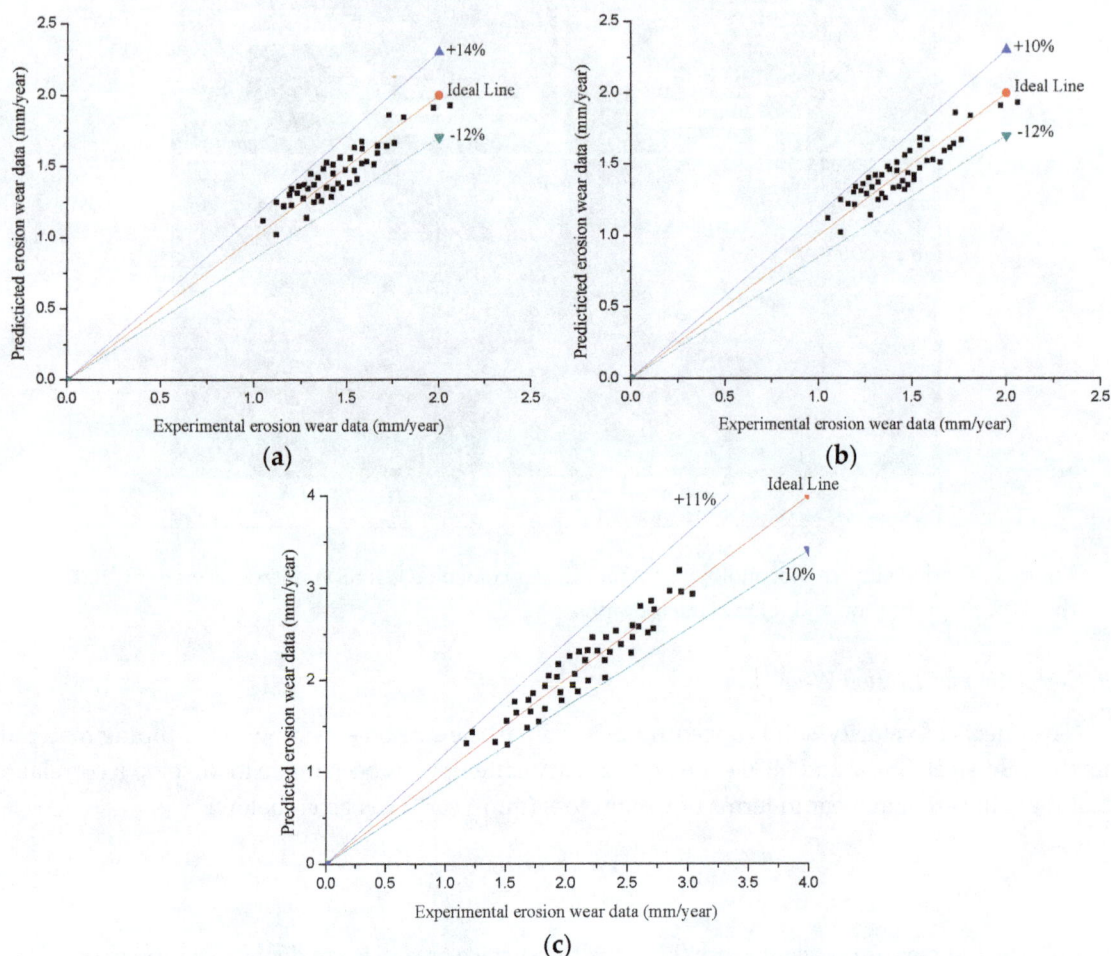

Figure 9. Variation of erosion wear for experimental and theoretical values: (**a**) mild steel; (**b**) SS202 and (**c**) SS304.

4. Conclusions

Present study was conducted with the motivation of measuring average erosion wear on specimens placed in a horizontal direction. The erosion wear on three different slurry pipe materials, namely, mild steel, SS202 and SS304 due to slurry mixture flow has been investigated using a slurry erosion pot tester. Erosion tests were conducted to establish the influence of rotational speed, concentration, and time period. On the basis of the experimental investigation, the following concise and precise forms of outcomes are found: SS202 shows better erosion resistance as compared with mild steel and SS304 under all circumstances. Rotational speed is found to be a highly influencing parameter for the erosion wear rate as compared to solid concentration, time duration, and weighted mean diameter. WC-$_{10}$Co$_4$Cr HVOF coating improved the erosion resistance of piping materials up to 3.5 times. The predicted erosion wear is in agreement with the experimental data and found within a deviation of ±12%.

Acknowledgments: The authors would like to inform that this work did not receive any grant from any funding agency either from public or commercial sector.

Author Contributions: Kaushal Kumar and Satish Kumar conceived and designed the experiments; Kaushal Kumar and Gurprit Singh performed the experiments; Kaushal Kumar, Jashanpreet Singh and Jatinder Pal Singh analyzed the data; Kaushal Kumar and Jashanpreet Singh wrote the paper; Satish Kumar supervised the whole work. All the authors have read and approved the final manuscript.

Conflicts of Interest: The authors declare no conflict of interest.

References

1. Kumar, S.; Mohapatra, S.K.; Gandhi, B.K. Effect of addition of fly ash and drag reducing on the rheological properties of bottom ash. *Int. J. Mech. Mater. Eng.* **2013**, *8*, 1–8.

2. Gandhi, B.K.; Singh, S.N.; Seshadri, V. Study of the parametric dependence of erosion wear for the parallel flow of solid–liquid mixtures. *Tribol. Int.* **1999**, *32*, 275–282. [CrossRef]

3. Gandhi, B.K.; Singh, S.N.; Seshadri, V. A study on the effect of surface orientation on erosion wear of flat specimens moving in a solid-liquid suspension. *Wear* **2003**, *254*, 1233–1238. [CrossRef]

4. Chandel, S.; Singh, S.N.; Seshadri, V. An experimental study of erosion wear in a centrifugal slurry pump using coriolis wear test rig. *Particul. Sci. Technol.* **2012**, *30*, 179–195. [CrossRef]

5. Kosa, E.; Göksenli, A. Effect of impact angle on erosive abrasive wear of ductile and brittle materials. *Int. J. Mech. Aerosp. Ind. Mechatron. Manuf. Eng.* **2015**, *9*, 1638–1642.

6. Modi, O.P.; Dasgupta, R.; Prasad, B.K. Erosion of a high carbon steel in coal and bottom ash slurries. *J. Mater. Eng. Perform.* **2000**, *9*, 522–529. [CrossRef]

7. Desale, G.R.; Gandhi, B.K.; Jain, S.C. Slurry erosion of ductile materials under normal impact condition. *Wear* **2008**, *264*, 322–330. [CrossRef]

8. Prasad, B.K.; Jha, A.K.; Modi, O.P. Effect of sand concentration in the medium and travel distance and speed on the slurry wear response of a zinc-based alloy alumina particle composite. *Tribol. Lett.* **2004**, *17*, 301–309. [CrossRef]

9. Shitole, P.P.; Gawande, S.H.; Desale, G.R. Effect of impacting particle kinetic energy on slurry erosion wear. *J. Bio- Tribo-Corros.* **2015**, *1*, 29. [CrossRef]

10. Kulu, P.; Käerdi, H.; Surzenkov, A. Recycled hard metal-based powder composite coatings: Optimisation of composition, structure and properties. *Int. J. Mater. Prod. Technol.* **2014**, *49*, 180–202. [CrossRef]

11. Xie, Y.; Pei, X.; Wei, S. Investigation of erosion resistance property of WC-Co coatings. *Int. J. Surf. Sci. Eng.* **2016**, *10*, 365–374. [CrossRef]

12. Cho, T.Y.; Yoon, J.H.; Kim, K.S. A study on HVOF coatings of micron and nano WC-Co powders. *Surf. Coat. Technol.* **2008**, *202*, 5556–5559. [CrossRef]

13. Ma, H.R.; Li, J.W.; Jiao, J. Wear resistance of Fe-based amorphous coatings prepared by AC-HVAF and HVOF. *Mater. Sci. Technol.* **2017**, *33*, 65–71. [CrossRef]

14. Valentinelli, L.; Valente, T.; Casadei, F. Mechanical and tribocorrosion properties of HVOF sprayed WC–CO coatings. *Corros. Eng. Sci. Technol.* **2004**, *39*, 301–307. [CrossRef]

15. Gupta, R.; Singh, S.N.; Seshadri, V. Prediction of uneven wear in a slurry pipeline on the basis of measurements in a pot tester. *Wear* **1995**, *184*, 169–178. [CrossRef]

16. Desale, G.R.; Gandhi, B.K.; Jain, S.C. Improvement in the design of a pot tester to simulate erosion wear due to solid–liquid mixture. *Wear* **2005**, *259*, 196–202. [CrossRef]

17. Goyal, D.K.; Singh, H.; Kumar, H.; Sahni, V. Slurry erosion behavior of HVOF sprayed WC-$_{10}$Co$_4$Cr and Al$_2$O$_3$ + 13TiO$_2$ coatings on a turbine steel. *Wear* **2012**, *289*, 46–57. [CrossRef]

18. Truscott, G.F. *A Literature Survey on Wear on Pipelines*; Publication TN 1295; BHRA Fluid Engineering: Cranfield, UK, 1975.

19. Bajracharya, T.R.; Acharya, B.; Joshi, C.B.; Saini, R.P.; Dahlhaug, O.G. Sand erosion of Pelton turbine nozzles and buckets: A case study of Chilime hydropower plant. *Wear* **2008**, *264*, 177–184. [CrossRef]

8

Investigation of Coating Performance of UV-Curable Hybrid Polymers Containing 1H,1H,2H,2H-Perfluorooctyltriethoxysilane Coated on Aluminum Substrates

Mustafa Çakır

Department of Metallurgical and Materials Engineering, Faculty of Technology, Marmara University, Goztepe 34722, Istanbul, Turkey; mcakir@marmara.edu.tr

Academic Editor: Rita B. Figueira

Abstract: This study describes preparation and characterization of fluorine-containing organic-inorganic hybrid coatings. The organic part consists of bisphenol-A glycerolate (1 glycerol/phenol) diacrylate resin and 1,6-hexanediol diacrylate reactive diluent. The inorganically rich part comprises trimethoxysilane-terminated urethane, 1H,1H,2H,2H-perfluorooctyltriethoxysilane, 3-(trimethoxysilyl) propyl methacrylate and sol–gel precursors that are products of hydrolysis and condensation reactions. Bisphenol-A glycerolate (1 glycerol/phenol) diacrylate resin was added to the inorganic part in predetermined amounts. The resultant mixture was utilized in the preparation of free films as well as coatings on aluminum substrates. Thermal and mechanical tests such as DSC, thermo-gravimetric analysis (TGA), and tensile and shore D hardness tests were performed on free films. Water contact angle, gloss, Taber abrasion test, cross-cut and tubular impact tests were conducted on the coated samples. SEM examination and EDS analysis was performed on the fractured surfaces of free films. The hybrid coatings on the aluminum sheets gave rise to properties such as moderately glossed surface; low wear rate and hydrophobicity. Tensile strength of free films increased with up to 10% inorganic content in the hybrid structure and this increase was approximately three times that of the control sample. As expected; the % strain value decreased by 17.3 with the increase in inorganic content and elastic modulus values increased by a factor of approximately 6. Resistance to ketone-based solvents was proven and an increase in hardness was observed as the ratio of the inorganic part increased. Samples which contain 10% sol–gel content were observed to provide optimal properties.

Keywords: fluoropolymer coating; organic-inorganic hybrid; UV curable

1. Introduction

Aluminum and its alloys have superior properties as they are lightweight and have high thermal and electrical conductivity. Aluminum and aluminum alloys are of remarkable industrial and economic importance [1–4]. However, surfaces of unalloyed aluminum sheets have low scratch and abrasion resistance. Polymer-based coatings such as organic–inorganic hybrids are the proposed solutions for elimination of these shortcomings without losing the natural appearance. Hybrids originating from epoxy and urethane systems have been developed in recent years [5–8].

Organic–inorganic hybrid coating materials have become increasingly important materials due to their extraordinary properties such as resistance to scratching and abrasion, thermal stability, and higher chemical stability [9,10]. Hybrid materials derive these advantages from the synergy between the properties of each individual component [11]. Generally, the sol–gel route is utilized for the preparation of organic–inorganic hybrid materials. Various organo-metal precursors based on silicone, titanium, aluminum and zirconium have been used in sol–gel processes for synthesis of the inorganic part [12,13]. Organic–inorganic hybrid materials are cured generally thermally. However, the UV curing process has been used as an alternative to thermal curing in the preparation of hybrid materials for decades [14–16].

UV-curable coatings have various advantages such as higher chemical stability, lower harm for environment, lower process costs and higher cure rates. These types of coatings have been widely used on a variety of substrates such as paper, metal, plastic and wood since the curing process is achieved using low temperatures. UV-curable coating systems can be modified by incorporating different monomeric groups. Utilization of fluorosilane structures is one of the effective methods for modification of UV-curable systems, especially for the improvement of surface characteristics [17–21].

The use of fluorinated monomers and oligomers for coatings has attracted considerable attention owing to the peculiar characteristics provided by the presence of fluorine atoms. Fluorinated coatings have hydrophobicity, chemical stability, weathering resistance, low refractive indices, good release properties, low coefficients of friction, water impermeability, low surface free energies and non-stick properties [22–24].

In this study, organic–inorganic hybrid formulations were employed, for which the inorganic part consisted of silane-terminated diethylene glycol, 1H,1H,2H,2H-perfluorooctyl triethoxysilane and 3-(trimethoxysilyl)propyl methacrylate. The organic part consisted of bisphenol-A glycerolate (1 glycerol/phenol) diacrylate and 1,6 hexanediol diacrylate. In the organic part, 3-(trimethoxysilyl) propyl methacrylate has two different roles in the hybrid formulation; on one hand, it links covalently to functional groups of silane-terminated diethylene glycol by means of Si–O–Si bonds and on the other hand it provides methylacrylate groups that allow co-polymerization with the acrylic monomer during a UV-curing process [25]. Another component present in the sol–gel precursors is 1H,1H,2H,2H-perfluorooctyltriethoxysilanes. The main role of that component in the structure is to make the organic part hydrophobic. UV-curable hybrid materials were characterized using various methods such as gloss, cross-cut, water contact angle, hardness, Taber abrasion test, tubular impact test and tensile test. Morphology and thermal properties were also investigated.

2. Experimental

2.1. Materials

Bisphenol-A glycerolate (1 glycerol/phenol) diacrylate (Bis-GA), 1.6 hexanediol diacrylate (HDDA), 3-isocyanatopropyl trimethoxysilane 95% (ICPTMS), diethylene glycol $\geq$ 99.5% (DEG), dibutyltin dilaurate, 3-(trimethoxysilyl) propyl methacrylate 98% (MEMO), 1-hydroxy-cyclohexyl-phenyl-ketone and p-toluenesulfonic acid (PTSA) were all obtained from Sigma-Aldrich. In addition, 1H,1H,2H,2H-perfluorooctyltriethoxysilane 97% (PTES) was purchased from ABCR. Anode-oxidized aluminum sheets (100 mm × 50 mm × 1 mm) were used as substrates in all coating applications.

2.2. Synthesis of Silane-Terminated Diethylene Glycol

Firstly, 5 g of diethylene glycol was charged to a 100-mL three-necked round bottom flask which was fitted with a thermometer pocket, water condenser and a magnetic stirrer. Then, 23.3094 g of 3-isocyanatopropyl trimethoxysilane was slowly added to the reaction flask. Dibutyltin dilaurate was added to the reaction flask as a catalyst at a concentration of 0.5%. The temperature was then raised to 60 °C and the mixture was stirred for 8 h. Completion of the urethane reaction was confirmed by the disappearance of the characteristic isocyanate (NCO) peak at 2275 cm^{-1} in the FT-IR spectrum.

2.3. Modification of Silane-Terminated Diethylene Glycol

A 250-mL three necked round bottom flask, equipped with a magnetic stirrer, a dropping funnel and a nitrogen inlet, was filled with 24.0461 g of 1H,1H,2H,2H-perfluorooctyl triethoxysilane (PTES),

35.0918 g of 3-(trimethoxysilyl)propyl methacrylate (MEMO) and 28.3094 g of silane-terminated diethylene glycol (STDEG). Ethanol was added to the reaction mixture. The mixture was then homogenized by mixing with a magnetic stirrer. The water/silicone ratio was calculated as $r = 3$ and the calculated amount of water was added to the mixture. Paratoluenesulfonic acid corresponding to 0.1 wt % of the above mixture was added as catalyst. The mixture was stirred for 4 h at room temperature with a magnetic stirrer. A schematic representation of the modification of STDEG is given in Figure 1. In this formulation, 3-(trimethoxysilyl) propyl methacrylate in the inorganic part has two different roles. In one of the roles, it links covalently to hydroxyl groups of STDEG by means of Si–O–Si bonds, the other role; it provides methylacrylate groups that allows co-polymerization with the acrylic monomer during a UV-curing process. Another component contained in the sol–gel precursors is 1H,1H,2H,2H-perfluorooctyltriethoxysilane. The main role of that component in the structure is to make the organic part hydrophobic.

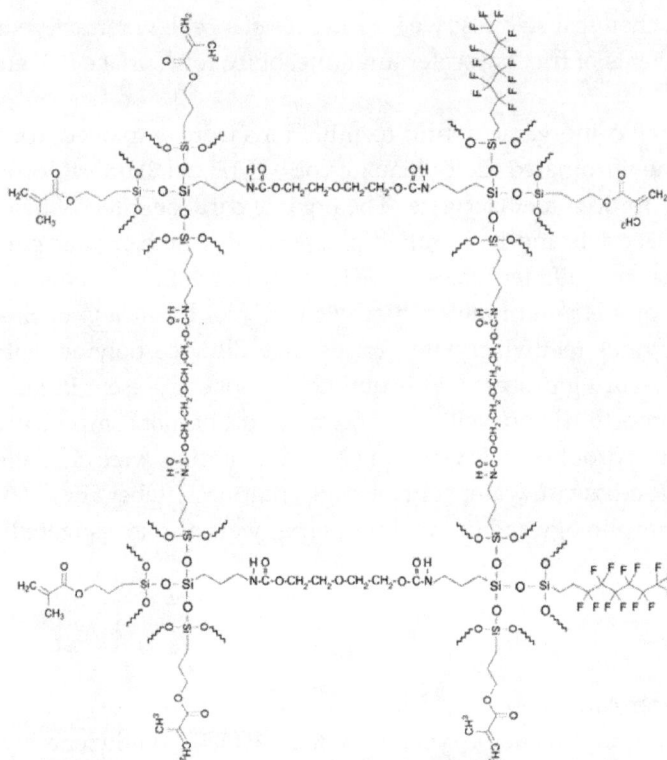

Figure 1. Schematic representation of silane-terminated diethylene glycol (STDEG) modified by both 1H,1H,2H,2H-perfluorooctyl triethoxysilane (PTES) and 3-(trimethoxysilyl)propyl methacrylate (MEMO).

2.4. Preparation of Hybrid Coatings

Hybrid coating mixtures were prepared by adding pre-hydrolyzed PTES, MEMO and STDEG to the organic part in predetermined (2.5 wt %, 5 wt %, 7.5 wt %, 10 wt %, 15 wt % and 20 wt %) ratios. The organic part of formulation comprised a UV-curable resin (Bis-GA) and a reactive diluent (HDDA). Compositions of hybrid coatings are presented in Table 1. A 3 wt % this photoinitiator was added to each of the hybrid compositions. Each composition was stirred until all the components were miscibilised completely. The efficiency of mixing was assessed by the amount of cloudiness visible during the mixing. Hybrid compositions were heated at 35 °C in a vacuum oven for about 10 min to remove the trapped air formed during the mixing process. These solutions were applied onto aluminum sheets by using a wire wound bar applicator to obtain a layer thickness of 30 μm. The obtained coatings were cured by a Raycon UV drying system with a medium pressure mercury lamp (150 W/cm, λ_{max}: 320–390 nm), whereas the speed of conveyor belt was 5 m/min. Free films were prepared by pouring the UV curable hybrid formulations into a mold made out of Teflon® (supplied by DuPont, İstanbul, Turkey) with dimensions 10 mm × 100 mm × 1 mm. The free films were cured by passing the Teflon mold through the UV drying system under similar curing conditions applied to that of coatings.

Table 1. Composition of hybrid coatings. Bis-GA: bisphenol-A glycerolate (1 glycerol/phenol) diacrylate; HDDA: 1.6 hexanediol diacrylate.

Notation	Organic Part		Sol–Gel Precursors			Photoinitiator
	Bis-GA (wt %)	HDDA (wt %)	PTES (wt %)	MEMO (wt %)	STDEG (wt %)	Irgacure 184 Total (wt %)
S-0	80	20	–	–	–	3
S-2.5	78	19.5	0.6874	1.0032	0.8093	3
S-5	76	19	1.3748	2.0064	1.6186	3
S-7.5	74	18.5	2.0623	3.0096	2.4279	3
S-10	72	18	2.7497	4.0129	3.2373	3
S-15	68	17	4.1246	6.0193	4.8559	3
S-20	64	16	5.4995	8.0258	6.4746	3

2.5. Characterization

The FT-IR spectrum of the silane-terminated diethylene glycol was recorded on a Shimadzu 8300 FT-IR spectrophotometer (Shimadzu, Kyoto, Japan). The soluble gel content of the UV-cured free films was determined by soxhlet extraction method for 8 h using pure acetone. Insoluble gel fraction was dried in vacuum oven at 80 °C to constant weight and the soluble gel content was calculated. Thermo-gravimetric analysis (TGA) of the UV-cured free films was performed by using a Perkin Elmer thermo-gravimetric analyzer (Perkin Elmer, Waltham, MA, USA). The tests were run from 30 to 800 °C with a heating rate of 20 °C/min under an air atmosphere. DSC analysis of the hybrid films was performed by using Perkin Elmer DSC equipment (Perkin Elmer, Waltham, MA, USA). The tests were run from 20 to 220 °C with a heating rate of 10 °C/min under a N_2 atmosphere. The fracture surface of the free films was examined on JOEL JSM-5910 LV SEM equipment (Joel, Tokyo, Japan). Elemental concentrations of silicon, fluorine, carbon and oxygen atoms were determined by means of an Oxford Instruments-INCA energy dispersive spectrum (EDS) system (Oxford Instruments, Oxfordshire, UK). Water contact angles were measured by using a Krüss DSA-2 goniometer (Krüss, Hamburg, Germany). The volume of droplets was controlled to be about 3 μL. Gloss property of the coated surfaces was measured by using a BYK-Gardner Glossmeter (BYK-Gardner, Geretsried, Germany) at an angle of 60°. Mechanical properties of the specimens were determined by standard tensile test to measure the modulus (E), tensile strength (σ) and elongation at break (ε). Tensile experiments were performed at room temperature on a materials testing system Zwick Z010, using a crosshead speed of 5 mm/min. The impact resistance was assessed by falling weight method in a tubular impact tester (Sheen Instruments, Surrey, UK) (ISO 6272), 908 g weight, 25″ Tube BS 6496-84. The solvent resistance of coatings was examined by performing the methyl ethyl ketone (MEK) rubbing test, according to ASTM D-5402. Adhesion properties of the coated aluminum sheets were determined by cross-cut test. For this purpose, two sets of 6 cuts with a spacing of 1 mm were made perpendicular to each other, thus making a lattice of 25 small blocks. Adhesion of coatings was determined through a tape adhesion test according to ASTM D-3359 standard by using a TESA™ 4124 tape. Then, a standardized tape was stuck on the lattice and pulled off with a constant force. The number of blocks removed was an indication of the adhesion. Hardness of free films was measured using a Zwick hardness tester (Zwick, Ulm, Germany). This test was performed according to ASTM D-2240 using Shore-D scale. The abrasion resistance of the coatings was determined by using a standardized Taber abrasion tester (Taber Industries, New York, NY, USA) ASTM D4060. CS-10 wheels were used with a 500-g load on each wheel, and the substrates were abraded at 100, 200 and 300 cycles.

3. Results and Discussion

Completion of the urethane reaction was confirmed by the disappearance of the characteristic NCO peak at 2275 cm^{-1}, 3339 cm^{-1} stretching N–H, 1692 cm^{-1}, C=O associated urethane and isocyanurate ring stretch in the FT-IR spectrum, as shown in Figure 2.

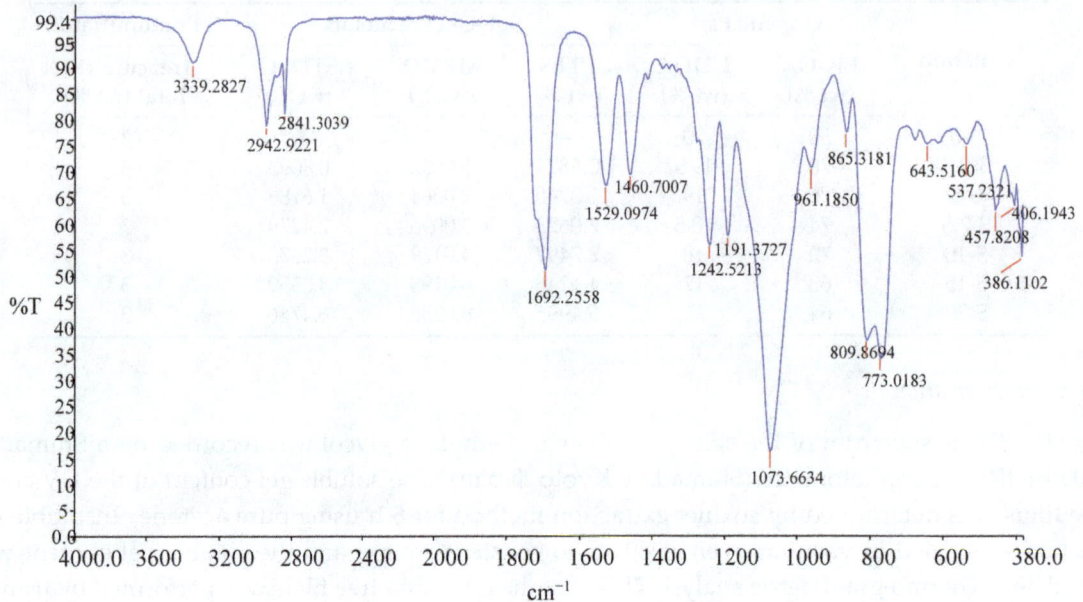

Figure 2. FT-IR spectrum of silane terminated diethylene glycol.

Figure 3 shows the contact angles of UV-cured coatings as a function of sol–gel content in the coating formulation. The contact angles of the hybrid coatings increased with the increase in sol–gel content, because the sol–gel formulation contained a high proportion of fluorine which was added to increase the hydrophobicity of the coating. The water contact angle of the control sample was 56.86°. With the incorporation of only 2.5 wt % sol–gel content, the contact angle increased to 85.97°. Only a very small amount of fluorine content significantly affected the contact angle. The water contact angle reached 98.71° at 15% sol–gel content. These results suggested that the CF$_3$ groups in hybrid coating increased the contact angle values. When the contact angle value of the surface is greater than 90°, the surface can be called hydrophobic [26,27]. Therefore coatings having sol–gel content greater than 5% can be considered within the class of hydrophobic coatings.

Gloss measures the amount of light reflected from a surface at a given angle and it is a significant coating property when the primary concern is aesthetics or decorative appearance. Surface roughness in clear coating influences gloss to a significant extent [28]. Besides surface roughness, gloss is also affected by the topographical orientation of the microfacets and refractive index of the coatings. The higher refractive index is, the higher surface gloss generally becomes [29]. The gloss values of the coatings were measured at 60° according to ASTM D523. The gloss values of control samples (S-0) were measured as 98.8. Increases in gloss were observed with increasing sol–gel content in the coating formulation. Among the previous studies, Tang et al [30] stated that increasing fluorine rate led to a rise in the value of gloss. Gloss value reached up to 109 in the sample with 20% sol–gel content. All gloss values can be seen in Table 2. Mulazim et al. [20] used PTES in their formulation. While they found the highest value at 107 for gloss 60°, the highest value for gloss 20° was found to be 86.

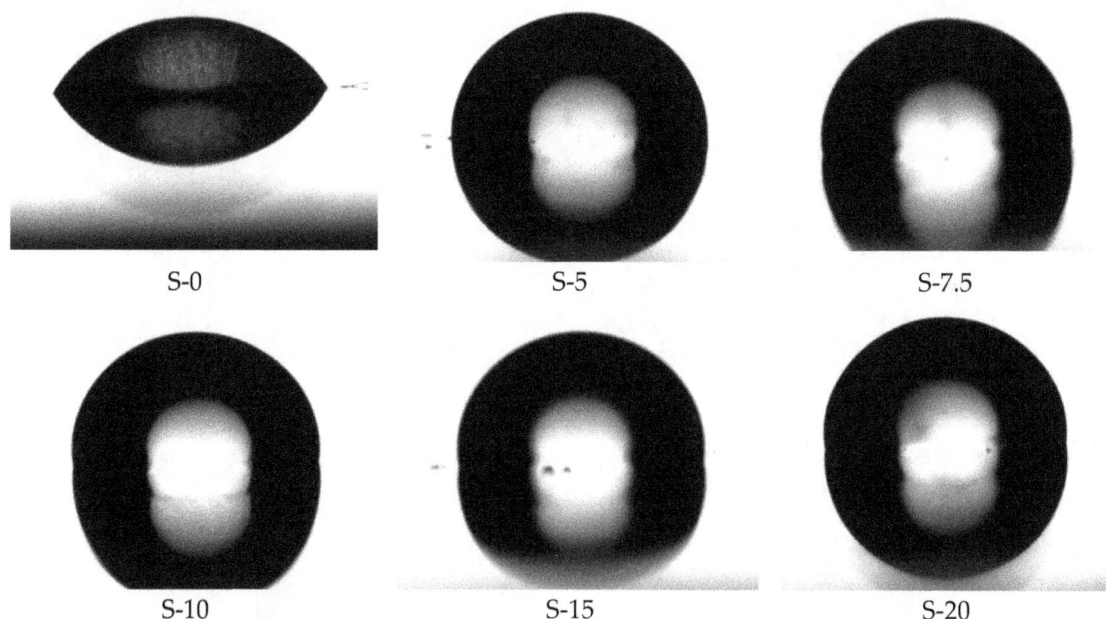

Figure 3. Contact angles of the UV-cured coatings depending on sol–gel content.

Table 2. Some of the physical properties of coatings.

Notation	Water Contact Angle	Gloss 60°	Cross-Cut ASTM D-3002	Impact Test (inch)	Hardness Shore-D	Soxhlet Extraction (% solid)	MEK Double Rub Test
S-0	56.86	98.2	5B	2	68	99.7	200
S-2.5	85.97	102.6	5B	2	71	99.8	200
S-5	95.22	103.5	5B	3	73	99.8	250
S-7.5	97.72	104.8	5B	3	73	99.6	300
S-10	98.65	107.5	4B	2	73	99.5	300
S-15	98.71	108.9	4B	1	73	99.8	300
S-20	98.10	109	3B	1	74	99.8	300

Cross-cut adhesion test was performed on the coatings according to ASTM D-3002 to determine adhesion properties of the coatings. Decreases in adhesion properties were observed with increasing ratios of sol–gel (Table 2). The reason for these decreases was the increased fluorine content. At first glance, it may be considered that adhesion properties can be improved by increasing the ratios of fluorosilane reacting with Al_2O_3 on aluminum sheets [31]. However wettability of hybrid coating formulation is decreased by increased fluorosilane content.

The ability of the coating to resist cracking led by rapid deformation was predicted by tubular impact resistance test. An indenter of fixed weight was dropped on the test surface through a graduated guided tube in this test, and a magnification lens was used to visually examine the cracks caused by rapid impact. An indenter with a hemispherical head and a diameter of 15.9 mm and 908-g load was used to conduct the tubular impact resistance test. Figure 4 shows the images of this test. At a height of one inch, only S-15 and S-20 coatings experienced cracks. Cracks formed at S-0, S-2.5 and S-10 coatings at the two-inch height impact test. S-5 and S-7.5 coating had cracks at the impact test with a height of three inches. Coatings cracked at different heights because increasing inorganic part content caused a decrease in the toughness of coatings, leading to easily-occurring cracks. Increase of sol–gel content in S-5 and S-7.5 coatings may explain the increased adhesion strength without causing decrease in toughness.

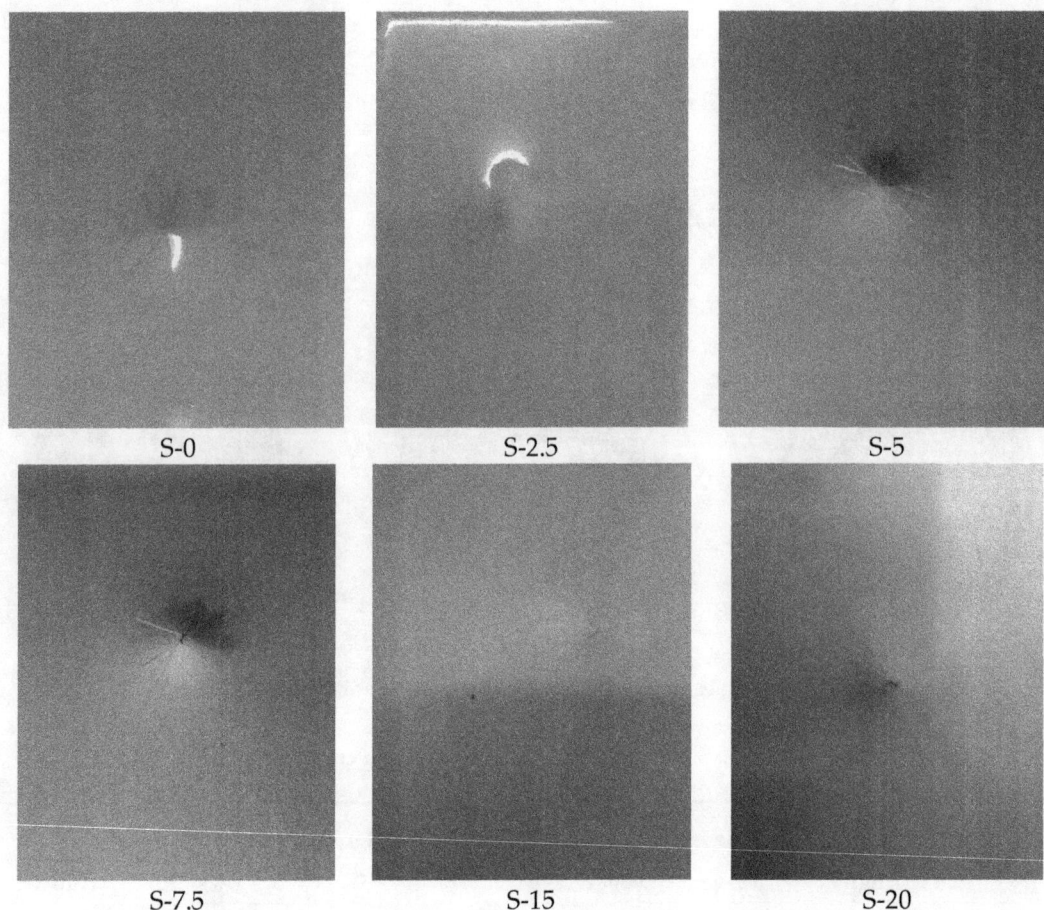

Figure 4. Picture of cracks that occurred on coatings after tubular impact test.

According to shore-D hardness test performed on free film, free film hardness values did not show significant increases by increasing sol–gel content (Table 2). The Shore-D hardness value of the S-0 sample was 68, whereas the Shore-D hardness value of the S-20 sample was 74.

UV-cured hybrid films were extracted with acetone in order to remove soluble ingredients. The weight loss of hybrid films was measured after 24 h in 80 °C in a drying process under vacuum. The solid content of the films was found to be between 99.6% and 99.8%. The solvent resistance of coatings was examined by performing MEK rubbing test, according to ASTM D-5402. MEK rubbing test values ranged from 200 to 300 cycles. Rub test values were observed to increase with increasing sol–gel content.

The sample films were prepared by free film manufacturing process in Teflon® mold. Their mechanical properties were evaluated and tabulated in Table 3. According to tensile test results, maximum tensile strength was increased with increasing sol–gel content. Tensile strength value of control sample (S-0) was 12.14 MPa. In contrast, the tensile strength value of the sample S-10 reached 31.21 MPa. An approximately threefold increase was observed in tensile strength. It was expected that elongation of coatings would decrease depending on sol–gel content.

As the fluorine-containing sol–gel content increased, % strain of the material decreased as expected. This is because inorganic materials have a lower % strain in comparison to organic materials. As the inorganic part in the hybrid structure increased, there was a decrease in % strain and it became rigid. When we examine the elastic modulus values of the samples, it is possible to see that the elasticity modulus increases with the increase of the sol–gel content. This increase in the elastic modulus provides more evidence of the hardening of the material due to the increases in sol–gel content.

Table 3. Mechanical properties of free films.

Notation	Maximum Tensile Strength (σ_M, MPa)	Strain at F_{max} (ε_M, %)	E-Modulus (E_t, MPa)
S-0	12.14	3.05	316.49
S-2.5	16.69	2.38	651.71
S-5	21.73	2.73	881.19
S-7.5	29.51	2.89	1127.96
S-10	31.21	2.85	1278.76
S-15	26.81	2.79	1427.67
S-20	19.34	2.52	1821.05

Figure 5 shows the results of the Taber abrasion test carried out using CS-10 wheels. A significant decrease was observed in weight loss due to the increases in the proportion of the fluorine-containing inorganic component. This decrease might be explained by the reduction of friction resistance due to the increase in the amount of fluorine. Previous studies have reported high abrasion resistance for coatings that have low friction resistance. The studies by McGrath et al. [32] also show decreases in friction resistance with the increasing fluorine content in the polymeric coatings.

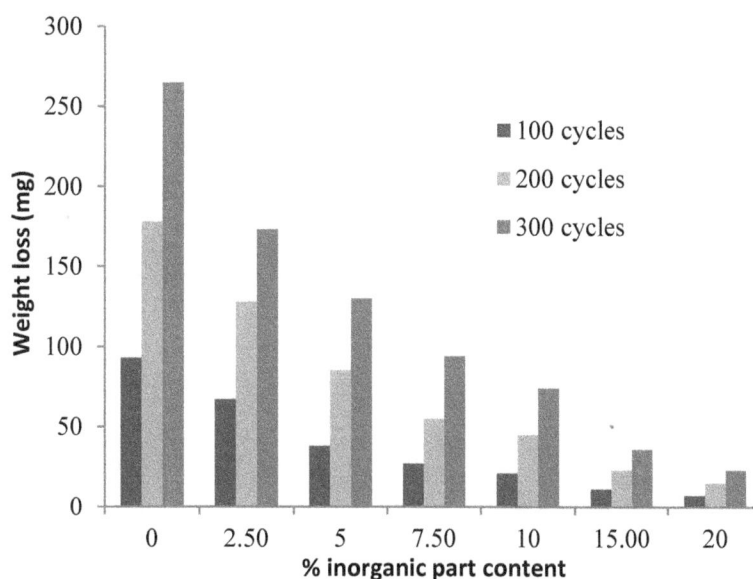

Figure 5. Abrasion resistance of coatings as a function of % inorganic part content.

Thermal properties of the hybrid films were evaluated by TGA and DSC analyses. When we examine the TGA thermograms of the hybrid films given in Figure 6, it can be seen that 10% weight loss occurs in the range of 358.5–393 °C (Table 4). 50% weight loss was at about 430 °C. Char yield increased with increasing sol–gel content as expected. However, char yield did not increase linearly with increasing inorganic part that was observed.

DSC thermograms of the samples of free films are given in Figure 7. It may be observed in Table 4 that the S-0 sample had a higher glass transition temperature (T_g = 69.27 °C) than the S-20 sample (T_g = 68.63 °C). The values of glass transition temperature (T_g) of the S-0 sample decreased with the increasing fluorine content in the inorganic part. Increases in fluorine content caused a decrease in T_g values [17,29].

SEM images of fracture surfaces of free films of the S-0, S-7.5 and S-20 samples are presented in Figure 8a–c. It is possible to see from the S-0 sample images that the unstable crack propagation took place. However, unstable crack lines were not observed on surface fracture of the S-7.5 and S-20 samples. The fracture of these samples showed a brittle tendency due to the inorganic part. It is also possible to see that phase separation does not occur when the fractured surfaces in Figure 8b,c are examined. The inorganic phase does not precipitate as a separate region in the organic phase.

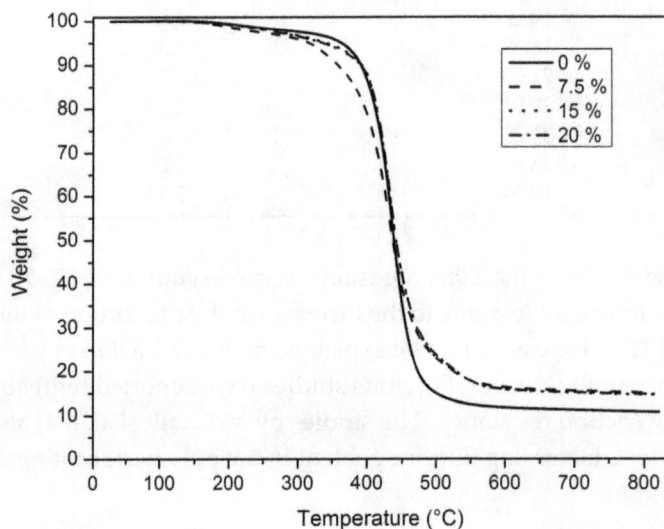

Figure 6. Thermo-gravimetric analysis (TGA) thermograms of the hybrid films.

Table 4. Thermal analysis results of UV curable hybrid films.

Notation	10% Weight Loss Temperature (°C)	50% Weight Loss Temperature (°C)	Final Weight Loss Temperature (°C)	Char Yield (%)	Glass Transition Temperature (T_g, °C)
S-0	391	438	680	10.8	69.27
S-7.5	358.5	429	685	14.66	66.49
S-15	393	434	685	14.80	68.82
S-20	391.3	427	705	15.13	68.63

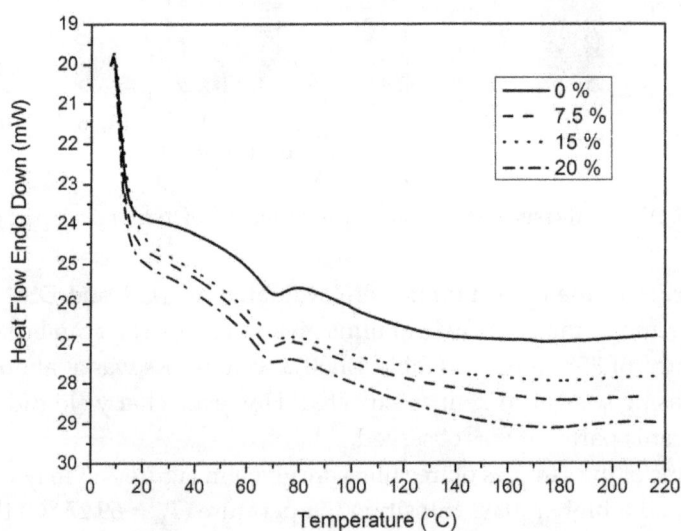

Figure 7. DSC curves of UV-curable hybrid films.

Elemental analysis of the S-20 sample was carried out by SEM-EDS analysis. SEM-EDS spectrum taken from fractured surface of the S-20 sample revealed the presence of 2.56 wt % silicon. This result confirms the effective incorporation of the inorganic precursor to the system.

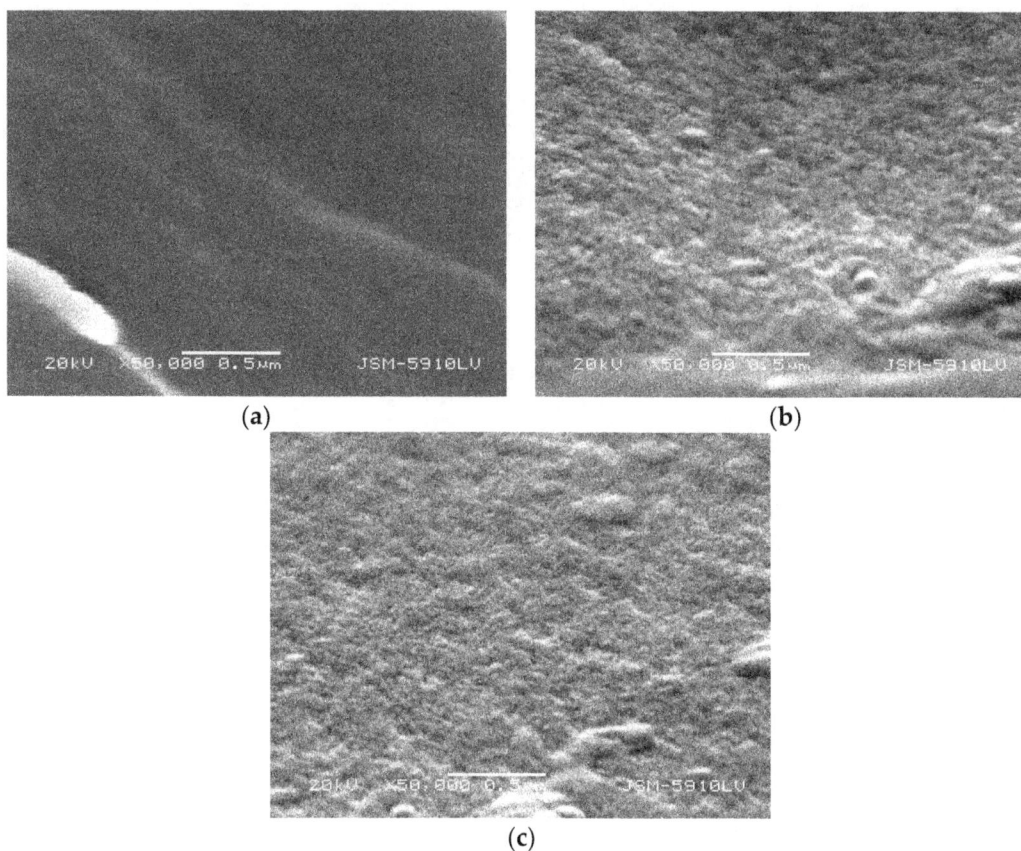

Figure 8. SEM images of the fracture surface of (a) S-0, (b) S-7.5, and (c) S-20 samples.

4. Conclusions

This study aimed to improve surface properties of aluminum sheets using fluorine-containing UV-curable hybrid coating formulations. Thus, hybrid coating mixtures were prepared by adding pre-hydrolyzed PTES, MEMO and STDEG to the organic part in predetermined ratios. The organic part of formulation comprised a UV-curable resin (Bis-GA), a reactive diluent (HDDA) and a photo-initiator. This prepared solution was applied onto the surface of aluminum sheets using a wire wound applicator. The coatings were then cured by passing through a UV dryer.

A significant increase was observed in the contact angles. The surfaces gained hydrophobic characteristics due to the increases in contact angles. Gloss 60° measurement values increased in parallel to the increases in the fluorine-containing inorganic part. Wear resistance of aluminum was improved without losing its natural appearance and gloss. Its resistance to ketone-based solvents was proven and an increase in hardness was observed as the ratio of the inorganic component increased. As expected, the tubular impact test values decreased with the increase in inorganic content. The adhesion of the coating to the aluminum surface was observed to be good up to an inorganic ratio of 10%, but it decreased after this ratio was exceeded. Tensile strength of free films increased with up to 10% inorganic content, and this increase reached approximately three times that of the pure sample at 31.21 MPa. As expected, the % strain value decreased by 17.3 with the increase in inorganic content and elastic modulus values increased by a factor of approximately 6. An approximately 1 °C decrease was observed in the T_g value by the increase in fluorine-containing sol–gel mixtures. Likewise, char efficiency in TGA analysis values rose from 10.8 to the value of 15.13. Considering all values mentioned above, the S-10 sample containing 10% sol–gel content was observed to provide optimal properties. It may be suggested that the formulation of the S-10 sample is suitable for coatings used for protecting the original appearance of aluminum sheets.

Conflicts of Interest: The author declares no conflict of interest.

References

1. Despic, A.; Parkhutik, V.P. Electrochemistry of Aluminum in Aqueous Solutions and Physics of Its Anodic Oxide. In *Modern Aspects of Electrochemistry*; Bockris, J.O.M., Conway, B.E., White, R.E., Eds.; Plenum Publishing: New York, NY, USA, 1989; No. 20; pp. 401–503.

2. Sheffer, M.; Groysman, A.; Mandler, D. Electrodeposition of sol–gel films on Al for corrosion protection. *Corros. Sci.* **2003**, *45*, 2893–2904. [CrossRef]

3. Ashassi-Sorkhabi, H.; Ghasemi, Z.; Seifzadeh, D. The inhibition effect of some amino acids towards the corrosion of aluminum in 1 M HCl + 1 M H_2SO_4 solution. *Appl. Surf. Sci.* **2005**, *249*, 408–418. [CrossRef]

4. Wernick, S.; Pinner, R.; Sheasby, P.G. The surface treatment of aluminum and its alloys. *ASM Int.* **1987**, *1987*, 1264.

5. Du, Y.J.; Damron, M.; Tang, G.; Zheng, H.; Chu, C.-J.; Osborne, J.H. Inorganic/organic hybrid coatings for aircraft aluminum alloy substrates. *Prog. Org. Coat.* **2001**, *41*, 226–232.

6. Liu, H.; Zheng, S.; Nie, K. Morphology and thermomechanical properties of organic–inorganic hybrid composites involving epoxy resin and an incompletely condensed polyhedral oligomeric silsesquioxane. *Macromolecules* **2005**, *38*, 5088–5097. [CrossRef]

7. Xu, J.; Pang, W.; Shi, W. Synthesis of UV-curable organic–inorganic hybrid urethane acrylates and properties of cured films. *Thin Solid Films* **2006**, *514*, 69–75. [CrossRef]

8. Pathak, S.S.; Sharma, A.; Khanna, A.S. Value addition to waterborne polyurethane resin by silicone modification for developing high performance coating on aluminum alloy. *Prog. Org. Coat.* **2009**, *65*, 206–216. [CrossRef]

9. Wen, J.; Wilkes, G.L. Organic/inorganic hybrid network materials by the sol-gel approach. *Chem. Mater.* **1996**, *8*, 1667–1681. [CrossRef]

10. Zandi-zand, R.; Ershad-langroudi, A.; Rahimi, A. Silica based organic–inorganic hybrid nanocomposite coatings for corrosion protection. *Prog. Org. Coat.* **2005**, *53*, 286–291. [CrossRef]

11. Sanchez, C.; Soler-Illia, A.; Ribot, F.; Lalot, T.; Mayer, C.R.; Cabuil, V. Designed hybrid organic-inorganic nanocomposites from functional nanobuilding blocks. *Chem. Mater.* **2001**, *13*, 3061–3083. [CrossRef]

12. Karatas, S.; Kayaman-Apohan, N.; Turunc, O.; Gungor, A. Synthesis and characterization of UV-curable phosphorus containing hybrid materials prepared by sol–gel technique. *Polym. Adv. Technol.* **2011**, *22*, 567–576. [CrossRef]

13. Tshabalala, M.A.; Sung, L.-P. Wood surface modification by in-situ sol-gel deposition of hybrid inorganic–organic thin film. *J. Coat. Technol. Res.* **2007**, *4*, 483–490. [CrossRef]

14. Zhang, L.; Zeng, Z.; Yang, J.; Chen, Y. Structure–property behavior of UV-curable polyepoxy-acrylate hybrid materials prepared via sol–gel process. *J. Appl. Polym. Sci.* **2003**, *87*, 1654–1659. [CrossRef]

15. Soppera, O.; Croutxe, C. Real-time fourier transform infrared study of free-radical UV-induced polymerization of hybrid sol–gel. I. Effect of silicate backbone on photopolymerization kinetics. *J. Polym. Sci. Part A Polym. Chem.* **2003**, *41*, 716–724. [CrossRef]

16. Kahraman, M.V.; Kugu, M.; Menceloglu, Y.; Kayaman-Apohan, N.; Gungor, A. The novel use of organo alkoxy silane for the synthesis of organic–inorganic hybrid coatings. *J. Non-Cryst. Solids* **2006**, *352*, 2143–2151. [CrossRef]

17. Beytut, H.; Cakir, M.; Kartal, I.; Sengul, M.; Tutak, D. Investigation of properties of fluorine containing hybrid coatings intended for use in printing. *Asian J. Chem.* **2015**, *27*, 3854–3860. [CrossRef]

18. Wang, T.; Isimjan, T.T.; Chen, J.; Rohani, S. Transparent nanostructured coatings with UV-shielding and superhydrophobicity properties. *Nanotechnology* **2011**, *22*, 265708. [CrossRef] [PubMed]

19. Yavas, H.; Öztürk, S.C.D.; Özhan, A.E.S.; Durucan, C. A parametric study on processing of scratch resistant hybrid sol–gel silica coatings on polycarbonate. *Thin Solid Films* **2014**, *556*, 112–119. [CrossRef]

20. Mulazim, Y.; Cakmakcı, E.; Kahraman, M.V. Photo-curable highly water-repellent nanocomposite coatings. *J. Vinyl Addit. Technol.* **2013**, *10*, 31–38. [CrossRef]

21. Park, S.; Nam, S.; Kim, L.; Park, M.; Kim, J.; An, T.K.; Yun, W.M.; Jang, J.; Hwang, J.; Park, C.E. Synthesis and characterization of a fluorinated oligosiloxane-containing encapsulation material for organic field-effect transistors, prepared via a non-hydrolytic sol–gel process. *Org. Electron.* **2012**, *13*, 2786–2792. [CrossRef]

22. Sangermano, M.; Bongiovanni, R.; Malucelli, G.; Priola, A.; Pollicino, A.; Recca, A. Fluorinated epoxides as surface modifying agents of UV-curable systems. *J. Appl. Polym. Sci.* **2003**, *89*, 1524–1529. [CrossRef]

23. Bongiovanni, R.; Malucelli, G.; Sangermano, M.; Priola, A. Properties of UV-curable coatings containing fuorinated acrylic structure. *Prog. Org. Coat.* **1999**, *36*, 70–78. [CrossRef]

24. Lin, Y.-H.; Liao, K.-H.; Huang, C.-K.; Chou, N.-K.; Wang, S.-S.; Chu, S.-H.; Hsieh, K.-H. Superhydrophobic films of UV-curable fluorinated epoxy acrylate resins. *Polym. Int.* **2010**, *59*, 1205–1211. [CrossRef]

25. Chang, C.-C.; Lin, Z.-M.; Huang, S.-H.; Cheng, L.-P. Preparation of highly transparent 13F-modified nano-silica/polymer hydrophobic hard coatings on plastic substrates. *J. Appl. Sci. Eng.* **2015**, *18*, 387–394.

26. Lau, K.K.S.; Bico, J.; Teo, K.B.K.; Chhowalla, M.; Amaratunga, G.A.J.; Milne, W.I.; McKinley, G.H.; Gleason, K.K. Superhydrophobic carbon nanotube forests. *Nano Lett.* **2003**, *3*, 1701–1705. [CrossRef]

27. Bhushan, B.; Jung, Y.C. Wetting study of patterned surfaces for superhydrophobicity. *Ultramicroscopy* **2007**, *107*, 1033–1041. [CrossRef] [PubMed]

28. Decker, C.; Keller, L.; Zahouily, K.; Benfarhi, S. Synthesis of nanocomposite polymers by UV-radiation curing. *Polymer* **2005**, *46*, 6640–6648. [CrossRef]

29. Kahraman, M.V.; Bayramoğlu, G.; Boztoprak, Y.; Güngör, A. Synthesis of fluorinated/methacrylated epoxy based oligomers and investigation of its performance in the UV curable hybrid coatings. *Prog. Org. Coat.* **2009**, *66*, 52–58. [CrossRef]

30. Tang, C.; Liu, W.; Ma, S.; Wang, Z.; Hu, C. Synthesis of UV-curable polysiloxanes containing methacryloxy/fluorinated side groups and the performances of their cured composite coatings. *Prog. Org. Coat.* **2010**, *69*, 359–365. [CrossRef]

31. Saleema, N.; Sarkar, D.K.; Gallant, D.; Paynter, R.W.; Chen, X.-G. Chemical nature of superhydrophobic aluminum alloy surfaces produced via a one-step process using fluoroalkyl-silane in a base medium. *ACS Appl. Mater. Interfaces* **2011**, *3*, 4775–4781. [CrossRef] [PubMed]

32. Kim, Y.-S.; Lee, J.-S.; Ji, Q.; McGrath, J.E. Surface properties of fluorinated oxetane polyol modified polyurethane block copolymers. *Polymer* **2002**, *43*, 7161–7170. [CrossRef]

9

Application of High-Velocity Oxygen-Fuel (HVOF) Spraying to the Fabrication of Yb-Silicate Environmental Barrier Coatings

Emine Bakan [1,*], Georg Mauer [1], Yoo Jung Sohn [1], Dietmar Koch [2] and Robert Vaßen [1]

[1] Forschungszentrum Jülich GmbH, Institute of Energy and Climate Research (IEK-1), 52425 Jülich, Germany; g.mauer@fz-juelich.de (G.M.); y.sohn@fz-juelich.de (Y.J.S.); r.vassen@fz-juelich.de (R.V.)

[2] German Aerospace Center, Institute of Structures and Design, 70569 Stuttgart, Germany; Dietmar.Koch@dlr.de

* Correspondence: e.bakan@fz-juelich.de

Academic Editor: Yasutaka Ando

Abstract: From the literature, it is known that due to their glass formation tendency, it is not possible to deposit fully-crystalline silicate coatings when the conventional atmospheric plasma spraying (APS) process is employed. In APS, rapid quenching of the sprayed material on the substrate facilitates the amorphous deposit formation, which shrinks when exposed to heat and forms pores and/or cracks. This paper explores the feasibility of using a high-velocity oxygen-fuel (HVOF) process for the cost-effective fabrication of dense, stoichiometric, and crystalline $Yb_2Si_2O_7$ environmental barrier coatings. We report our findings on the HVOF process optimization and its resultant influence on the microstructure development and crystallinity of the $Yb_2Si_2O_7$ coatings. The results reveal that partially crystalline, dense, and vertical crack-free EBCs can be produced by the HVOF technique. However, the furnace thermal cycling results revealed that the bonding of the $Yb_2Si_2O_7$ layer to the Silicon bond coat needs to be improved.

Keywords: environmental barrier coatings (EBC); ytterbium silicate; high-velocity oxygen-fuel (HVOF)

---

1. Introduction

Silicon carbide (SiC) fiber reinforced SiC ceramic matrix composites (CMCs), having a relatively low density and higher temperature capability compared to their metallic superalloy counterparts, are notable materials for the future of gas turbine engine technology. A significant enhancement in gas turbine efficiency leading to lower fuel consumption and a lower emission of NOx are anticipated with the implementation of CMCs in the gas turbine engines [1–5].

However, the lack of durable CMCs in combustion environments interferes with this implementation, as water vapor reacts with the silica scale that forms on the CMCs, leading to the formation of gaseous reaction products such as $Si(OH)_4$ [4,6]. An enhanced recession rate of the CMCs in high pressure and high gas velocity combustion atmospheres reveals the requirement of protective environmental barrier coatings (EBCs) for their long-term use under these conditions [7–9]. The establishment of environmentally stable (i) and well-adhering coatings on the CMC substrate (ii) that have a chemical compatibility within the layers (iii), and that develop minor stresses as a result of thermal expansion mismatch and phase transformations etc; (iv) are desired for this purpose [10].

EBC systems consisting of a silicon bond coat and a rare-earth (RE) silicate (e.g., $Yb_2Si_2O_7$, $Lu_2Si_2O_7$) top coat layer were successfully tested in gas turbine engines [10–14]. In these EBC systems, RE-silicates with a high water vapor stability impede the diffusion of oxygen and water vapor to the substrate. In the meantime, the silicon bond coat plays an important role in the oxidation protection of the substrate, as well as for the adhesion of the coating on the substrate. While the RE-silicates were shown to be promising, the fabrication of crystalline and single-phase silicate coatings with a dense microstructure has been troublesome when using the conventional plasma spraying process. Modification in the plasma spraying process was thus proposed, suggesting the utilization of a furnace in which the substrate to be deposited is placed and heated up to 1200 °C [15]. By this adjustment, the deposition of dense and crystalline RE-silicate coatings was found to be ensured, yet, secondary phases in the coatings as a result of the evaporation of Si-containing species could not be avoided [16]. Furthermore, the feasibility and adaptation of such a process involving a furnace for the big and complex-shaped components are questionable. In a recent study [17], it was reported, based on initial trials, that the HVOF process can be an alternative method to achieve dense $Yb_2Si_2O_7$ coatings with desired properties. This is because, in contrast with the APS process, HVOF offers a low heat transfer to the particles (flame temperatures up to 3000 °C), reducing the molten fraction of the sprayed powders which is expected to be favorable for obtaining crystalline and stoichiometric coatings. Moreover, the high particle velocities achieved in HVOF typically provide well-bonded particles resulting in high-density coatings. To this end, we further investigated the HVOF processing of $Yb_2Si_2O_7$ coatings and address our findings in this study in terms of process optimization, microstructure, composition, and crystallinity. The crystallinity, as well as the adhesion of the $Yb_2Si_2O_7$ coatings on the Si bond coat, are discussed in relation to the feedstock particle size and the furnace thermal cycling behavior of a $Si/Yb_2Si_2O_7$ EBC system is reported.

2. Experimental Procedure

2.1. Materials and Process

Commercially available Si and $Yb_2Si_2O_7$ powders provided by Oerlikon Metco (US) Inc. (New York, NY, USA) were used in this study. The details about the two kinds of powders are given in Table 1. The elemental analysis of the Si powder, determined by inductively coupled plasma optical emission spectroscopy (ICP-OES, TJA-Iris-Intrepid, Thermo Scientific GmbH, Kleve, Germany), revealed 98.7 wt % $\pm$ 0.7 wt % purity. The phase composition of the $Yb_2Si_2O_7$ feedstock was determined with an X-ray diffractometer (XRD, D4 Endeavor, Bruker AXS GmbH, Karlsruhe, Germany) (Cu Kα radiation, operating voltage 40 kV, current 40 mA, step size 0.02°, step time 0.75 s) over a 2θ-range of 10°–80°. The deconvolution of the XRD peaks for quantitative phase analysis was performed using Rietveld analysis (TOPAS Software, Bruker AXS GmbH, Karlsruhe, Germany). The XRD analysis of the $Yb_2Si_2O_7$ powder, as shown in Figure 1, yielded a crystalline structure with the presence of monoclinic $Yb_2Si_2O_7$ (C2/m, JCPDS No 01-082-0734) and secondary monoclinic Yb_2SiO_5 (I2/a, JCPDS No 00-040-0386) phases. According to the Rietveld analysis, the powder was found to contain 88 wt % and 12 wt % $Yb_2Si_2O_7$ and Yb_2SiO_5, respectively.

Diamond Jet 2700 high-velocity oxy-fuel spraying equipment (Oerlikon-Metco, Wohlen, Switzerland) and a three-cathode TriplexPro-210 APS gun (Oerlikon-Metco, Wohlen, Switzerland) manipulated with a six-axis robot (IRB 2400, ABB, Zürich, Switzerland) in a Multicoat facility (Oerlikon-Metco, Wohlen, Switzerland) were used to fabricate the $Yb_2Si_2O_7$ and Si coatings, respectively. The HVOF burner was fitted with a convergent cylindrical design nozzle (Type 2705) which yields lower particle velocities and thus longer dwell times in comparison to that of a convergent-divergent nozzle [18]. Methane has been used as the fuel gas in the HVOF process. For the optimization of the HVOF spraying conditions, the methane/oxygen ratios and stand-off distances (SOD) were varied. The flow rate of the methane was selected to be 190, 200, 215 slpm, whereas the oxygen flow rate was kept constant, at its maximum possible value of 395 slpm for the equipment. The more detailed spraying parameters used in the present study are listed in Tables 2 and 3. Ferritic stainless steel (20 × 20 mm^2) substrates were used for the initial spraying optimization of the $Yb_2Si_2O_7$ coatings. The substrates were alumina grit blasted with an average particle size of 45–125 μm using 3 bar pressure for 20 s to reach an average roughness of 3.9 $\pm$ 0.1 μm. For the calculation of the area specific weight (g/m^2) of the $Yb_2Si_2O_7$ coatings, the area of each substrate was measured and the weight of each sample was recorded before and after spraying (using five spraying cycles). In order to evaluate the splat morphologies, single splats of the $Yb_2Si_2O_7$ were produced via a wipe test on single-side polished silicon wafers. The surface morphology of the splats was afterward measured by a 3D laser confocal microscope (Keyence VK-9700, Keyence, Osaka, Japan). For furnace cycling, SiC/SiCN CMC substrates provided by DLR Stuttgart [19] were coated with the optimized $Si/Yb_2Si_2O_7$ EBC system.

Table 1. Manufacturing method and particle size of the feedstock powders measured with laser diffraction (LA-950-V2, Horiba Ltd., Tokyo, Japan).

Powder	Manufacturing Method	Particle Size (µm)		
		d_{10}	d_{50}	d_{90}
Si	Fused-crushed	28	40	59
$Yb_2Si_2O_7$	Agglomerated-sintered	22	34	52

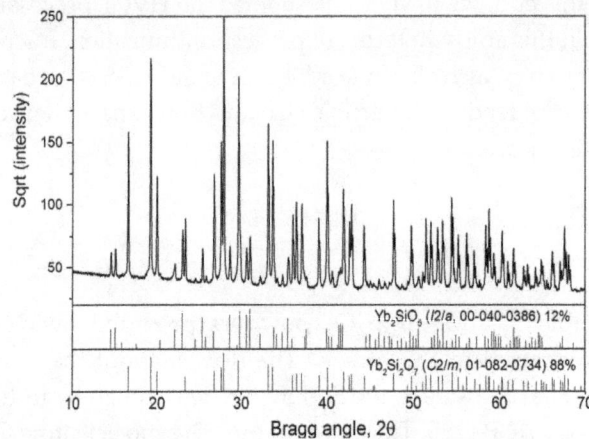

Figure 1. X-ray diffractogram of the $Yb_2Si_2O_7$ feedstock.

Table 2. HVOF spray parameters for processing $Yb_2Si_2O_7$ coatings.

Number of Spray Sets	Oxygen Flow (slpm)	Methane Flow (slpm)	Equivalence Ratio (φ) *	Air Flow (slpm)	Powder Feed Rate (%)	Spray Distance (mm)	Robot Velocity (mm/s)	Substrate Temperature (°C)
#1	395	190	0.96	250	20	350	1200	135
#2	395	200	1.01	250	20	350	1200	150
#3	395	215	1.08	250	20	350	1200	155
#4	395	200	1.01	250	20	375	1200	130
#5	395	200	1.01	250	20	400	1200	125

* φ = (fuel/oxygen)/(fuel/oxygen)stoichiometric.

Table 3. APS spray parameters for Si deposition.

Spray Current (A)	Ar Flow Rate (slpm)	Feed Rate (%)	Spray Distance (mm)	Robot Velocity (mm/s)	Substrate Temperature (°C)
450	50	30	100	500	200

2.2. Calculations of Gas Temperature and Velocity in the HVOF Nozzle

Chemical Equilibrium and Applications (CEA) [20] software was used to calculate the gas temperature and velocity in the nozzle throat using an infinite-area combustion chamber model. In this one-dimensional model, the combustion conditions are obtained with the assumption of the chemical equilibrium of the combustion products. Further details of the model's assumptions, parameters, and procedures for obtaining the combustion and throat conditions can be found in the report of Gordon and McBride [20].

The calculations were made for three different oxygen/methane flow rate combinations (#1, #2 and #3 in Table 2). The individual nozzle entry pressure input for each oxygen/methane combination was located by iteration, according to the condition of the constant mass flow rate ($\dot{m}$, kg/s),

$$\dot{m} = \rho A u \qquad (1)$$

where ρ, A, and u are the gas density (kg/m^3), nozzle cross-sectional area (m^2), and gas velocity (m/s), respectively. $\dot{m}$ of the gas mixture and A of the nozzle throat ($\varnothing$ = 7 mm) were known in this equation, while ρ and u of the gas mixture at the throat of the nozzle were calculated by the software for a given chamber pressure. After the nozzle entry pressures were correctly located for each mixture, the gas velocity and temperatures were finally calculated at the throat of the nozzle, which corresponds to the nozzle exit conditions.

2.3. Characterization of the Coatings

The microstructures of the as-sprayed and thermally cycled coatings were analyzed using a scanning electron microscope (SEM, TM3030, Hitachi High-Technologies, Chiyoda, Tokyo, Japan) coupled with an energy dispersive X-ray spectrophotometer (EDX, Quantax70, Bruker Nano GmbH, Berlin, Germany); secondary electron (SE) and backscattered electron (BSE) signals were collected. Acquired BSE-SEM images were also employed to assess the volume fractions of pores in the coatings by image analysis, using an image thresholding procedure with the analySIS pro software (Olympus Soft Imaging Solutions GmbH, Münster, Germany). The analysis was performed on 10 SEM micrographs ($\times 2000$ magnification) per sample, each having a resolution of 1280×1100 pixels and covering a horizontal field width of 126 μm. A metallographic cross-section of an as-sprayed coating was also investigated with FEG-SEM (LEO Gemini 1530 from Carl Zeiss NTS GmbH, Oberkochen, Germany) equipped with an electron backscatter diffraction (EBSD) detector (NordlysNano, Oxford Instruments, Oxfordshire, UK) to demonstrate the amorphous and crystalline areas, as well as to generate a phase map of the crystalline regions.

The phase composition of the coatings was determined with the same measurement parameters for the feedstock given above (D4 Endeavor, Bruker AXS GmbH, Karlsruhe, Germany). The PONKCS (partial or no known crystal structure) method was used to determine the amorphous content of the as-sprayed coatings using the XRD data, as described in the previous work [17]. Further details about the procedure can also be found in [21].

The roughness (arithmetic mean deviation of the surface, R_a) of the substrates and the as-sprayed coatings was measured with an optical profilometer using a chromatic white light sensor (CHR 1000) (CyberScan CT 300, CyberTechnologies, Ingolstadt, Germany). An evaluation of the roughness measurements was made in accordance with the recommendations of the standards DIN EN ISO 4288 [22] and DIN EN ISO 3274 [23]. Accordingly, the cut-off wavelength (λ_c) was selected, depending on the expected roughness values. At the same time, the evaluation length (In) and the corresponding traverse length (It) were defined, according to the standards.

2.4. Furnace Thermal Cycling Test

The Si/Yb$_2$Si$_2$O$_7$ EBCs deposited on the SiC/SiCN substrates were placed in a muffle furnace and heated in ambient atmosphere to 1200 °C, with a rate of 10 K/min. The sample was held for 20 h at this high temperature and then cooled to room temperature with a rate of 10 K/min, and this thermal cycle was repeated four times.

3. Results and Discussion

3.1. Effect of Methane/Oxygen Flow Rates and Stand-Off Distance

For the assessment of the effect of the chosen oxygen/methane flow rates (#1, #2, and #3 in Table 2), the area specific weight (ASW) and average roughness (R_a) of the coatings were evaluated.

A correlation was found between the ASW and the roughness values of the as-sprayed coatings; the maximum ASW was calculated for the coating with the minimum R_a and vice versa, and was employed as an indirect method for optimization (Figure 2a,b). It is reasonable to ascribe the high ASW—low roughness combination of the coating to the well-molten, flattened, and bonded particles, which can be translated to the high particle temperature and velocities delivered by the combustion of the fuel/oxygen mixture. The particles will otherwise remain in the non-molten/partially-molten state and they will either bounce-off the substrate resulting in a reduction in the ASW or will stick to it, but the latter will lead to a rougher surface finish. For ease of understanding, a comparison of the smooth and rough surface morphologies of HVOF sprayed molten and non-molten/partially molten single $Yb_2Si_2O_7$ splats, respectively, is given in Figure 3a,b.

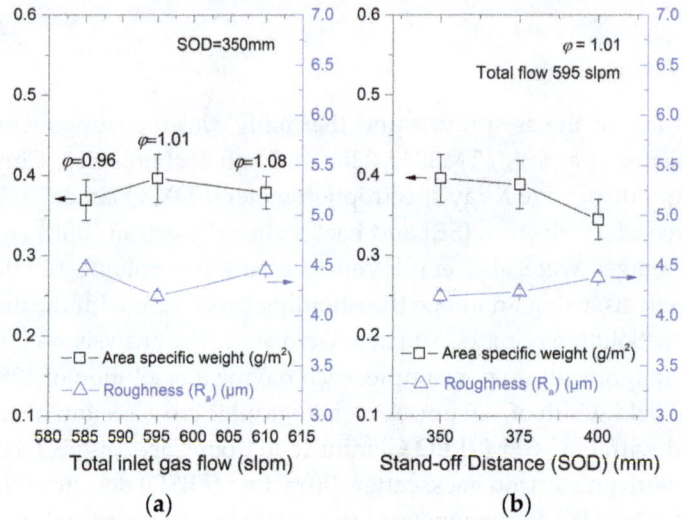

(a)   **(b)**

Figure 2. ASW and roughness of the HVOF sprayed $Yb_2Si_2O_7$ coatings as a function of the total inlet gas flow (**a**) and (**b**) the stand-off distance. Error bars of the ASW values calculated from the standard deviation of four substrate area measurements and error bars of the roughness show the standard deviation of 10 profile measurements ($\lambda_c = 2.5$ mm).

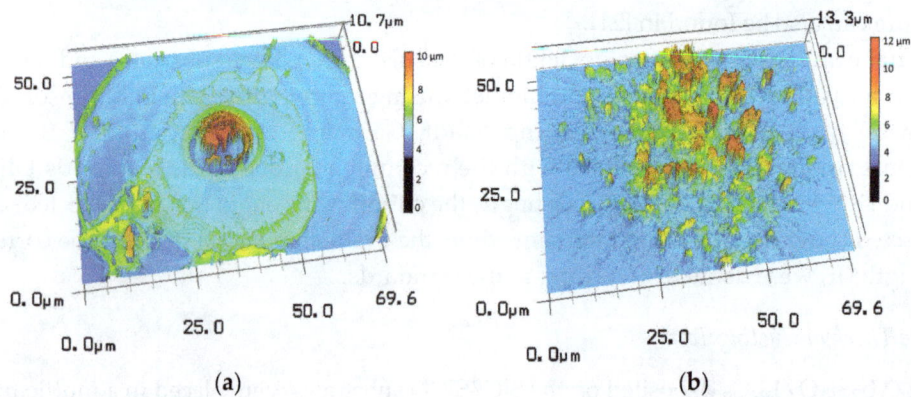

(a)   **(b)**

Figure 3. 3D-surface morphology of HVOF-sprayed (**a**) molten ($\phi = 1.01$) and (**b**) non-molten ($\phi = 0.96$) $Yb_2Si_2O_7$ single splats on Si wafer. The images were produced using a confocal laser microscope.

Based on these assumptions, it can be interpreted from Figure 2a that the maximum particle temperature and velocities were reached at the total gas flow of 595 slpm, while the increased total flow rate (610 slpm) induced a reduction. The equivalence ratio of the fuel/oxygen mixtures, which is described as the actual fuel/oxygen ratio divided by the stoichiometric fuel/oxygen ratio ($\phi = $ (fuel/oxygen)/(fuel/oxygen)$_{stoichiometric}$, fuel rich if $\phi > 1$, fuel lean if $\phi < 1$ and stoichiometric if $\phi = 1$) is also noted on the graph, indicating that the selected fuel/oxygen ratios are close to a stoichiometric mixture. The stoichiometric mixture ($\phi = 1$) is suggested by the combustion thermodynamics to obtain a maximum adiabatic flame temperature from the combustion of a hydrocarbon fuel at a given chamber pressure [24]. Furthermore, the relevance of the stoichiometric fuel/oxygen ratios (or slightly fuel-rich mixtures) to the maximum particle temperature and velocities in the HVOF process has been revealed in the literature [25–27]. On the other hand, no general agreement was found regarding the effect of the total inlet gas flow on the particle temperature and velocities. Zhao et al. and Turunen et al. [28,29] reported an increase in both the particle temperature and velocities with increasing total gas flow rates associated with higher burning enthalpy and gas expansion. However, Picas et al. [26] showed a decrease in particle temperatures when fuel rates exceeded a certain level. Moreover, Guo et al. [30], who worked with methane fuel similar to the current study, recorded increasing particle velocities with rising fuel flow rates and yet they found that the particle temperatures gradually reduced at the same conditions. The decreasing temperatures of the particles at higher fuel flow rates were thus attributed to their reducing dwell time in the flame, which can also explain the abovementioned changes in the ASW and roughness values at the highest total flow rate. To further check this, the gas temperature and velocities were calculated for the chosen constant oxygen and increasing methane flow rates (#1, #2, and #3 in Table 2) in this study and the results are shown in Figure 4. Accordingly, the highest gas velocity and the highest gas temperature are found at the highest total gas flow ($\phi = 1.08$). It can be expected that the temperature would drop again if ϕ was further increased. Considering that the optimum roughness and ASW values are obtained applying a gas mixture of an almost stoichiometric composition ($\phi = 1.01$) in this work, it is clear that the maxima of the roughness and ASW are seen at specific particle velocities and temperatures, and beyond these, begin to decrease. Thus, the highest temperature and velocity values are not the optimum parameters in this respect.

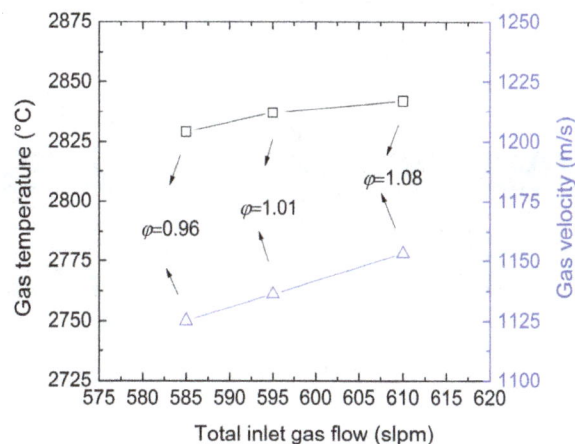

Figure 4. Calculated gas temperature and velocities in the throat of the nozzle at different total gas flow rates. The oxygen flow is constant at 395 slpm.

Using this optimum fuel/oxygen mixture, the $Yb_2Si_2O_7$ coatings were also sprayed at different stand-off distances (SODs) (Figure 2b). The SOD < 350 mm yielded lower ASW values of the coatings and thus these results are not shown in the graph. Furthermore, almost no difference was observed by increasing the SOD from 350 mm to 375 mm. However, a slight decrease in the ASW and a simultaneous increase in the roughness were obtained at 400 mm SOD. The effect of the SOD will be further discussed below.

3.2. Microstructure, Crystallinity and Phase Composition

Figure 5a–e shows the low magnification cross-section microstructures of five $Yb_2Si_2O_7$ coatings sprayed with different total flow rates and SODs, and the measured porosity of these coatings via image analysis are given in. The microstructures of all coatings were found to be fairly dense and similar. Nevertheless, the lowest porosity content was measured at $\phi = 1.01$ with 350 mm SOD (#2). This finding confirms the prior discussions regarding the highest ASW and the lowest roughness recorded due to well-molten particles with the same spray parameters. On the other hand, the highest porosity was measured at $\phi = 1.01$ with 400 mm SOD (#5). Obviously, particles start to significantly cool down/slow down beyond 375 mm and more particles bounce off the substrate. Supportively, Figure 5e reveals a thinner coating with a rougher surface finish due to the presence of a high amount of non-molten or partially molten particles in the microstructure.

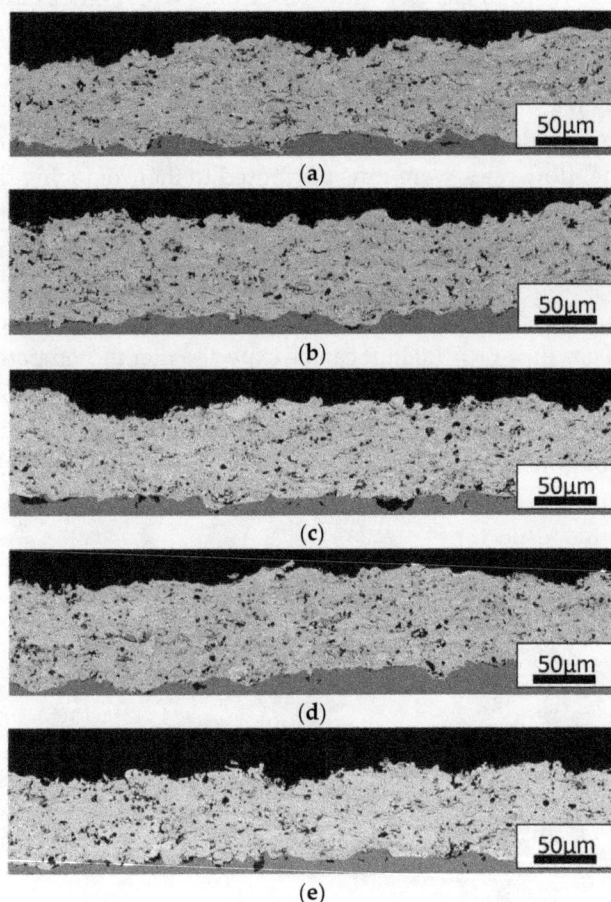

Figure 5. Microstructure of HVOF sprayed $Yb_2Si_2O_7$ coatings with the spray settings of (**a**) #1; (**b**) #2; (**c**) #3; (**d**) #4; (**e**) #5.

Figure 6 shows the XRD patterns of the HVOF sprayed $Yb_2Si_2O_7$ deposits. The indexed peaks in the patterns suggest the presence of monoclinic $Yb_2Si_2O_7$ (C2/m, JCPDS No 01-082-0734) and Yb_2SiO_5 (I2/a, JCPDS No 00-040-0386) phases in the feedstock. The broad humps at 2θ angle ranges of $24°-38°$ and $40°-70°$ imply that the as-sprayed deposit contains amorphous phases, as expected and discussed in the previous study [17]. A quantitative comparison of the amorphous contents of the as-sprayed coatings determined by the PONKCS method is shown in Table 4. The results were found to be quite similar to each other within the estimated margin of error of 5%, as well as higher than in the earlier work (48% amorphous content at 325 mm SOD [17]). This is probably due to the smaller particle size distribution of the feedstock in the current work, as well as the longer SOD, which is selected to be a minimum of 350 mm to obtain higher ASWs.

The crystalline/non-crystalline areas in the coatings were also investigated by an EBSD analysis. The analysis was performed on the cross-section of the HVOF sprayed coating #2 as an example (Figure 7a,b). The non-molten/partially molten particles embedded in a molten and well-bonded coating matrix are observed in Figure 7a, and Figure 7b shows a phase map of the same coating area.

The latter clearly demonstrates that the molten matrix is non-crystalline and hence yields no diffraction (black regions), whereas the non-molten particles preserved their crystallinity. The $Yb_2Si_2O_7$ and Yb_2SiO_5 phases were indexed for these non-molten particles, supporting the XRD analysis results.

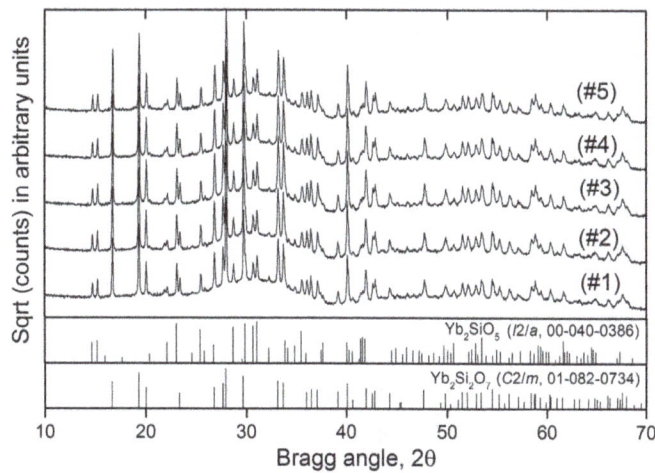

Figure 6. XRD patterns of HVOF deposited coatings with the given process parameters (#1, #2, #3, #4, #5).

Table 4. Measured porosity and amorphous content of the $Yb_2Si_2O_7$ coatings.

Number of Spray Sets	Porosity (%) $\pm$ SD	Amorphous Content (wt %)
#1	8.0 ± 0.8	72 ± 5
#2	7.7 ± 0.8	73 ± 5
#3	8.3 ± 0.6	69 ± 5
#4	8.1 ± 0.5	70 ± 5
#5	9.1 ± 1.2	65 ± 5

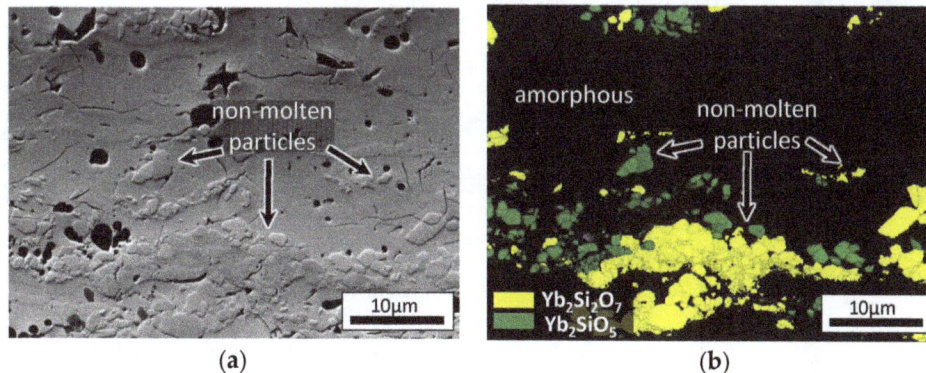

Figure 7. EBSD analysis results of the $Yb_2Si_2O_7$ coating (#2): (**a**) Forescatter diode (FSD) image and (**b**) phase map.

3.3. Adhesion of HVOF Sprayed Top Coat on Si Bond Coat

The Si coatings with a thickness of 60 µm were sprayed by the APS with the given parameters in Table 3 on SiC/SiCN CMC substrates. Afterward, the $Yb_2Si_2O_7$ was sprayed using the HVOF parameters (#2) on the Si bond coat and Figure 8a shows the cross-section micrograph of this trial. A decreased thickness of the Si bond coat and the $Yb_2Si_2O_7$ coated/uncoated patches on top of it clearly suggest that the sprayed $Yb_2Si_2O_7$ particles eroded the brittle Si layer. A deposition could be achieved later by reducing the particle size of the feedstock, due to the ease of melting fine powders, which results in a superior wetting behavior (Figure 8b). A higher density of the coating is also evident from the microstructure. However, as shown with the XRD patterns in Figure 8c, reducing the particle size is also accompanied by a decrease in the coating crystallinity. According to the PONCKS analysis, this coating has a more than 90% amorphous content and yet no vertical crack formation was observed in the as-sprayed state, as well as the after thermal cycling, as shown in Figure 9a. This behaviour was found to be different to that of the APS deposited highly amorphous $Yb_2Si_2O_7$ coatings, as they are vertically cracked, even after the deposition (such a microstructure can be seen in [17]). The cracking in the APS coating is attributed to its higher thermal expansion (particularly if the Yb_2SiO_5 is present) than the substrate, which results in the development of the tensile stresses in the coating during cooling from high deposition temperatures (above 500 °C) to room temperature. Possible reasons preventing the similar crack formation in the HVOF sprayed coatings in the as sprayed state and after the thermal cycling are: (i) a higher porosity in the HVOF coatings than the APS coatings (in the range of 5–8% higher in the former, based on the image analysis), which provides the HVOF coatings with some strain tolerance; (ii) lower HVOF deposition temperatures (150–170 °C) which diminishes the magnitude of stresses due to the reduced cooling rates; (iii) the possibly lower content of Yb_2SiO_5 with a higher thermal expansion in the HVOF coatings than the APS coatings due to lower deposition temperatures, which hinders the evaporation of Si-bearing species; and finally (iv) the possible compressive stress growth in the HVOF coatings as a result of the peening effect of the particles on the substrate or on the previously deposited layer.

Nevertheless, although the HVOF coating remained attached to the bond coat after cycling and built no vertical cracks (Figure 9a), in a few locations, where a particularly loose non-molten particle is in contact with the bond coat, partial delaminations were observed (Figure 9c). It is probable that the reduced interfacial bonding strength by the presence of these loosely bonded particles eases the crack propagation and hence the delamination.

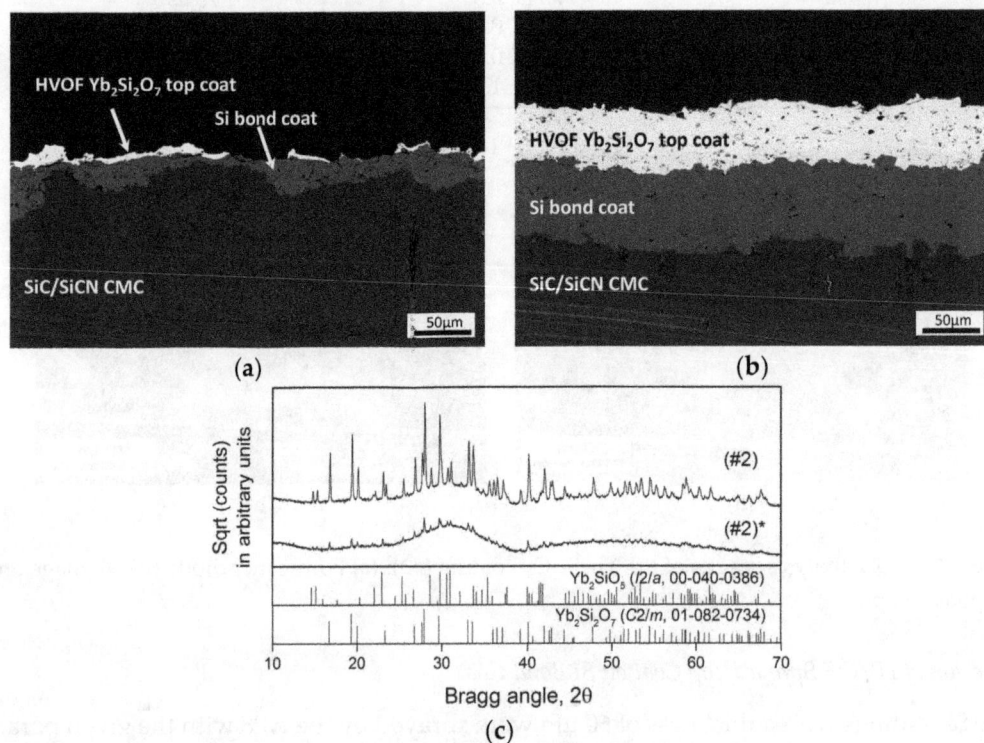

(a)   (b)

(c)

Figure 8. (a,b) Backscattered SEM micrographs of EBC cross-sections. (a) HVOF sprayed $Yb_2Si_2O_7$ (#2) on Si bond coat; (b) HVOF sprayed $Yb_2Si_2O_7$ (#2) using reduced particle size ($d_{10} = 17$ μm, $d_{50} = 25$ μm, $d_{90} = 34$ μm) on Si bond coat. (c) Comparison of the XRD patterns of the $Yb_2Si_2O_7$ coatings deposited using same spray parameters (#2) but different particle sizes. (#2) and (#2) * denote original and reduced particle sizes, respectively.

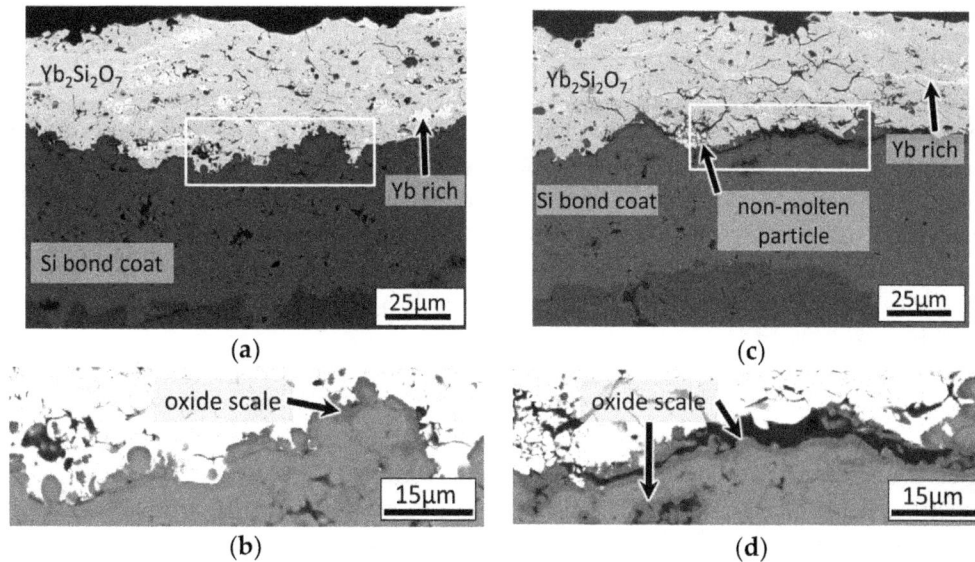

Figure 9. Backscattered SEM micrographs of EBC cross-section after five cycles between 1200 °C and room temperature with 20 h high temperature dwell time. Images (**a**,**c**) are taken from the different regions of the same sample. White rectangles in (**a**,**c**) indicate the region of high-magnification images shown below (**b**,**d**), in which oxidation of the silicon coating at the Si/Yb$_2$Si$_2$O$_7$ interface as well as on the pore surfaces is clearly visible via darker image contrast.

4. Conclusions

The HVOF process was used for the deposition of the Yb$_2$Si$_2$O$_7$ coatings, and the effects of the oxygen/methane ratio and stand-off distance on the coating microstructure and crystallinity were discussed. The concurrent changes in the area specific weight and arithmetic mean roughness associated with the presence of molten/non-molten particles in the coatings at different process conditions (e.g., fuel lean or rich conditions) were interpreted. Accordingly, an equivalence ratio of 1.01 was found to yield the highest area specific weight when it was combined with a stand-off distance of 350–375 mm. The amorphous content of the HVOF sprayed coatings was found to be in the range of 65%–73% by PONKCS analysis, and the EBSD analysis clearly demonstrates that the non-molten particles explain this partial crystallinity. However, despite the fact that the depositions successfully worked on the metallic substrates, under the same deposition conditions, the silicon bond coat layer was found to be eroded. It was shown that this could be avoided by decreasing the particle size of the Yb$_2$Si$_2$O$_7$ feedstock; however, by doing that, the amorphous content of the coating is further increased. Nevertheless, no vertical cracks were observed in this highly amorphous and dense as-deposited coating, revealing the advantage of HVOF deposition over the APS, which was elaborated in the results and discussion section. The thermal cycling test of this highly amorphous coating, however, suggested that non-molten particles at the Si/Yb$_2$Si$_2$O$_7$ interface may be unfavorable for the adhesion, as their presence at the interface coincided with the delamination cracks. Consequently, if the bonding of the HVOF top coat to the Si bond coat can be improved, the conventional HVOF method can be used for the manufacturing of silicate EBCs without requiring very high deposition temperatures as is the case in the plasma spraying process. Current investigations are looking at different methods to develop a well-adhered interface and steam cycling tests are planned for the validation of the EBCs.

Acknowledgments: The authors thank K.H.R., R.L. and F.K. (all in IEK-1, JÜLICH) for their support on manufacturing the coatings. Special thanks to E. Wessel (IEK-2, JÜLICH) for performing EBSD analysis.

Author Contributions: E.B., G.M., and R.V. conceived and designed the experiments, D.K. contributed the materials, E.B. performed the experiments, E.B., G.M., and Y.S. analyzed the data, and E.B. wrote the paper.

Conflicts of Interest: The authors declare no conflict of interest.

References

1. Gupta, A.K.; Lilley, D.G. Combustion and environmental challenges for gas turbines in the 1990s. *J. Propuls. Power* **1994**, *10*, 137–147. [CrossRef]
2. Jacobson, N.S.; Smialek, J.L.; Fox, D.S.; Opila, E.J. *Durability of Silica-Protected Ceramics in Combustion Atmospheres*; American Ceramic Society: Westerville, OH, USA, 1995.
3. Ohnabe, H.; Masaki, S.; Onozuka, M.; Miyahara, K.; Sasa, T. Potential application of ceramic matrix composites to aero-engine components. *Compos. Part A Appl. Sci. Manuf.* **1999**, *30*, 489–496. [CrossRef]
4. Bansal, N.P. *Handbook of Ceramic Composites*; Bansal, N.P., Ed.; Kluwer: Boston, MA, USA, 2005.
5. Bansal, N.P.; Lamon, J. *Ceramic Matrix Composites: Materials, Modeling and Technology*; Wiley: Hoboken, NJ, USA, 2014.
6. Opila, E.J.; Hann, R.E. Paralinear oxidation of CVD SiC in water vapor. *J. Am. Ceram. Soc.* **1997**, *80*, 197–205. [CrossRef]
7. Robinson, R.C.; Smialek, J.L. SiC recession caused by SiO_2 scale volatility under combustion conditions: I, experimental results and empirical model. *J. Am. Ceram. Soc.* **1999**, *82*, 1817–1825. [CrossRef]
8. Smialek, J.L.; Robinson, R.C.; Opila, E.J.; Fox, D.S.; Jacobson, N.S. SiC and Si_3N_4 recession due to SiO_2 scale volatility under combustor conditions. *Adv. Compos. Mater.* **1999**, *8*, 33–45. [CrossRef]
9. Opila, E.J.; Smialek, J.L.; Robinson, R.C.; Fox, D.S.; Jacobson, N.S. SiC recession caused by SiO_2 scale volatility under combustion conditions: II, thermodynamics and gaseous-diffusion model. *J. Am. Ceram. Soc.* **1999**, *82*, 1826–1834. [CrossRef]
10. Lee, K.N. Current status of environmental barrier coatings for Si-based ceramics. *Surf. Coat. Technol.* **2000**, *133*, 1–7. [CrossRef]
11. Eaton, H.E.; Linsey, G.D.; More, K.L.; Kimmel, J.B.; Price, J.R.; Miriyala, N. EBC Protection of SiC/SiC Composites in the Gas Turbine Combustion Environment. In Proceedings of the ASME Turbo Expo 2000: Power for Land, Sea, and Air, Munich, Germany, 8–11 May 2000.
12. Eaton, H.E.; Linsey, G.D.; Sun, E.Y.; More, K.L.; Kimmel, J.B.; Price, J.R.; Miriyala, N. EBC Protection of SiC/SiC Composites in the Gas Turbine Combustion Environment: Continuing Evaluation and Refurbishment Considerations. In Proceedings of the ASME Turbo Expo 2001: Power for Land, Sea, and Air, New Orleans, LA, USA, 4–7 June 2001.
13. Lee, K.N.; Fox, D.S.; Eldridge, J.I.; Zhu, D.; Robinson, R.C.; Bansal, N.P.; Miller, R.A. Upper temperature limit of environmental barrier coatings based on mullite and BSAS. *J. Am. Ceram. Soc.* **2003**, *86*, 1299–1306. [CrossRef]
14. Lee, K.N.; Fox, D.S.; Bansal, N.P. Rare earth silicate environmental barrier coatings for SiC/SiC composites and Si_3N_4 ceramics. *J. Eur. Ceram. Soc.* **2005**, *25*, 1705–1715. [CrossRef]
15. Richards, B.T.; Zhao, H.; Wadley, H.N.G. Structure, composition, and defect control during plasma spray deposition of ytterbium silicate coatings. *J. Mater. Sci.* **2015**, *50*, 7939–7957. [CrossRef]
16. Richards, B.T.; Young, K.A.; de Francqueville, F.; Sehr, S.; Begley, M.R.; Wadley, H.N.G. Response of ytterbium disilicate–silicon environmental barrier coatings to thermal cycling in water vapor. *Acta Mater.* **2016**, *106*, 1–14. [CrossRef]
17. Bakan, E.; Marcano, D.; Zhou, D.; Sohn, Y.J.; Mauer, G.; Vaßen, R. $Yb_2Si_2O_7$ coatings deposited by various thermal spray techniques: A preliminary comparative study. *J. Therm. Spray Technol.* **2017**, submitted.
18. Korpiola, K.; Hirvonen, J.P.; Laas, L.; Rossi, F. The influence of nozzle design on HVOF exit gas velocity and coating microstructure. *J. Therm. Spray Technol.* **1997**, *6*, 469–474. [CrossRef]
19. Mainzer, B.; Friess, M.; Jemmali, R.; Koch, D. Development of polyvinylsilazane-derived ceramic matrix composites based on Tyranno SA3 fibers. *J. Ceram. Soc. Jpn.* **2016**, *124*, 1035–1041. [CrossRef]
20. Gordon, S.; MeBride, B.J. *Computer Program for Calculation of Complex Chemical Equilibrium Compositions and Applications. I. Analysis*; NASA Reference Publications; Lewis Research Center: Cleveland, OH, USA, 1994; pp. 25–32.
21. Scarlett, N.V.Y.; Madsen, I.C. Quantification of phases with partial or no known crystal structures. *Powder Diffr.* **2006**, *21*, 278–284. [CrossRef]
22. *ISO 4288:1996–Geometrical Product Specifications (GPS)–Surface Texture: Profile Method–Rules and Procedures for the Assessment of Surface Texture*; International Organization for Standardization: Geneva, Switzerland, 1996.

23.   *ISO 3274:1996–Geometrical Product Specifications (GPS)–Surface Texture: Profile Method–Nominal Characteristics of Contact (Stylus) Instruments*; International Organization for Standardization: Geneva, Switzerland, 1996.

24.   Glassman, I. *Combustion*; Elsevier Science: Amsterdam, The Netherlands, 1997.

25.   Li, M.; Christofides, P.D. Multi-scale modeling and analysis of an industrial HVOF thermal spray process. *Chem. Eng. Sci.* **2005**, *60*, 3649–3669. [CrossRef]

26.   Picas, J.A.; Punset, M.; Baile, M.T.; Martín, E.; Forn, A. Effect of oxygen/fuel ratio on the in-flight particle parameters and properties of HVOF WC-CoCr coatings. *Surf. Coat. Technol.* **2011**, *205*, S364–S368. [CrossRef]

27.   Cheng, D.; Xu, Q.; Tapaga, G.; Lavernia, E.J. A numerical study of high-velocity oxygen fuel thermal spraying process. Part I: Gas phase dynamics. *Metall. Mater. Trans. A* **2001**, *32*, 1609–1620. [CrossRef]

28.   Zhao, L.; Maurer, M.; Fischer, F.; Lugscheider, E. Study of HVOF spraying of WC–CoCr using on-line particle monitoring. *Surf. Coat. Technol.* **2004**, *185*, 160–165. [CrossRef]

29.   Turunen, E.; Varis, T.; Hannula, S.P.; Vaidya, A.; Kulkarni, A.; Gutleber, J.; Sampath, S.; Herman, H. On the role of particle state and deposition procedure on mechanical, tribological and dielectric response of high velocity oxy-fuel sprayed alumina coatings. *Mater. Sci. Eng. A* **2006**, *415*, 1–11. [CrossRef]

30.   Guo, X.; Planche, M.-P.; Chen, J.; Liao, H. Relationships between in-flight particle characteristics and properties of HVOF sprayed WC-CoCr coatings. *J. Mater. Process. Technol.* **2014**, *214*, 456–461. [CrossRef]

10

Coating Qualities Deposited Using Three Different Thermal Spray Technologies in Relation with Temperatures and Velocities of Spray Droplets

Yasuyuki Kawaguchi [1,*], Fumihiro Miyazaki [1], Masafumi Yamasaki [1], Yukihiko Yamagata [2], Nozomi Kobayashi [2] and Katsunori Muraoka [1]

[1]   Plazwire Co., Ltd., 2-3-54 Higashi-naka, Hakata-ku, Fukuoka 812-0829, Japan;
      f-miyazaki@plazwire.co.jp (F.M.); m-yamasaki@plazwire.co.jp (M.Y.); k-muraoka@plazwire.co.jp (K.M.)

[2]   Interdisciplinary Graduate School of Engineering Sciences, Kyushu University, Kasuga-koen, Kasuga, Fukuoka 816-8580, Japan; yamagata@asem.kyushu-u.ac.jp (Y.Y.); pq18fragment@gmail.com (N.K.)

*   Correspondence: Y-Kawaguchi@plazwire.co.jp

Academic Editor: Yasutaka Ando

Abstract: Three guns based on different thermal spray technologies—namely, gas flame spray, wire arc spray, and wire plasma spray—were operated at each best cost–performance condition, and the resulting spray droplets and deposited coating qualities were investigated. For the former, a simple optical monitoring system was used to measure temperatures and velocities of spray droplets ejected from the guns. On the other hand, for the latter, qualities of coating layers on substrates—namely, surface roughness, atomic composition, hardness, adhesive strength, and porosity—were characterized. Then, these coating qualities were discussed with respect to the measured temperatures and velocities of spray droplets, which revealed novel features in the coatings that have not been seen before, such as atomic composition and hardness strongly dependent on temperature and environments of droplets towards the substrates, and porosity on velocity of droplets impinging onto the substrates.

Keywords: spray droplet from wire; temperature; velocity; gas flame spray; wire arc spray; wire plasma spray; coating quality

---

1. Introduction

Thermal spray technologies have been used for the last several decades for various applications, such as for anti-corrosion of electricity transmission towers and energy generation, etc. [1], together with bridges [2], subsea rises/pipelines [3], and anti-wear for various contacting components [4]. As was pointed out by Fauchais and his coauthors [5], company workers of plasma spray have been trying to find good coating services (corrosion and wear resistance, thermal protection, etc.) and coating repeatability, reproducibility, and reliability for applications in the fields. This endeavor has to be carried out at the best cost–performance; that is, acceptable coating qualities for various purposes have to be pursued at reasonably competitive costs compared with other competitors/methods/techniques. On the other hand, researchers in laboratories tend to seek to understand coating processes and resulting coating microstructure and properties [5].

This article is a result of the joint effort of field workers in a thermal spray company who have been routinely employing thermal spray for more than a decade at the above-mentioned "best cost–performance", with university researchers who have been trying to narrow the existing gap of their effort with industry by supplying the most pertinent data for understanding and optimizing thermal spray processes. In the process of trying to find operating conditions of various thermal spray guns at the best cost–performance in the fields, wire arc spray has been observed the to yield, for example, fast deposition at relatively porous and soft coating layers, while gas flame spray to yield hard coating layers of low porosity at relatively low adhesive strength. On the other hand, coauthors of the present article at the University have believed that temperature and velocity of spray droplets to have profound effects on these characteristics/qualities of coating layers, and have spent much effort to realize a simple monitoring system for the purpose [6].

In this background, a joint work has been started to investigate the relationship of the qualities of coating layers deposited using three spray guns based on the different thermal spray technologies operated at their best cost–performances, in order to closely look at them with respect to temperatures and velocities ejected from the guns and impinging onto substrates. As the measures of characteristics/qualities of thus-deposited coating layers, the measured surface roughness, atomic composition, hardness, porosity, and adhesive strength were compared on the same comparative basis for the coating layers deposited over such wide ranges of operating parameters. As far as the present authors are aware, there has not been any work of this kind carried out and reported on in open literature in the past. Therefore, it is their belief that the results revealed novel features, such as composition and hardness heavily dependent on the temperature and production environment, and porosity on the velocity of droplets impinging on substrates. These results, together with other details, are described.

This article is structured as follows: Section 2 describes the feed stock and deposited material, methods, and equipment; Section 3 presents the measured results of droplet temperatures and velocities together with those of coating qualities; and Section 4 draws observations among these two values and the resultant discussions. Section 5 summarizes the results.

2. Material, Methods, and Equipment

2.1. Material and the Method of Surface Treatment for Thermal Spray Coating

Substrates on which coating layers were deposited were JISG3101 SS400 ($75 \times 150 \times 6$ mm^3), which were first blast-treated following the procedures described in ISO8501-1 Sa3.0 (surface roughness to be more than Ra 8.0 μm and Rz 50.0 μm). The substrates were blasted with F-24 Al$_2$O$_3$ grit (NANIWA ABRASIVE MFG. Co., Ltd., Osaka, Japan) at a pressure of 0.6 MPa using a nozzle diameter of 6 mm at a standoff distance of 100 mm and an angle of attack of 90°. The type A-3R (ATSUCHI TEKKO Co., Ltd., Osaka, Japan) was used as the blasting equipment.

2.2. Productions of Spray Droplets Using Al/5Mg Wire and Spray Parameters for Coating Productions

As the material for the droplet formation of thermal spray, the present work employed an aluminum-5% magnesium alloy A5056 (hereafter referred to as Al/5Mg). This is widely used for anti-corrosion of iron and other surfaces. Table 1 shows the Al/5Mg wire composition. The gas flame gun used the wire with a diameter of 3.2 mm, while wire arc and wire plasma guns used that of 1.6 mm (two of this for the wire arc gun).

Table 1. Composition of the coating material (wt.%).

Material	Wire Diameter (mm)	Si	Fe	Cu	Mn	Mg	Cr	Zn	Ti	Al
Al/5Mg	3.2	0.09	0.19	0.01	0.09	4.53	0.07	0.00	0.04	Balance
wire	1.6	0.13	0.17	0.00	0.05	4.83	0.06	0.00	–	Balance

As the methods of wire thermal spray, the three most widely used types of guns were employed in the present study—namely, gas flame, wire arc, and wire plasma. Figure 1 shows a cross-section of the "plazwire" gun head, which is a type of wire plasma system, and has been most extensively used in fields of the present authors' company [7]. The other two guns are of the conventional types—namely, a gas flame [8] and a wire arc [9]—and are not described any further here.

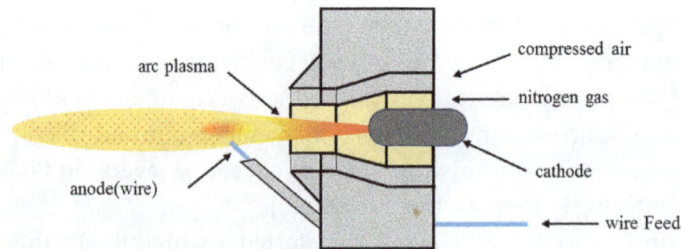

Figure 1. A cross-section of the "plazwire" system.

Table 2 gives the spray parameters of the three guns used for the coating productions.

Table 2. Spray parameters used for coating productions.

Method	Equipment	Standard Spray Distance (mm)	Wire Feed Rate (kg/h)
Gas flame	Metco14E	150	4.0
Wire arc	Metallisation 150/s450	150	3.2
Wire plasma	PW-120	150	3.0

2.3. The Methods and Arrangements of Equipment and Instruments for Measuring Temperature and Velocity of Spray Droplets

The method of measuring the temperature of spray droplets is well-known as a "two-color" thermometry, in which an intensity ratio at two wavelengths of an optical emitter is compared. For this purpose, the Planck emission formula [10] is used, which is written as follows:

$$I(\lambda, T) = \varepsilon(\lambda, T) \frac{2\pi hc^2}{\lambda^5} \frac{1}{\exp\left(\frac{hc}{\lambda kT}\right) - 1} \tag{1}$$

Here, $I(\lambda, T)$ is an intensity of radiation emitted at a wavelength of λ from a body surface at a temperature of T; $\varepsilon(\lambda, T)$ is an emissivity which is dependent on the material and its surface conditions (such as surface roughness and the degree of oxidation), together with wavelength and the temperature of the emitting surface; h is the Planck's constant; c is the speed of light; and k is the Boltzmann constant.

This method is usually applied at visible and near-infrared regions of electromagnetic wave radiations. In order to use this method in practice, one has to know values of emissivity of the emitter surfaces at hand at the two wavelengths by referring to suitable references [11]. In order to implement the method, the emitted radiation from the surface is collected through two narrow bandpass filters, which are electronically measured. Thus, it is only a matter of calibrating the two measured intensities against temperature by correlating them using the Planck's formula and the surface emissivity.

The method of velocity measurement is also well known as a "time-of-flight (TOF)" technique, in which a time difference Δt of a passage of a radiating body between well-defined positions having a distance of Δx is measured to yield the velocity $v = \xi \Delta x / \Delta t$, where ξ is a magnification or reduction ratio of a collection optics of a radiating body onto a recording surface. Because the passage of an image of a radiating body on the recording surface is very brief, it has to be confirmed whether the time response of each pixel element of the recording surface is fast enough to catch a real-time history, which necessitates a calibration procedure.

Figure 2 shows the arrangement of the simple monitoring system, consisting of a sensor unit (a), a control board unit (b), and a PC for data analysis (c). The sensor head consisted of a light collecting lens followed by a part for separating the light into two wavelengths, with the light of each wavelength guided through an interference filter (central wavelengths of 650 nm and 750 nm with a bandwidth of 50 nm each) into a fiber-coupled silicon photo-diode (Si-PD) to be electronically detected.

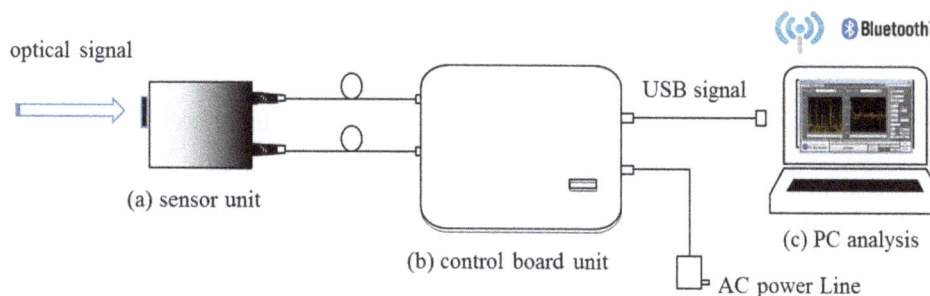

Figure 2. Arrangement of the simple monitoring system.

For the present purposes of obtaining relationships between measured temperture and velocity of spray droplets with coating qualities, discussions are concentrated on the results at the measurement position of temperature and velocity at the central position of substrates for the deposition of coating layers, namely 150 mm downstream and on the center-line of the spray guns.

2.4. Methods of Coating Depositions and Employed Techniques of Measuring Coating Qualities

Coating layers were deposited using three thermal spray technologies of combustion gas flame, wire arc, and wire plasma spraying, using Al/5Mg wires shown in Table 1 to coating thicknesses of approximately 100 μm.

The methods of measuring coating qualities are as described below, with their validations mostly approved by the JIS (Japanese Industrial Standards) and the ISO (International Organization for Standardization), as shown below with respective numbers.

First, the surface roughness was measured using Surftest SJ-201 (Mitutoyo Corporation, Kanagawa, Japan) following the JISB0601 "2001" instruction.

Atomic composition was observed using a scanning electron microscope (SEM) instrument for the cross-sectional view using 15.0 kV × 100 times magnification, and an electron probe micro analyzer (EPMA) instrument (JXA-8500F, JEOL Ltd., Tokyo, Japan) for atomic concentration with their semi-quantitative analysis. During the analyses and measurements, "the instructions for users" were rigorously followed.

Hardness was measured using a Micro-hardness tester (MVK-G1, Mitutoyo Corporation, Kanagawa, Japan) at a load of 100 g for sample numbers of $n = 10$. During the analyses and measurements, "the instructions for users" were rigorously followed.

Adhesive strength was measured following the JISH8300 instructions (in accordance with ISO2063) using Techno-tester2000LD (SANKO-TECHNO Co., Ltd., Chiba, Japan).

Porosity was measured following the JISR1643 instructions, where an open-pore measurement technique was employed, namely by measuring weight changes using distilled water in vacuum, among sample numbers of $n = 3$.

3. Results of the Measurements

3.1. Temperature and Velocity of Spray Droplets

Table 3 summarizes the measured values of temperatures and velocities. It is to be noted that there is an uncertainty of up to 20% overestimate in these temperatures [6] and 10% underestimate in these velocities. The overestimates come from uncertainty in the emissivity value [11] of droplets, and the underestimates come from the necessity of subtracting the contributions of low temperature droplets existing at the peripheries of the droplet flow which are inevitably observed for this kind of line-of-sight measurements (this subtraction was not carried out in the present values).

Table 3. Measured temperatures and velocities of spray droplets ejecting from three different thermal spray guns.

Method	Temperature (K)	Velocity (m/s)
Gas flame	1900	200
Wire arc	2900	60
Wire plasma	3000	90

It is to be noted that these measured values are broadly in agreement with those obtained using commercially available instruments [12,13]. However, the results shown in Table 3 were obtained after careful calibration of the present instrument [6] with confirmed accuracies with two significant digits. In addition, the above uncertainties in temperature and velocity were only available after obtaining raw data by the present authors.

3.2. Qualities of Coating Layers

The three photographs in Figure 3 show the cross-sectional views as observed using the SEM method. These results do not show any marked difference for coating layers deposited using the three thermal spray guns. The red circles in the figures show the positions of semi-quantitative analyses for the EPMA method. From the atomic compositions, one recognizes an almost complete disappearance of Mg for the wire arc spray case, while the Mg compositions were almost the same for both gas flame spray and wire plasma spray cases.

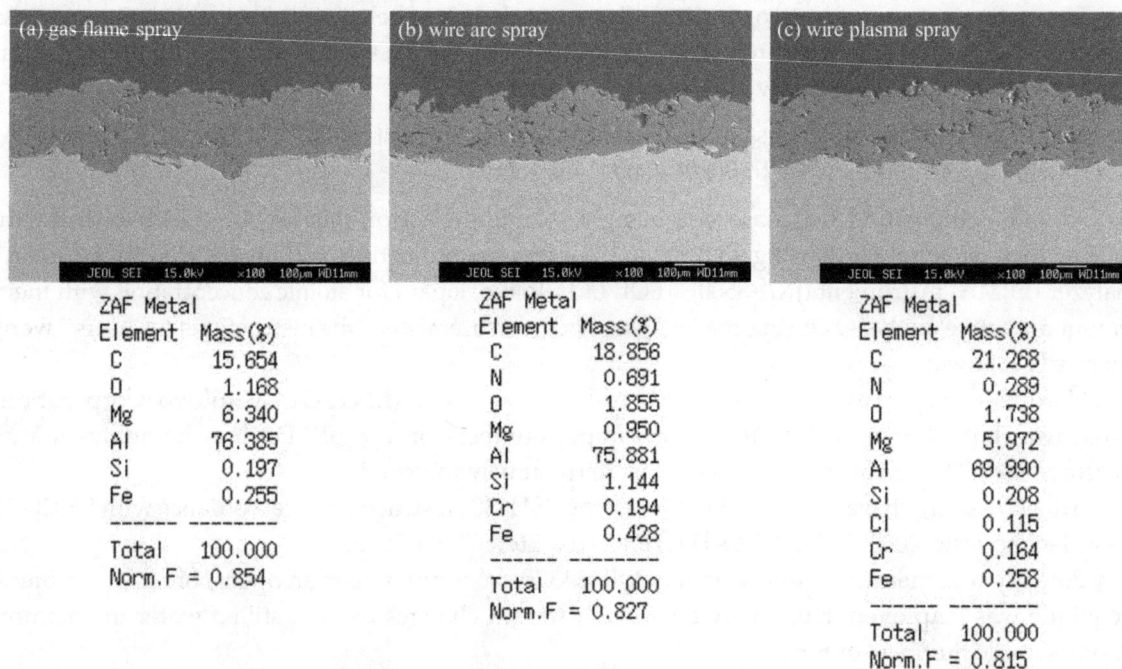

ZAF Metal

Element	Mass(%)
C	15.654
O	1.168
Mg	6.340
Al	76.385
Si	0.197
Fe	0.255

Total 100.000
Norm.F = 0.854

ZAF Metal

Element	Mass(%)
C	18.856
N	0.691
O	1.855
Mg	0.950
Al	75.881
Si	1.144
Cr	0.194
Fe	0.428

Total 100.000
Norm.F = 0.827

ZAF Metal

Element	Mass(%)
C	21.268
N	0.289
O	1.738
Mg	5.972
Al	69.990
Si	0.208
Cl	0.115
Cr	0.164
Fe	0.258

Total 100.000
Norm.F = 0.815

Figure 3. Cross-sectional views obtained using SEM (above) and the results of the semi-quantitative analyses (following the instruction commonly referred to as the "ZAF" corrections) using an electron probe micro analyzer (EPMA) for coating layers deposited using the three thermal spray guns (below, which are direct-printouts pasted here from the EPMA instrument). (**a**) Gas flame spray; (**b**) Arc spray; (**c**) Plasma spray.

Table 4 shows the values of the measured surface roughnesses, hardnesses, adhesive strengths and porosities obtained for coating layers deposited using the three thermal spray guns. The surface roughnesses were R_z 80–100 μm for both combustion gas flame and wire plasma cases, were a little smoother for the former (gas flame case), and R_z 150 µm for the wire arc case. The mesured average hadnesses were, in the harder order, HV = 69.5 for wire plasma spray, 65.3 for gas flame spray, and 41.4 for wire arc spray. Measured average adhesive strengths were, in the stronger order, 14 MPa for wire arc spray, 7.5 MPa for wire plasma spray, and 6.6 MPa for gas flame spray. Finally, the measured porosities were, in the more porous order, 24% for wire arc spray, 19% for wire plasma spray, and 14% for gas flame spray.

Table 4. Results of the measured roughnesses, hardnesses, adhesive strengths, and porosities obtained for coating layers deposited using the three thermal spray guns. In this table, "–" indicates the upper and lower values of two measurements, and "±" indicates the ranges of more than three measurements with their average values at each front.

Method	Surface Roughness R_z (µm)	Hardness (HV)	Adhesive Strength (MPa)	Porosity (%)
Gas flame	80–100	65.3 ± 15.0	6.61–6.65	14 ± 1.0
Wire arc	130–150	41.1 ± 8.0	13.4–14.5	24 ± 2.0
Wire plasma	80–100	69.5 ± 14.0	7.50–7.85	19 ± 2.0

4. Discussions

4.1. Power Balances in the Three Thermal Spray Guns

It is of interest to know how much power is dissipated in a thermal spray gun, and how much of it is used to heat spray droplets impinging onto substrates. For this purpose, measured values of temperatures were used together with known input powers. The heat loss from a gun head to the surrounding air was calculated to be negligible from the measured surface temperatures of the gun head and the surrounding ambient air.

The powers to gun heads are obtainable from electrical input powers for plasma- and arc-heated guns, while from a gas flow-rate and a heat of combustion for a gas-heated gun where complete burning of the supply gas is assumed.

On the other hand, the powers to spray droplets are obtainable by referring to Figure 4; from feed-rates of wires multiplied by (1), a sensible heat to raise temperatures of wires from an ambient temperature to the melting point of 933 K for Al (neglecting the contribution of 5% Mg), (2) a latent heat at the melting point (397 kJ/kg), and (3) a sensible heat to raise temperatures of wires from the melting point to measured values. The power efficiencies are obtainable from η = ((1) + (2) + (3))/ (input power).

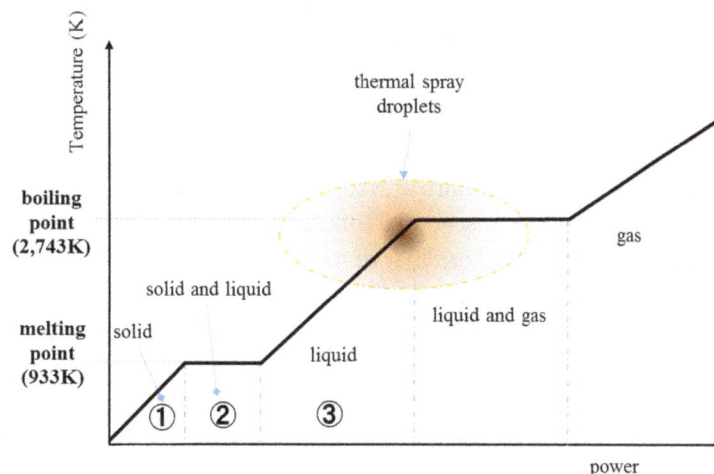

Figure 4. Schematic diagram showing how an input power to a gun is used to heat up spray droplets (abscissa is in a relative unit). The melting and boiling points in the ordinate is for Al.

The obtained results are shown in Table 5. As seen from this table, efficiencies are, in the order of higher efficiency, 45% for arc-heated, 23% for plasma-heated, and 9% for gas-heated guns. The implications of this result are discussed in the latter half of Section 4.3.

Table 5. Powers to guns and various heating powers to spray droplets, together with power efficiencies for three thermal spray guns.

Method	Input Power (kW)	Input Power to Thermal Spray Droplets (kW)			Total Power into Spray Droplets (kW)	Efficiency (%)
		Step (1)	Step (2)	Step (3)		
Gas flame	18.0	0.51	0.35	0.75	1.61	9
Wire arc	7.3	0.71	0.44	2.13	3.28	45
Wire plasma	10.0	0.48	0.33	1.53	2.34	23

4.2. Coating Qualities in Relation with Temperatures and Velocities of Spray Droplets

4.2.1. Gas Flame Spray Gun

Gas flame sprayed droplets are characterized by a low temperature (about 1900 K) and a high velocity (about 200 m/s), as seen in Table 3. The first evidence observed in Figure 3 is that the atomic composition of coating layers has hardly changed from that of the supplied wire, which is believed to be due to almost no loss of Mg due to the low temperature of droplets. The high hardness of coating layers as shown in Table 4 is explained by this maintenance of the Mg composition. The low adhesive strength may have resulted from the low temperature of spray droplets. The low porosity of the coating layers in the same table is also explained by the high velocity of spray droplets impinging onto a substrate.

4.2.2. Wire Arc Spray Gun

Wire arc sprayed droplets are characterized by a high temperature (about 2900 K) and a low velocity (about 60 m/s), as seen in Table 3. As evidently observed in Figure 3, the Mg in the coating layers has almost disappeared due to its low boiling temperature of 1364 K, and a small amount of nitrogen has appeared. The very low hardness of coating layers—almost similar to Al, as shown in Table 4—is believed to be due to the almost complete loss of Mg by the high temperature of droplets. The high adhesive strength may be due to the high temperature of the spray droplets. The high porosity of the coating layers may have resulted from the low velocity of spray droplets impinging onto the substrate.

4.2.3. Wire Plasma Spray Gun

Wire plasma sprayed droplets are characterized by a high temperature (about 3000 K) and an intermediate velocity (about 90 m/s), as seen in Table 3. The atomic composition of coating layers as shown in Table 4 has not changed much from that of the supplied wire in spite of the high temperature of spray droplets, similar to arc-heated droplets. This evidence, together with the high hardness of coating layers (e.g., Table 4), may have resulted from the presence of a shielding gas supplied around and along the flow of the spray droplets, which is intended to suppress mixing of spray droplets with the surrounding ambient air. The observed fact of lower adhesive strength of coating layers in the same table, compared with those of arc-heated droplets—despite the similar high temperatures for both cases—may have resulted from other unidentified cause(s). The porosity of the plasma-heated case (also from the same table) is between the other two cases, which may be due to the intermediate values of the temperature and velocity.

4.3. Overall Assessment of the Obtained Results

There have been extensive studies and resulting literatures for obtaining good coating layers by thermal spray processes (a good overview of a history up to the present time is given in [5], together with what have been carried out from the standpoint of monitoring thermal spray processes) and for practical applications in various fields [1–4]. These studies are mostly concerned with particular types of guns for understanding thermal spray processes with its optimization and/or for specific applications. This article is distinctly different from these past studies in that three guns of completely different technologies were employed. Their dimensions and operating conditions were optimized, as described in Section 1, for "the best cost–performance" after much trial-and-error, and are now routinely in use for depositions of corrosion- and wear-resistant layers in field works of the present coauthors' company. The wide ranges of temperatures and velocities as shown in Table 3 together with wide ranges of coating qualities as evident from Table 4 and of power efficiency as shown in Table 5 are only possible through the present approach. It is to be noted that the values of coating qualities shown in Table 4 may not be as good as one may hope, but it is the result of "the best cost–performance" choice, as one may obtain better qualities at the expenses of other factors, such as cost and/or coating time.

The observed results of coating qualities in relation with temperature and velocity of spray droplets were described above in Section 4.2, and a few words may be of value on the power efficiencies shown in Table 5. It is to be noted from Table 5 that the power efficiency of the wire arc spray gun is the highest at 45% among the three guns, followed by the wire plasma spray gun at 23%, then by the gas flame spray gun at 9%. These results are well understandable, as follows: power to the wire arc spray is supplied directly outside the gun, unhindered much from the compressed air sent to the arc region to transfer spray droplets to substrates, with the resultant power to the wire approaching almost half of the supplied amount. On the other hand, power to the wire plasma spray is first used to ionize the supplied nitrogen gas, and thus-formed plasma reaches the anode, which is in fact the wire to become spray droplets after melting. In this process, almost three-quarters of the supplied power is taken away by the ionized gas (plasma), and about one quarter remains in the droplets. Finally, power to the gas flame spray is only through heat transfer from burnt gas to the wire, which is very slow and of indirect process compared with electric discharges towards the wire electrodes, resulting in only less than 10% of the supplied power transferred to formed droplets for this case.

These features yield the following obvious outcomes: fast production of spray droplets and coating layers deposited by the wire arc spray, with rather rough surface and high porosity due to low velocity of the droplets resulting from their poor acceleration by the compressed air. High velocity of spray droplets in gas flame spray is the result of their good acceleration in a well-constricted nozzle, where the constricted nozzle shape is necessary to make as much thermal contact of burnt gas with the central wire. This high velocity yields low porosity of the coating layer. For the case of the wire plasma spray, the nozzle is constricted as shown in Figure 1 in order to yield good acceleration there, but not to the extent of the small annular structure for the gas flame spray. Additionally, a compressed air is supplied (as shown in Figure 1) to shield and prevent the outside air from mixing and decelerating the spray droplets' flow in the central core. The authors believe this feature to have played an important role in keeping the atomic composition as shown in Figure 3 and the resulting hardness as shown in Table 4.

The results of power efficiencies shown in Table 5 and the resulting spray qualities shown in Figure 3 and Table 4 are summarized as the wire plasma spray to be placed in between the gas flame spray and the arc wire spray in almost every aspect. Therefore, these characteristics are to be considered when considering which technology is to be selected for a particular application.

5. Conclusions

From the comparative studies of coating qualities deposited using three thermal spray guns with the measured temperatures and velocities of spray droplets impinging onto the substrates for the three cases, the following conclusions were obtained, summarized in terms of coating qualities, rather than from the viewpoints of various guns as described in Section 4.

- **Atomic Composition.** This is believed to be decided by the atmosphere when spray droplets are formed, and by their temperatures. Spray droplets produced by the wire arc spray-heating are formed in an open atmosphere and at a high temperature, resulting in a deposited atomic composition much altered from that of the supplied wire. On the other hand, discharges in the wire plasma spray-heating are maintained in nitrogen gas, and the flow of thus-formed droplets are surrounded by a shielding gas, with the result that there is little composition change from that of the wire. Spray droplets produced by the gas flame spray-heating are formed at a low temperature, such that the deposited atomic composition is also hardly changed from that of the supplied wire.

- **Hardness.** This is believed to be mostly dictated by the atomic compositions of formed coating layers. In particular, the coating layers deposited using Al/5Mg is decided by the Mg composition, as evident for coating layers using all three different thermal spray guns.

- **Adhesive Strength.** This is believed to be decided by the temperatures of spray droplets, resulting in high adhesive strengths at high temperatures. However, the reason for different adhesive strengths deposited using wire arc spray- and wire plasma spray-heating—despite their similar temperatures—is not known at the present time. A weak adhesive strength deposited using gas flame spray-heating is well understood by a low temperature of spray droplets.

- **Porosity.** This is believed to be mostly dictated by the velocities of spray droplets, resulting in low porosities at high velocities and vice versa.

Acknowledgments: This work has been carried out by the Grant-in-Aid "Seed-utilization business study" from the Japanese Ministry of Economy, Trade and Industry (METI), and the authors wish to thank their financial support. The Grant does not cover the costs to publish in open access.

Author Contributions: Yasuyuki Kawaguchi conceived and designed the experiments, together with writing the paper, with an advice and support of Fumihiro Miyazaki; Masafumi Yamasaki performed the experiments; Yukihiko Yamagata and Nozomi Kobayashi also joined the experiments, together with analyzing the data; Katsunori Muraoka contributed the overall strategy of this work and wrote the paper.

Conflicts of Interest: The authors declare no conflict of interest.

References

1. Vardelle, A.; Moreau, C.; Akedo, J.; Ashrafizadeh, H.; Berndt, C.C.; Berghaus, J.O.; Boulos, M.; Brogan, J.; Bourtsalas, A.C.; Dolatabadi, A. The 2016 thermal spray roadmap. *J. Thermal Spray Technol.* **2016**, *25*, 1376–1440. [CrossRef]

2. *Steel Bridge Design Handbook: Corrosion Protection of Steel Bridges*; Federal Highway Administration, Department of Transportation: Washington, DC, USA, 2015.

3. Ce, N.; Paul, S. Thermally Sprayed Aluminum Coatings for the Protection of Subsea Risers and Pipelines Carrying Hot Fluids. *Coatings* **2016**, *6*, 58. [CrossRef]

4. Fauchais, P.; Vardelle, A. Thermal spray coatings used against corrosion and corrosive wear. In *Advanced Plasma Spray Application*; Jazi, H.S., Ed.; INTECH Open Access Publisher: Rijeka, Croatia, 2012; pp. 3–38.

5. Fauchais, P.; Vardelle, M.; Vardelle, A. Reliability of plasma-sprayed coatings: Monitoring the plasma spray process and improving the quality of coatings. *J. Phys. D Appl. Phys.* **2013**, *46*, 1–16. [CrossRef]

6. Kawaguchi, Y.; Miyazaki, F.; Yamasaki, M.; Yamagata, Y.; Muraoka, K. The first results of an optical monitoring system for optimization of thermal plasma droplets. *J. Instrum.* **2015**, *10*, 1–8. [CrossRef]

7. Muraoka, K.; Kawaguchi, Y.; Miyazaki, F.; Nagayama, K.; Koso, T. Plazwire technology and the use of laser-aided diagnostics for its future evolution. *J. Instrum.* **2013**, *8*, 1–11. [CrossRef]

8. Oerlikon Metco Homepage. Available online: https://www.oerlikon.com/ (accessed on 15 December 2015).

9. Metallisation Ltd. Homepage. Available online: http://www.metallisation.com/ (accessed on 15 December 2015).

10. Bekefi, G.; Barrett, A. Sources of Radiation. In *Electromagnetic Vibrations, Waves, and Radiation*; The MIT Press: Cambridge, MA, USA, 1977; pp. 301–312.

11. Brewster, M.Q. Radiative Properties and Simple Transfer. In *Thermal Radiative Transfer and Properties*; John Wiley: New York, NY, USA, 1992; pp. 55–60.

12. AccuraSpray-G3C. Tecnar Automation Ltée Homepage. Available online: http://tecnar.com/ (accessed on 15 December 2015).

13. Spray watch. Oseir Ltd. Homepage. Available online: http://www.oseir.com/ (accessed on 15 December 2015).

11

Assessment of Environmental Performance of TiO$_2$ Nanoparticles Coated Self-Cleaning Float Glass

Martina Pini [1,*], Erika Iveth Cedillo González [2,3], Paolo Neri [1], Cristina Siligardi [2] and Anna Maria Ferrari [1]

[1] Department of Sciences and Engineering Methods, University of Modena and Reggio Emilia, Via Amendola, 2, 42100 Reggio Emilia, Italy; paolo.neri@unimore.it (P.N.); annamaria.ferrari@unimore.it (A.M.F.)

[2] Department of Engineering "Enzo Ferrari", University of Modena and Reggio Emilia, Via Vignolese, 905/A, 41125 Modena, Italy; erikaiveth.cedillogonzalez@unimore.it (E.I.C.G.); cristina.siligardi@unimore.it (C.S.)

[3] Facultad de Ciencias Químicas, Universidad Autónoma de Nuevo León, Guerrero y Progreso s/n Col. Treviño, Monterrey 64570, Mexico

* Correspondence: martina.pini@unimore.it

Academic Editor: Alessandro Lavacchi

Abstract: In recent years, superhydrophilic and photocatalytic self-cleaning nanocoatings have been widely used in the easy-to-clean surfaces field. In the building sector, self-cleaning glass was one of the first nanocoating applications. These products are based on the photocatalytic property of a thin layer of titanium dioxide (TiO$_2$) nanoparticles deposited on the surface of any kind of common glass. When exposed to UV radiation, TiO$_2$ nanoparticles react with the oxygen and water molecules adsorbed on their surface to produce radicals leading to oxidative species. These species are able to reduce or even eliminate airborne pollutants and organic substances deposited on the material's surface. To date, TiO$_2$ nanoparticles' benefits have been substantiated; however, their ecological and human health risks are still under analysis. The present work studies the ecodesign of the industrial scale-up of TiO$_2$ nanoparticles self-cleaning coated float glass production performed by the life cycle assessment (LCA) methodology and applies new human toxicity indicators to the impact assessment stage. Production, particularly the TiO$_2$ nanoparticle application, is the life cycle phase most contributing to the total damage. According to the ecodesign approach, the production choices carried out have exacerbated environmental burdens.

Keywords: self-cleaning; life cycle assessment; titanium dioxide nanoparticles; ecodesign; scale-up; float glass

1. Introduction

Since Fujishima and Honda discovered the photo-splitting of water in a titanium dioxide (TiO$_2$) anode photochemical cell in 1972 [1], research in the self-cleaning field based in photocatalytic nanoparticles has continuously grown. Among all the various metal oxides that have been tested for photocatalytic applications, TiO$_2$ has received the most attention because of its chemical stability and high reactivity under ultraviolet (UV) light irradiation. When a TiO$_2$ particle absorbs a photon with $hv \geq E_g$ (E_g = band gap of TiO$_2$ = 3.2 eV) [2], an electron is transferred from the valence band to the conduction band (e^-), leaving behind a positive hole (h^+). If the e^--h^+ pair interacts with adsorbed species, it forms radicals capable of oxidizing a wide range of organic pollutants into H$_2$O and CO$_2$ [3]. This property of TiO$_2$ can be used to impart the self-cleaning functionality to a variety of materials including tiles, glass, plastic coatings, panels, wallpapers, window blinds, paints, tunnel walls and road blocks to name a few [4–6]—and the field is still growing. Indeed, according to the BCC Research Advanced Materials Report AVM069B, the total market for photocatalyst products is forecasted to grow over the next five years, and is estimated to be valued at nearly $2.9 billion by 2020 [7].

Although the self-cleaning property that photocatalytic TiO_2 nanoparticles can impart to common materials is promising, the unexpected growth of nanotechnology is raising several concerns about the potential negative impacts that these new materials could cause on human health and the environment. The release of nanoparticles into environmental matrices could occur during different stages of their life cycles [8,9]. Therefore, considerable efforts should be made to assess the toxicity of nanoparticles, first on humans and then—though no less important—on the environment. The European Commission encouraged the life cycle approach to assess the sustainability of nanoproducts [10]. Life cycle assessment (LCA) is the most adequate methodology for determining the potentially adverse effects on human health and the environment of a product, process or service. It has thus been recognized as a useful tool to assess the environmental performance of nanoproducts [11].

Hischier et al. [12] investigated numerous review articles about the use of LCA in the nanotechnology field [13–19]. A key and open issue addressed in these reviews is the human toxicity and ecotoxicity characterization factors (CFs) for nanomaterials [12]. Thus far, CFs for a toxicity assessment have been published for two nanoparticles only, namely carbon nanotubes (CNT) [20] with graphene oxide [21] and TiO_2 nanoparticles (nanoTiO$_2$) [22,23].

The present work studies the ecodesign of the industrial scale-up of nanoTiO$_2$ self-cleaning coated float glass production performed by LCA methodology, focusing on the assessment of both human health effects and environmental loads of the entire life cycle of this new nanomaterial. Therefore, previously developed frameworks [23,24] established to evaluate the potential human toxicity impacts of nanoTiO$_2$ have been implemented in the impact assessment stage. This study was a part of an Italian project named "ARACNE" [25]. The main aim of this project is to study and ecodesign eco-friendly building materials with higher technological properties. In addition to the present LCA study, several LCA case studies of building nanomaterials have been carried out within ARACNE [24,26–28].

Over the last several years, few LCA studies that deal with releases of nanoparticles have been carried out. In particular, these studies are analyses of nanoTiO$_2$ [12,26–30], silver nanoparticles [31], CNT [20,32] and silica [33]. Nevertheless, only five LCA studies [26–30] were implemented in the life cycle impact assessment (LCIA) phase with the preliminary human toxicity factors calculated following the Ecoindicator 99 framework for carcinogenic substances [24], and only two of these [28,30] further applied the human CFs to a nanoTiO$_2$ analysis performed with the USEtox™ (version 2.0, Lyngby, Denmark) framework [23]. Moreover, the study of Hischier et al. applied only the latter CFs in the LCIA [12].

This work, together with two Pini et al. studies [28,30] (belonging to the ARACNE project and concerning different building materials, i.e., enameled steel panels and porcelain stoneware tiles), are the first LCA case studies assessing the nanoparticles released during the building nanomaterial life cycles, subsequently using the LCIA for all human toxicity factors performed by two different frameworks before analyzing the obtained results. Again, in accordance with the ecodesign approach, the production choices carried out have led to concerns about environmental burdens and safety of human health. Finally, the benefit derived from the nanoTiO$_2$ application of was also assessed considering toluene and NO$_x$ abatements.

2. Materials and Methods

2.1. Ecodesign of an Industrial Scale Process

In this work, a modified coating method [34], consisting first of a decrease in initial substrate roughness with acetic acid and then dip-coating of the softened glass into a TiO_2 acid nanosuspension, was used with the aim of producing films with enhanced adhesion to the substrate. This coating method was optimized thanks to experimental tests carried out in a chemical lab. The research continued with the intent to design an industrial scale-up of the developed coating method. Nevertheless, when a technology is not ready for the commercial scale, which is often the case with emerging technologies, sufficient data is scarcely available and so the environmental performance evaluation is based on incomplete information [35]. Therefore, LCA analysis of a production process at a laboratory scale should not be considered since the LCA results do not necessarily represent the environmental burdens which would be caused after scaling up to typical mass production [36–39]. The reasons are:

- There might be changes due to scale up in process yield as well as in energy efficiency of the process; these can influence the environmental burdens, as these affect the material and energy use as well as the amount of emissions and waste.
- There might be changes in technology and in the material or energy supplies.
- In LCA analysis of pilot/laboratory plants, processes are often seen as isolated or independent from each other. The effects due to changes in plant utilization are not considered sufficiently.

Gavankar et al. [36] studied the role that scale and technology maturity play in LCA of new technologies, e.g., nanotechnologies. They stated, "the magnitude of environmental impacts of emerging technologies at their mass production scale can be significantly smaller than a linear extrapolation of early LCAs may suggest".

In this work, starting from laboratory data, the best environmental performance of the industrial scale-up process of nanoTiO$_2$ self-cleaning coated float glass was evaluated. Here, the authors adopted a first linear extrapolation to convert lab-scale data into industrial-scale data. Future steps would be to include more elaborate up-scaling schemes.

To ecodesign the industrial-scale process, it was necessary to consider literature data and databases included in SimaPro 8 software [40] (e.g., ecoinvent v2 database [41] was used to model the float glass process), since the laboratory scale does not give meaningful information about plants, equipment, internal transports, nor about ordinary maintenance operations of equipment and machineries. In addition, no data related to the installation, use and end-of-life stages of nanoTiO$_2$ self-cleaning coated float glass have been provided by the laboratory.

2.2. Goal and Scope Definition

The goal of the study is to assess the environmental impacts of a nanoTiO$_2$ self-cleaning coated float glass over its entire life cycle in order to identify the hot spots of the system during the entire life cycle. The system studied is a self-cleaning glass coated with nanoTiO$_2$ film to create a surface that remains cleaner for longer than conventional glass. Titanium dioxide incorporation in building materials and its activation by the near-UV fraction of incident solar irradiation offers promising potential, namely the reduction of organic and inorganic pollutants. Therefore, the benefits derived from its application have been considered, i.e., the abatement of inorganic and organic substances (e.g., NO$_x$ and toluene emissions). In particular, an abatement of 4.01 mg/h·m^2 for NO$_x$ substance (studied by Chen and Poon [4]) and a reduction of 100 mg/h·m^2 for toluene emission (proposed by Demeestere [42]) were taken into account. To evaluate the reduction in concentration of these substances in the LCA studies, negative values were considered as input data.

The function of self-cleaning is applications in private buildings, such as traditional windows and curtain walls as well as glazing. 1 m^2 of nanoTiO$_2$ self-cleaning coated float glass is analyzed. The system boundaries cover the entire life cycle of the system analyzed, following the LCA approach. The analysis includes the supply of all raw materials involved in the coating process, packing, installation and end of life (Figure 1). The production, maintenance and disposal of facilities as well as the environmental burdens related to the production of chemicals, packaging and other auxiliary materials are also included in the present study. Emissions into the air and water, as well as the solid waste produced in each step are taken into account. The transportation to a treatment facility of the solid waste is also considered.

Starting from laboratory data, the best environmental performance of the industrial scale-up process of nanoTiO$_2$ self-cleaning coated float glass was evaluated. Moreover, because of the limited knowledge currently available regarding the effects nanoTiO$_2$ may have on the environment or human health [43], safe behavior was adopted for all life cycle steps in which workers may come into contact with or inhale nanoparticles released by a nanocoating surface. The following assumptions have been made:

- HEPAs (high efficiency particulate air filters), possessing 99.97% efficiency, were installed during cutting, soaking in acetic acid and coating steps.
- Use of PPE (personal protective equipment), particularly the face mask with its 95% efficiency [44] in protecting workers from dust and nanoparticles inhalation during coating, installation, use and end-of-life steps was implemented.
- A closed manufacturing system was designed.
- Use of specific packaging to limit the release of nanoparticle emissions during transportation was used.
- Transport distances of facilities, raw material, chemicals, materials for packaging from supplier to the production site have been assumed equal to 100 km, as required by the environmental product declaration (EPD) certification [45].
- Italian mixed-electric energy obtained by non-renewable sources (the electricity type mainly used in Italy) and created by ecoinvent was assumed. Obviously, adopting renewable energy such as photovoltaic energy, would enhance the environmental performance. In particular, environmental damage associated with the use of renewable sources can decrease by more than 87%. Nevertheless, this study is part of a regional Italian project, so its production must be located in the Italian territory.

2.3. Impact Assessment

Life cycle impact assessment (LCIA) results were modeled by a modified IMPACT 2002+ v2.10 [46] method as described below and successively by a modified USEtox™ method v1.03 [47] in order to consider the human health CFs for nanoTiO$_2$ in an indoor and outdoor environment as calculated by Pini et al. [23]. For a more representative index of the considered system, some additions and modifications were implemented in IMPACT 2002+, i.e., modification to the categories *Land use* (different types of land transformations were considered) and *Mineral extraction* (additional resources were added), as well as the *Radioactive waste* category (radioactive waste and its occupied volume was evaluated) [24,26].

Further, this study assesses the releases of nanoTiO$_2$ into the air (outdoor environment) and those inhaled by workers. Therefore, human toxicity of nanoTiO$_2$ for the outdoor environment and that breathed in by workers were calculated as reported in Ferrari et al. [26] and Pini, [24] and then incorporated into the IMPACT 2002+ method.

The environmental benefits derived from nanoTiO$_2$ application were evaluated only by the IMPACT 2002+-modified method.

2.4. Life Cycle Inventory

The entire life cycle of a nanoTiO$_2$ self-cleaning coated float glass (shown in Figure 1) consists of four main steps: (1) production; (2) installation; (3) use and (4) end of life. The production step, in turn, is divided into: (a) cutting; (b) lapping; (c) ultrasonic cleaning; (d) soaking in acetic acid; and (e) dip-coating.

The present study considers the outdoor application of a self-cleaning float glass in a private building. Inventory data, related to the life cycle of the bottom-up hydrolytic synthesis of nanoTiO$_2$, is reported by Pini et al. in a previous work [29]. The synthesis procedure was patented and employed by Colorobbia Italia S.p.A. [48]. The entire production and the end of life are the main life cycle steps that require electric energy. The life cycle of nanoTiO$_2$ self-cleaning coated float glass is described below.

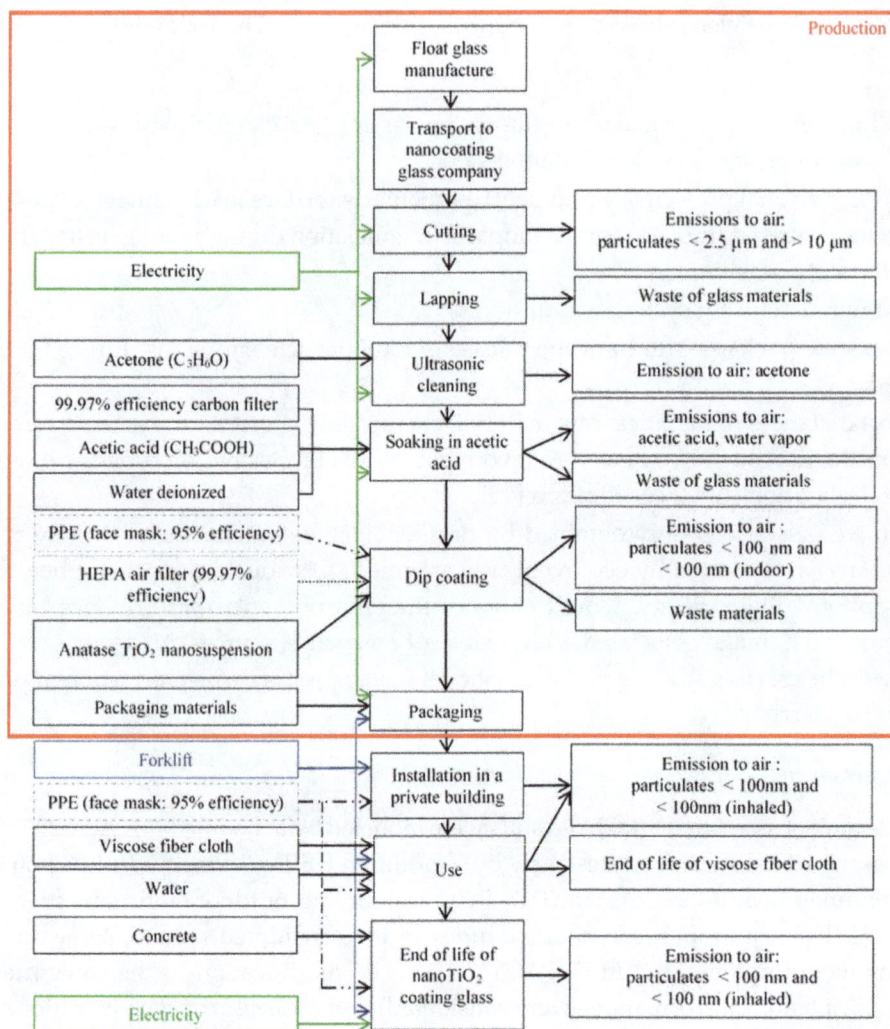

Figure 1. Flow chart of nanoTiO$_2$ (titanium dioxide) self-cleaning coated float glass.

2.4.1. Production

First, the Saint Gobain soda-lime float glass is cut into the customer's required size. The obtained glass is then polished to bevel the edges and corners. The successive ultrasonic cleaning step is a process that is able to clean the glass surface using ultrasound and acetone as solvent media. The clean glass is soaked in 96% CH$_3$COOH for 4 h to decrease the surface roughness of the substrate; the etched glass is subsequently coated with five layers of a nanoTiO$_2$ suspension at a coating rate of 85 mm/min. Finally, nanoTiO$_2$ coated float glasses are packed in a wooden box.

2.4.2. Installation and Use Phase

The nanoTiO$_2$ self-cleaning coated float glass was used for a private building as windows. In the installation step, the transport by lorry from the production company to the installation site and the handling of glasses from the lorry to the private building were evaluated. The installation of a single glass with nanoTiO$_2$ coating side oriented externally was considered.

In the use phase, nanoTiO$_2$ self-cleaning coated float glass was considered for applications such as windows, external windows, conservatories, etc. In accordance with Fujishima et al. [49], the duration of ten years of nanoTiO$_2$ coating effects was assumed. In the study, the heat reflected outside (thanks to the nanocoating) and the heat that transferred through the glass was assessed. Therefore, in summertime, the nanoTiO$_2$ coating kept the indoor room cooler thus obtaining a benefit.

On the contrary, in winter, this phenomenon meant that part of the solar heat did not pass through the glass windows, decreasing the radiation heat inside the room. Furthermore, the benefits of nanocoating such as the reduction of NO_x and VOCs concentrations was evaluated. Finally, annual maintenance of glazing with only water and viscose fiber cloth was included.

2.4.3. End of Life

To protect human health, and considering the uncertainty of the potential damage caused by nanoparticles after ten years (duration of nanoTiO$_2$ coating effects), making the glass inert through specific waste treatment was assumed; the waste glass was covered with concrete and then buried. Different glass lifetimes were evaluated in order to take into account the real lifetime of glass compared to that of the coating. Therefore, refunctionalization of glass after ten years was assumed. Considering a glass lifetime of 30 years and two functionalization treatments are needed. A final inertization treatment was considered.

The compilation of inventory data was carried out using databases included in SimaPro 8 software [40]. The ecodesign of industrial scale-up production of self-cleaning glass coated with nanoTiO$_2$ film was performed on lab data, carried out by the experiments to determinate the optimized coating method. The remaining data was obtained from specialized databases and literature such as devices, machineries, plants, internal transports, ordinary maintenance operations and all data regarding installation, use and end-of-life steps. A selection of important data used in the LCI (life cycle inventory) of nanoTiO$_2$ self-cleaning coated float is reported in Table 1.

Table 1. Inventory data of 1 m^2 of nanoTiO$_2$ self-cleaning coated float.

Category	Components	Quantity	Unit	Source
Energy input	Electricity consumption	244.4	kWh	Energetic process I/O data derived from ecoinvent database. Energy consumptions were supplied by the chemical lab and scaled up with linear rate
Materials I/O	Float glass uncoated	9.91	kg	Data supplied by the chemical lab and scaled up with linear rate and Colorobbia Italia SpA. for nanoTiO$_2$ suspension
	Tap water	52.77	L	
	Acetone	263.33	kg	
	Acetic acid	4.37	kg	
	Water deionized	2.39	kg	
	Compressed air	423.33	L	
	nanoTiO$_2$ suspension	5.84E−03	kg	
	Protection film (LDPE)	1.92E−02	kg	Data was supplied by one of the company leaders in glass production
	Viscose fiber cloth	0.13	kg	
	Concrete	0.24	m^3	
	Heat gain in summer season due to nanocoating	825.2	kW	Data supplied by the chemical lab and scaled up with linear rate
	Heat lost in winter season due to nanocoating	754.13333	kW	
Emissions to air	Particulates <2.5 µm	1.43E−02	kg	Data supplied by the chemical lab and scaled up with linear rate
	Particulates >10 µm	2.61E−02	kg	
	Particulates >2.5 µm and <10 µm	6.53	kg	
	Acetic acid	7.20E−02	kg	
	Water	1.29E−02	kg	
	Acetone	3.31E−06	kg	
	Particulates <100 nm in air	6.67E−03	kg	
	Particulates <100 nm inhaled	0.75	kg	
	NO_x	1.17E−01	kg	
	Nitric acid	2.40E−04	kg	
	Toluene	92E−03	kg	
	CO_2	3.92E−02	kg	
Transports	Road	85.49	tkm	

Table 1. *Cont.*

Category	Components	Quantity	Unit	Source
Waste to treatment	Disposal to residual landfill of nanoTiO$_2$ particulates captured by filter	4.01E−04	g	Waste quantities were given from the chemical lab while waste treatment statistics were derived from the ecoinvent process
	Acetone wastes captured by filter to residual landfill	5.05E−03	cm^3	
	Acetic acid wastes captured by filter to residual landfill	4.33	kg	
	Wastewater treatment (water used during the maintenance operations of equipment)	52.77	L	
	Disposal of particulates <2.5 μm and >10 μm dust captured by filter to residual landfill	1248.21	g	
	Disposal waste glass (inertization)	8.04	kg	End of life of functionalized glass was built ad hoc according to ecodesign approach. Data were appropriately assumed

3. Life Cycle Impact Assessment

3.1. The Modified IMPACT 2002+ Method

The environmental analysis of 1 m^2 of nanoTiO$_2$ self-cleaning coated float glass was conducted. Single score damage is equal to 25.22 mPt. The results of the analysis at mid-point level reported in Table 2 and Figure 2 show that the phases of the life cycle with the highest environmental burdens are the production (65.08%) and the use (28.16%) stages, followed by end of life (6.08%) and installation (0.67%).

Figure 3 highlights that the most significant contribution to the total damage is due to the *Non-renewable energy* impact category (37.89%), which is primarily affected by natural, in-ground gas (63.35%) due to the production phase (41.7%), in particular for electric energy consumption. Subsequently, the second major contribution to the total damage is generated by the *Global warming* impact category (34.49%), mainly due to fossil carbon dioxide (96.73%), which is caused by the production process (49.6%) and the use phase (46.68%), especially for glass manufacture and energy spent on air conditioning in the summer.

Table 2. Characterized LCIA results of 1 m^2 of nanoTiO$_2$ self-cleaning coated (*IMPACT 2002+ Method*).

Impact Category	Unit	Total	Production	Installation	Use Phase	End of Life
Carcinogens	kg C$_2$H$_3$Cl eq	6.35E−01	2.33E−03	3.37E−01	1.99E−02	6.35E−01
Non-carcinogens	kg C$_2$H$_3$Cl eq	6.14E−01	3.02E−03	5.15E−02	4.30E−02	6.14E−01
Respiratory inorganics	kg PM2.5 eq	4.60E−02	3.58E−04	−8.27E−03	3.11E−03	4.60E−02
Ionizing radiation	Bq C-14 eq	8.20E+02	4.17E+00	1.88E+02	6.35E+01	8.20E+02
Ozone layer depletion	kg CFC-11 eq	7.75E−06	1.01E−07	6.14E−06	3.51E−07	7.75E−06
Respiratory organics	kg C$_2$H$_4$ eq	2.13E−02	2.78E−04	−1.86E+00	2.69E−03	2.13E−02
Aquatic ecotoxicity	kg TEG water	5.46E+03	3.58E+01	9.13E+02	2.31E+02	5.46E+03
Terrestrial ecotoxicity	kg TEG soil	5.88E+02	8.05E+00	1.10E+02	6.83E+01	5.88E+02
Terrestrial acid/nutri	kg SO$_2$ eq	7.76E−01	9.62E−03	−4.79E−01	6.78E−02	7.76E−01
Land occupation	m2org.arable	5.31E−01	5.64E−03	5.08E−01	1.84E+00	5.31E−01
Aquatic acidification	kg SO$_2$ eq	2.29E−01	1.62E−03	8.40E−02	1.24E−02	2.29E−01
Aquatic eutrophication	kg PO$_4$ P-lim	7.15E−03	2.69E−05	1.58E−03	3.43E−04	7.15E−03
Global warming	kg CO$_2$ eq	4.26E+01	2.43E−01	4.03E+01	2.97E+00	4.26E+01
Non-renewable energy	MJ primary	8.31E+02	4.44E+00	5.83E+02	3.39E+01	8.31E+02
Mineral extraction	MJ surplus	2.87E+00	5.98E−03	1.89E−01	8.30E−02	2.87E+00
Radioactive waste	kg	3.76E+01	1.73E−01	7.59E+00	6.47E+01	3.76E+01
Carcinogens inhaled	kg	1.10E−03	3.82E−06	2.23E−04	8.17E−05	1.10E−03
Total	mPt (milli-point)	2.522E+01	1.641E+01	1.700E−01	7.100E+00	1.534E+00

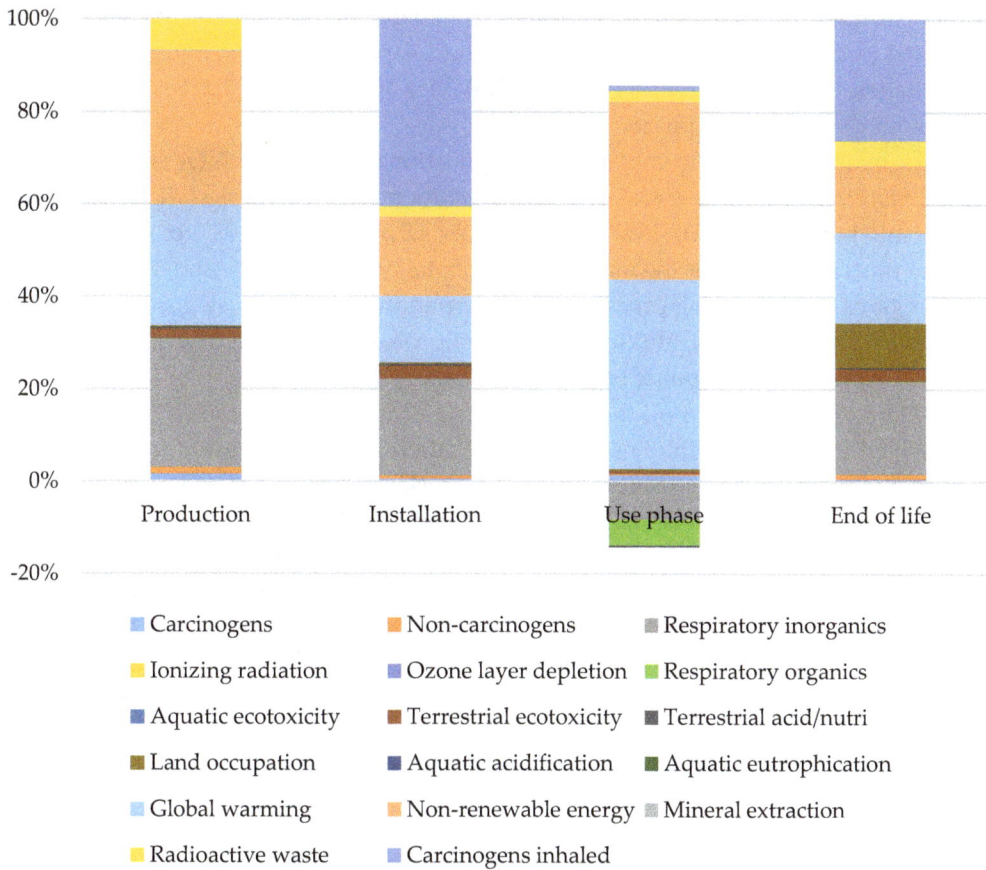

Figure 2. Evaluation by single score of 1 m^2 of nanoTiO$_2$ self-cleaning coated float glass.

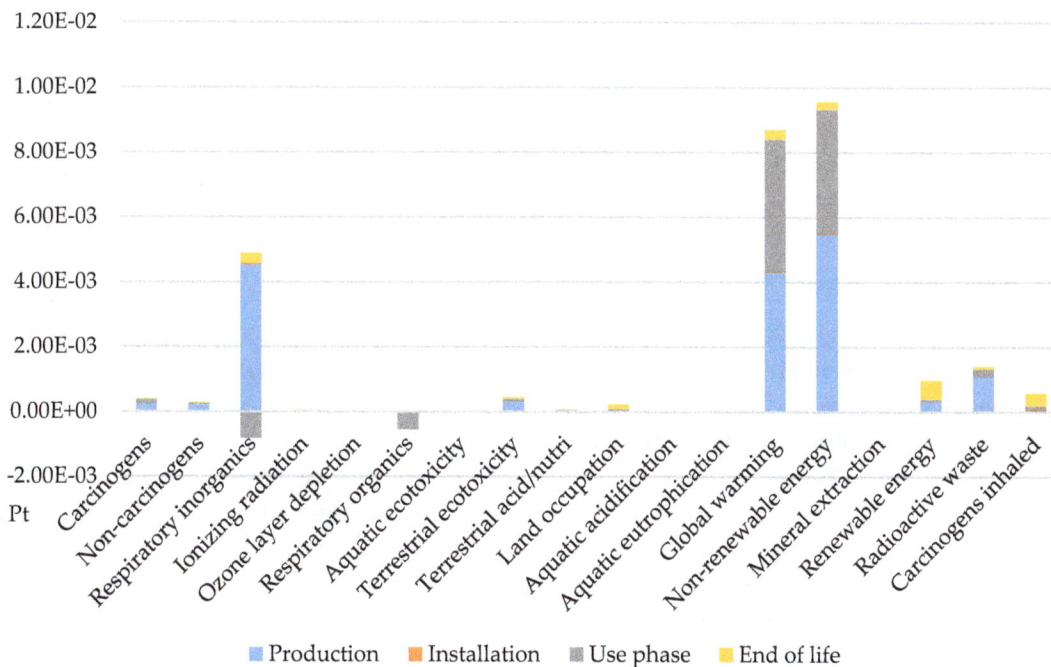

Figure 3. Weighted results by impact categories of 1 m^2 of nanoTiO$_2$ self-cleaning coated float glass.

The human toxicity effects generated by releases of nanoTiO$_2$ afflict the *Carcinogens* (outdoor environment) and *Carcinogens inhaled* (nanoparticles inhaled by worker) impact categories.

In *Carcinogens inhaled* (2.32%), the damage is entirely due to the releases of 7.45E−07 kg of particulates, <100 nm inhaled (anatase TiO_2 nanoparticles) by human, especially during end-of-life (68.76%) and use (19.45%) stages. *Carcinogens* impact category (1.56%) is affected by of 7.25E−4 kg of particulates <100 nm in the air during the use phase.

Finally, the benefits derived from nanoTiO_2 application (toluene and NO_x emission reductions) involve the *Respiratory inorganics* and *Respiratory organics* impact categories. *Respiratory inorganics* (16.12%) is mainly influenced by 37.33% of particulates, <2.5 µm, and 32.02% of sulfur dioxide, and the production process determines the main environmental burden (86.56% and 86.13% respectively), especially in regards to the lapping process and glass manufacture. This category is also affected by nitrogen oxides in the air (8.69%), and the production process determines the main environmental burden (385.35%) balance by use phase benefit (−331%).

In *Respiratory organics* (−2.18%), the reduction of −2.92 kg of toluene (VOC) emission to air (−100%) is derived from the benefit of nanoTiO_2 application in the use phase.

The impact of nanoTiO_2 release and inhaled by worker expressed in eco-point (Pt) is equal to 0.584 mPt. Conversely, the environmental benefit generated by toluene and NO_x abatement is equal to 1.77 mPt. The benefit derived from organic and inorganic emissions reduction counterbalances the negative impact of nanoTiO_2 releases; they differ in one order magnitude. However, the limited negative effect of nanoTiO_2 emissions depends on the safe choice defined in keeping with the ecodesign approach.

The endpoint analysis highlights (Table 3) that the total damage is affected by 16.74% to *Human health* (4.22E−3 Pt), 37.97% to *Resources* (9.57E−3 Pt), 34.49% to *Climate change* (8.69E−3 Pt), 2.89% to *Ecosystem quality* (7.29E−4 Pt), 5.59% to *Radioactive waste* (1.41E−3 Pt) and 2.31% to *Carcinogens inhaled* (5.84E−4 Pt).

Table 3. LCIA results at end-point level of 1 m^2 of nanoTiO_2 self-cleaning coated float glass.

Damage Category	Unit	Total	Production	Installation	Use Phase	End of Life
Human health	DALY	2.99E−05	3.59E−05	2.67E−07	−8.61E−06	2.37E−06
Ecosystem quality	PDF·m^2·year	9.99E+00	6.31E+00	8.17E−02	9.72E−01	2.63E+00
Climate change	kg CO_2 eq	8.61E+01	4.26E+01	2.43E−01	4.03E+01	2.97E+00
Resources	MJ primary	1.46E+03	8.34E+02	4.44E+00	5.83E+02	3.40E+01
Radioactive waste	kg	1.10E+02	3.76E+01	1.73E−01	7.59E+00	6.47E+01
Carcinogens inhaled	DALY	1.41E−03	1.10E−03	3.82E−06	2.23E−04	8.17E−05

Effects of Different Electricity Sources

The LCIA results highlight that the electric energy consumptions produce the main environmental loads. Therefore, a sensitivity analysis was conducted in order to assess the environmental improvement adopting renewable electricity, here represented by photovoltaic electricity mix, instead of the one derived from fossil fuel as required by the electric energy mix.

Table 4 and Figure 4 show the environmental performance enhancements of 41.51% (−1.047 mPt) when renewable energy is used instead of the electric energy mix generated mainly by fossil fuels. The world's trend is to increase renewable energy use. Therefore, the comparison between these two scenarios allows evaluating the environmental performance of an ideal situation, where the total electric energy mix is completely replaced by renewable sources, such as a photovoltaic mix. Nevertheless, today, the share of fossil fuels in the global mix is around 82% (the same as it was 25 years ago) and the contribution of renewable energy only reduces this to around 75% in 2035 [50]. This means the "nanoTiO_2 self-cleaning coated using electric energy mix" currently represents the real energy context. Finally, the LCIA results highlight that the benefit derived from nanoTiO2 application (1.77 mPt) has the same order of magnitude of the environmental improvement obtained by the use of renewable electricity (1.047 mPt).

Table 4. Environmental comparison between 1 m^2 of nanoTiO$_2$ self-cleaning coated glass using electric energy mix and 1 m^2 of nanoTiO$_2$ self-cleaning coated glass using renewable energy sources.

Impact Category	Unit	NanoTiO$_2$ Self-Cleaning Coated (Electric Energy Mix)	NanoTiO$_2$ Self-Cleaning Coated (Renewable Energy Source)
Carcinogens	kg C$_2$H$_3$Cl eq	9.95E−01	7.36E−01
Non-carcinogens	kg C$_2$H$_3$Cl eq	7.11E−01	8.78E−01
Respiratory inorganics	kg PM2.5 eq	4.12E−02	3.04E−02
Ionizing radiation	Bq C-14 eq	1.08E+03	1.08E+03
Ozone layer depletion	kg CFC-11 eq	1.43E−05	8.42E−06
Respiratory organics	kg C$_2$H$_4$ eq	−1.83E+00	−1.84E+00
Aquatic ecotoxicity	kg TEG water	6.64E+03	6.37E+03
Terrestrial ecotoxicity	kg TEG soil	7.75E+02	7.46E+02
Terrestrial acid/nutri	kg SO$_2$ eq	3.74E−01	7.79E−02
Land occupation	m2org.arable	2.89E+00	2.69E+00
Aquatic acidification	kg SO$_2$ eq	3.27E−01	2.40E−01
Aquatic eutrophication	kg PO$_4$ P-lim	9.11E−03	1.14E−02
Global warming	kg CO$_2$ eq	8.61E+01	4.45E+01
Non-renewable energy	MJ primary	1.45E+03	6.76E+02
Mineral extraction	MJ surplus	3.15E+00	3.89E+00
Renewable energy	MJ	1.10E+02	6.87E+02
Radioactive waste	kg	1.41E−03	1.40E−03
Carcinogens inhaled	kg	7.45E−07	7.45E−07
Total	mPt (milli-point)	2.522E+01	1.475E+01

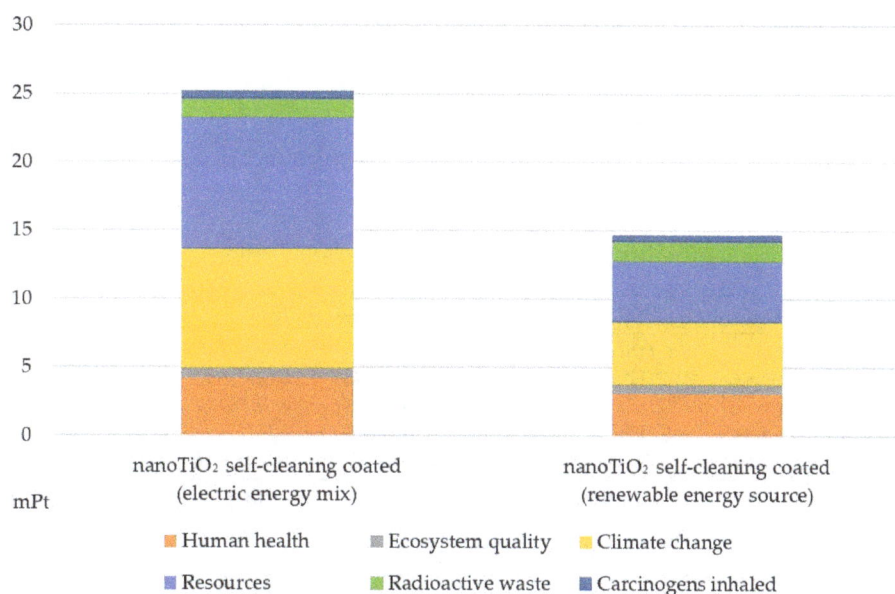

Figure 4. Environmental comparison between 1 m^2 of nanoTiO$_2$ self-cleaning coated glass using electric energy mix and 1 m^2 of nanoTiO$_2$ self-cleaning coated glass using renewable energy source.

3.2. The Modified USEtox™ Method

The results of the analysis at mid-point level reported in Figure 5 and Table 5 show that the life cycle phases with the highest environmental loads are the production stage, in particular due to the *Human toxicity*, *cancer* (85.5%), *Human toxicity*, *non-cancer* (80.6%) and *Ecotoxicity* (83.6%) impact categories and the end-of-life stage, specifically *Human toxicity*, *cancer*, *indoor* (68.8%) and *Human toxicity*, *non-cancer*, *indoor* (68.8%).

The total damage of *Human toxicity*, *cancer* and *Ecotoxicity* impact categories is mainly due to chromium VI in water (95.23% and 89.8%, respectively), which is caused by the production stage (86%), particularly the steel manufacture used to produce the air filter. Moreover, in *Human toxicity*, *non-cancer*, barium in water generates major environmental load (42%), specifically affected by the production stage (77.6%) producing the heavy fuel oil necessary for flat glass production. In *Human toxicity, cancer, indoor* and *Human toxicity, non-cancer, indoor* impact categories, the damage is completely caused by the releases of 7.45E−07 kg of particulates, <100 nm inhaled (anatase TiO_2 nanoparticles inhaled by people that are in the room) in indoor environment and is mainly due to end-of-life phase (68.76% for both impact categories). Releases of 2.6E−6 kg of particulates, <100 nm in the air affect *Human toxicity, cancer* by 0.261% and *Human toxicity, non-cancer* by 8.53E−2% and chiefly results from the installation and use phase (98.62% for both impact categories).

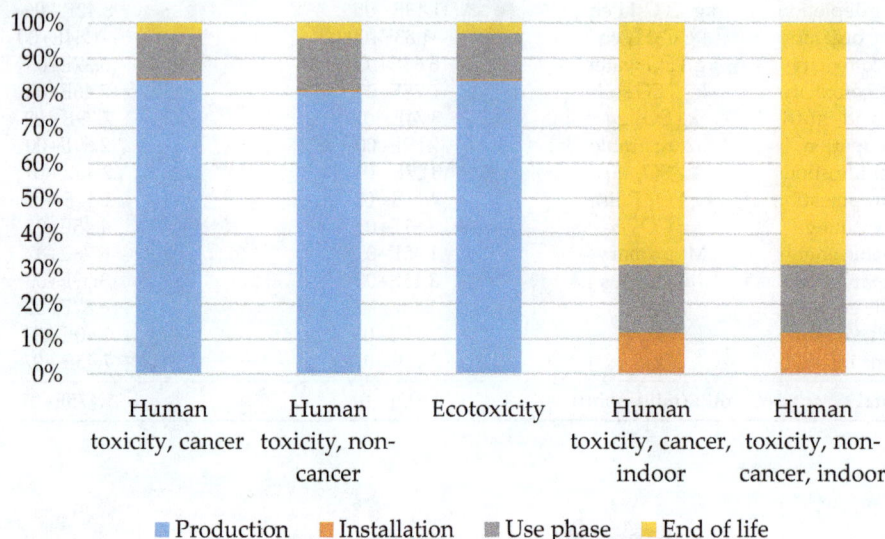

Figure 5. Environmental profile of 1 m^2 of nanoTiO$_2$ self-cleaning coated float glass (characterization results).

Table 5. Characterized LCIA results of 1 m^2 of nanoTiO$_2$ self-cleaning coated float glass.

Impact Category	Unit	Total	Production	Installation	Use Phase	End of Life
Human toxicity, cancer	CTUh [#]	4.401E−06	3.692E−06	1.497E−08	5.65E−07	1.297E−07
Human toxicity, non-cancer	CTUh [#]	1.565E−07	1.261E−07	6.657E−10	2.297E−08	6.797E−09
Ecotoxicity	CTUe [§]	46.236702	38.666071	0.1488913	6.0468404	1.3748995
Human toxicity, cancer, indoor	CTUh [#]	1.066E−08	6.344E−14	1.256E−09	2.073E−09	7.327E−09
Human toxicity, non-cancer, indoor	CTUh [#]	4.359E−13	2.595E−18	5.139E−14	8.479E−14	2.997E−13

[#] CTUh = cases/kg$_{emitted}$; [§] CTUe = PAF·m^3·year.

3.3. Comparison between the Environmental Performance NanoTiO$_2$ Functionalized Float Glass and the Conventional Ones

Finally, the study analyzes the different environmental performances determined by the nanoTiO$_2$ functionalized float glass (innovative building material) and a single float uncoated glass (conventional building material). For the latter building material, two different lifetime scenarios were considered. The first one considers that the float glass and the nanoTiO$_2$ coating have the same lifetime (10 years) (it is assumed that after 10 years the nanocoating no longer produces benefits). The second one considers that the float glass lifetime is equal to 30 years and the nanoTiO$_2$ coating lifetime equal to 10 years. Therefore, another two refunctionalization processes, after every 10 years, was needed in a period of 30 years. For both scenarios, the inertization process with concrete was taken into account as end-of-life treatment. The criteria followed to model the uncoated float glass are reported in the supplementary material (SM).

Figure 6 reports the LCIA results of the comparison, considering a lifespan of 30 years, among 1 m^2 of uncoated flat glass (conventional material), 3 m^2 of nanoTiO$_2$ coated float glass (10 years lifetime) and 1 m^2 of nanoTiO$_2$ coated float glasses (30 years lifetime) to be refunctionalized twice (innovative materials). LCIA was here performed by the modified IMPACT 2002+ method. The detailed environmental comparison results and the single LCIA results per glass are reported in the SM.

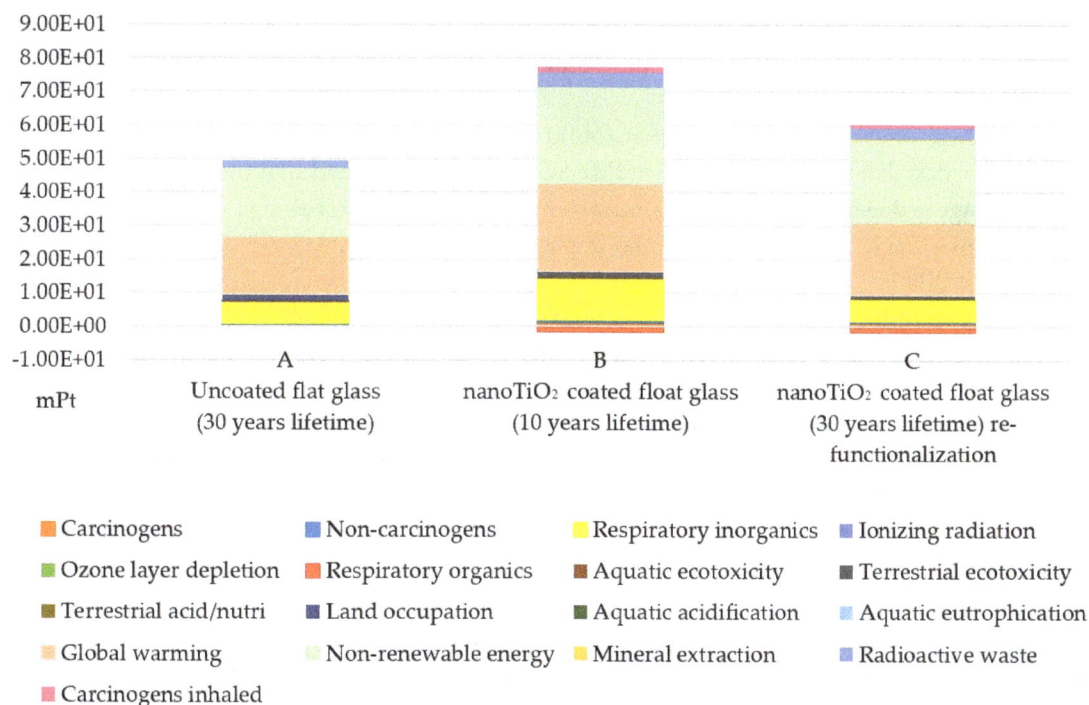

Figure 6. LCIA comparison by single score of 1 m^2 of conventional uncoated flat glass (conventional material) and 1 m^2 of nano-TiO$_2$ functionalized coated float glasses (innovative materials).

NanoTiO$_2$ functionalized float glass (scenario B) is the case study that produced the highest environmental damage (75.65 mPt), followed by scenario C (58.54 mPt) and finally scenario A (49.39 mPt). The impact categories that mainly determine the environmental loads on all analyzed case studies are **Non-renewable energy**, **Global warming** and **Respiratory inorganics**:

- In the **Non-renewable energy** impact category, case B determines the higher impact (28.7 mPt on the total damage) mainly due to *gas, natural, in-ground* emission generated by electric energy manufacture in the production process of nanoTiO$_2$ self-cleaning coated float glass;

- In **Global warming**, case B determines the higher impact (26.09 mPt on the total damage) mainly due to *carbon dioxide, fossil* emission generated by natural gas production used in the use phase for air conditioning.

- In **Respiratory inorganics** impact category, case B determines the higher impact (12.2 mPt on the total damage) mainly due to *particulates* <2.5 μm emission generated by the lapping process in the production stage. For innovative nanomaterials (case studies B and C), nitrogen oxide emissions in the air reduced by the photocatalytic activity of nanoTiO$_2$ coating generated a reduction of environmental load in this category.

Finally, Figure 6 shows that the **Respiratory organics** impact category determines an environmental benefit of 1.65 mPt for both B and C scenarios, specifically the reduction of toluene (VOC) emissions into the air.

4. Conclusions

Although the total market for photocatalytic products is estimated to be at $3 billion by 2020, and the most used photocatalyst is nanoparticled TiO$_2$, its ecological and human health risks are still under analysis. Therefore, in this work, the environmental sustainability of nanoTiO$_2$ functionalized coated float glass was performed with the life cycle assessment methodology. An ecodesign approach was followed in order to make the most appropriate choices for minimizing environmental loads and protecting human health. In this context, an industrial scale-up of the coating production and its successive application on the float glass were studied.

The analysis of results illustrates the same trend for both modified IMPACT 2002+ and USEtox™ methods.

The highest environmental burden is found to be the production phase of the life cycle of nanoTiO$_2$-functionalized coated float glass. IMPACT 2002+ determined that the main environmental load at this stage is due to the float glass manufacturing and acetic acid soaking processes. Furthermore, USEtox™ shows that the main environmental impact at the production stage is due to the steel used to produce the air filter.

An analysis of the benefits derived by nanoTiO$_2$ application by the modified IMPACT 2002+ method reveal a moderate gain in reducing airborne pollutants during the use phase, i.e., toluene (-2.92 kg) and NO$_x$ ($-9.3E-2$ kg) emissions for the *Respiratory organics* and *Respiratory inorganics* impact categories, respectively. However, it is necessary to point out that the data related to the nanoparticle emissions in all life cycle stages are not up to date and are still unknown. Therefore, scientific effort must be made to obtain adequate life cycle inventory (LCI) data on these new materials in order to ascertain the real sustainability of nanoparticle coatings for outdoor application [24].

In particular, precautions such as installation of high efficiency particulate air filters, closed systems for the production stage, protective equipment, and special end-of-life treatment in addition to guideline recommendations on how to treat nanoproducts throughout their entire life cycle will limit nanoparticle emissions into the air and/or inhaled by humans. In fact, the TiO$_2$ nanoparticles directly inhaled by humans is equal to $7.45E-07$ kg and affects *Carcinogens inhaled* (modified IMPACT 2002+ method), *Human toxicity, cancer* and *Human toxicity, non-cancer* impact categories (modified USEtox™ method) especially during end-of-life treatment (68.76%). In regards to TiO$_2$ nanoparticles released into the air, the quantity totals $7.26E-4$ kg and influences *Carcinogens* (modified IMPACT 2002+ method), *Human toxicity, cancer, indoor* and *Human toxicity, non-cancer, indoor* impact categories (modified USEtox™ method), especially during the use phase, by 98.62%.

The LCIA performed by the IMPACT 2002+ method highlighted that the benefit derived from organic and inorganic emissions reduction counterbalances the negative impact of nanoTiO$_2$ releases, differing by one order of magnitude. However, the limited negative effect of nanoTiO$_2$ emissions depends on the safe choice defined following the ecodesign approach. Therefore, if these choices change, the results could also vary.

The present work implements two preliminary LCIA frameworks (ecoinvent 99 and USEtox™) determined to quantify the potential human toxicity of an engineered nanoparticle (nanoTiO$_2$) using the LCA methodology [23,24].

The authors already discussed in Pini et al. [28,30] the limitations of applied LCIA frameworks. The fate module requires improvement by, for example, considering rate coefficients as descriptors for environmental fate processes. Moreover, as several gaps still exist in the toxicity assessment of nanomaterials, a database comprising the results of all the toxicological tests carried out thus far on these new materials is urgently required. As long as this data is unavailable, the effect analysis of these LCIA frameworks will suffer from lack of robustness. Therefore, the hereby presented environmental results must be updated as soon the weaknesses of the LCIA frameworks have been addressed. A future research step, then, might be the application of the preliminary human toxicity factors for nanoTiO$_2$ to already-existing LCA case studies that include nanoTiO$_2$ and that have not yet been investigated (i.e., functionalized building materials, synthesis processes, nanoparticle application, nanotechnologies production, etc.). The final aim is the validation of the preliminary LCIA frameworks for the assessment of human toxicity factors for nanoTiO$_2$ through their application to concrete LCA case studies. This allows a comparison of the obtained environmental results and their subsequent optimization. Future steps would be to include more elaborate up-scaling schemes.

In conclusion, the comparison analysis between nanoTiO$_2$ functionalized float glass and uncoated float glass showed that the latter building material causes higher environmental damage, mainly as a result of the higher solar factor value of uncoated glass compared to that of nanocoated glass.

Acknowledgments: Authors thank the financial support of the "ARACNE e Laboratorio Integrato Sviluppo Tecnologie Avanzate Materiali Innovativi per Costruzioni Ecosostenibili" through the Italian regional program, "Dai distretti produttivi ai distretti tecnologici".

Author Contributions: Martina Pini collected the data to carry out the life cycle inventory, performed the LCA study, implemented new toxicity factors for nanoTiO$_2$ in the life cycle impact assessment, interpreted the environmental results and wrote the manuscript. Anna Maria Ferrari contributed to the analysis of outcomes and drafted the final discussion. Cristina Siligardi and Erika Iveth Cedillo-González designed the laboratory experimental methodology and contributed to the discussion of the results and the writing of the paper. Erika Iveth Cedillo-González prepared the nanoTiO$_2$ functionalized float glass and performed the experiments, including the modified procedure for increasing the adhesion and the related tests.

Conflicts of Interest: The authors declare no conflict of interest and the founding sponsors had no role in the design of the study; in the collection, analyses, or interpretation of data; in the writing of the manuscript, and in the decision to publish the results.

References

1. Fujishima, A.; Honda, K. Electrochemical photolysis of water at a semiconductor electrode. *Nature* **1972**, *238*, 37–38. [CrossRef] [PubMed]
2. Ni, M.; Leung, M.K.H.; Leung, D.Y.C.; Sumathy, K. A review and recent developments in photocatalytic water-splitting using for hydrogen production. *Renew. Sust. Energ. Rev.* **2007**, *11*, 401–425. [CrossRef]
3. Akhavan, O.; Ghaderi, E. Self-accumulated Ag nanoparticles on mesoporous TiO$_2$ thin film with high bactericidal activities. *Surf. Coat. Technol.* **2010**, *204*, 3676–3683. [CrossRef]
4. Chen, J.; Poon, C. Photocatalytic construction and building materials: From fundamentals to applications. *Build. Environ.* **2009**, *44*, 1899–1906. [CrossRef]
5. Pichat, P. Self-cleaning materials based on solar photocatalysis. In *New and Future Developments in Catalysis: Solar Photocatalysis*; Suib, S.L., Ed.; Elsevier: Amsterdam, The Netherlands, 2013; pp. 167–190.
6. Raibeck, L.; Reap, J.; Bras, B. Investigating environmental burdens and benefits of biologically inspired self-cleaning surfaces. *CIRP J. Manuf. Sci. Technol.* **2009**, *1*, 230–236. [CrossRef]
7. Photocatalysts: Technologies and Global Markets; BCC Research Advanced Materials Report AVM069B. Available online: http://www.bccresearch.com/market-research/advanced-materials/photocatalysts-technologies-markets-report-avm069b.html (accessed on 3 September 2016).
8. Som, C.; Berges, M.; Chaudhry, Q.; Dusinska, M.; Fernandes, T.F.; Olsen, S.I.; Nowack, B. The importance of life cycle concepts for the development of safe nanoproducts. *Toxicology* **2010**, *269*, 160–169. [CrossRef] [PubMed]
9. Hsu, L.-Y.; Chein, H.-M. Evaluation of nanoparticle emission for TiO$_2$ nanopowder coating materials. In *Nanotechnology and Occupational Health*; Springer: Berlin, Germany, 2007; pp. 157–163.
10. United Nations Environment Programme (UNEP). *Global Guidance Principles for Life Cycle Assessment Databases. A Basis for Greener Processes and Products*; United Nations: Geneva, Switzerland, 2011.
11. Hischier, R. Framework for LCI modelling of nanoparticle releases along the life cycle. *Int. J. LCA* **2014**, *19*, 838–849. [CrossRef]
12. Hischier, R.; Salieri, B.; Pini, M. Most important factors of variability and uncertainty in an LCA study of nanomaterials – findings from a case study with nano titanium dioxide. *IMPACT* **2017**. Submitted.
13. Hischier, R.; Walser, T. Life cycle assessment of engineered nanomaterials: State of the art and strategies to overcome existing gap. *Sci. Total Environ.* **2012**, *425*, 271–282. [CrossRef] [PubMed]
14. Gavankar, S.; Suh, S.; Keller, A.F. Life cycle assessment at nanoscale: Review and recommendations. *Int. J. LCA* **2012**, *17*, 295–303. [CrossRef]
15. Gavankar, S.; Suh, S.; Keller, A.A. *Life Cycle Assessment of Engineered Nanomaterials*; Woodhead Publishing: Cambridge, UK, 2014.

16. Upadhyayula, V.K.K.; Meyer, D.E.; Curran, M.A.; Gonzalez, M.A. Life cycle assessment as a tool to enhance the environmental performance of carbon nanotube products: A review. *J. Clean Prod.* **2012**, *26*, 37–47. [CrossRef]

17. Kim, H.C.; Fthenakis, V. Life cycle energy and climate change implications of nanotechnologies. A critical review. *J. Ind. Ecol.* **2013**, *17*, 528–541. [CrossRef]

18. Lazarevic, D.; Finnveden, G. Life cycle aspects of nanomaterials. In *Environmental Strategies Research KTH*; Royal Institute of Technology: Stockholm, Sweden, 2013.

19. Miseljic, M.; Olsen, S.I. Life-cycle assessment of engineered nanomaterials: A literature review of assessment status. *J. Nanopart. Res.* **2014**, *16*, 1–33. [CrossRef]

20. Eckelman, M.J.; Mauter, M.S.; Isaacs, J.A.; Elimelech, M. New perspectives on nanomaterial aquatic ecotoxicity: Production impacts exceed direct exposure impacts for carbon nanotoubes. *Environ. Sci. Technol.* **2012**, *46*, 2902–2910. [CrossRef] [PubMed]

21. Deng, Y.; Li, J.; Qiu, M.; Yang, F.; Zhang, J.; Yuan, C. Deriving characterization factors on freshwater ecotoxicity of graphene oxide nanomaterial for life cycle impact assessment. *Int. J. LCA* **2016**. [CrossRef]

22. Salieri, B.; Righi, S.; Pasteris, A.; Olsen, S.I. Freshwater ecotoxicity characterisation factor for metal oxide nanoparticles: A case study on titanium dioxide nanoparticle. *Sci. Total Environ.* **2015**, *505*, 494–502. [CrossRef] [PubMed]

23. Pini, M.; Salieri, B.; Ferrari, A.M.; Nowack, B.; Hischier, R. Human health characterization factors of nano-TiO$_2$ for indoor and outdoor environments. *Int. J. LCA* **2016**, *21*, 1452–1462. [CrossRef]

24. Pini, M. Life Cycle Assessment of Nano-TiO$_2$ Functionalized Building Materials Extended to Historical Buildings. Ph.D. Thesis, University of Modena and Reggio Emilia, Modena, Italy, 2015.

25. Bando "Dai Distretti Produttivi ai Distretti Tecnologici"—DGR n. 1631/2009. Available online: http://www.innovazionefinanza.it/2009/emilia-romagna-distretti-tecnologici/ (accessed on 9 January 2017).

26. Ferrari, A.M.; Pini, M.; Neri, P.; Bondioli, F. Nano-TiO$_2$ coatings for limestone: Which sustainability for cultural heritage? *Coatings* **2015**, *5*, 232–245. [CrossRef]

27. Pini, M.; Gamberini, R.; Neri, P.; Rimini, B.; Ferrari, A.M. Life Cycle Assessment of a Self-Cleaning Coating Based on NanoTiO$_2$-Polyurea Resin Applied on an Aluminum Panel. Available online: http://digidownload. libero.it/giabon/tesi/pini/pres_pini_LCAResinaPoliurea_con_nanoTiO2.pdf (accessed on 9 January 2017).

28. Pini, M.; Bondioli, F.; Neri, P.; Montecchi, R.; Ferrari, A.M. Environmental and human health assessment of life cycle of nanoTiO$_2$ functionalized porcelain stoneware tile. *Sci. Total Environ.* **2017**, *577*, 113–121. [CrossRef] [PubMed]

29. Pini, M.; Rosa, R.; Neri, P.; Bondioli, F.; Ferrari, A.M. Environmental assessment of a bottom-up hydrolytic synthesis of TiO$_2$ nanoparticles. *Green Chem.* **2015**, *17*, 518–531. [CrossRef]

30. Pini, M.; Marinelli, S.; Gamberini, R.; Neri, P.; Rimini, B.; Ferrari, A.M. Life cycle assessment of a nano-TiO$_2$ functionalized enamel applied on a steel panel. *IJOQM* **2016**, *12*, 478–485.

31. Walser, T.; Demou, E.; Lang, D.J.; Hellweg, S. Prospective environmental life cycle assessment of nanosilver T-shirts. *Environ. Sci. Technol.* **2011**, *45*, 4570–4578. [CrossRef] [PubMed]

32. Hischier, R. Life cycle assessment study of a field emission display television device. *Int. J. LCA* **2015**, *20*, 61–73. [CrossRef]

33. Roes, A.L.; Tabak, L.B.; Shen, L.; Nieuwlaar, E.; Patel, M.K. Influence of using nanoobjects as filler on funtionality-based energy use of nanocomposites. *J. Nanopart. Res.* **2010**, *12*, 2011–2028. [CrossRef]

34. Cedillo-González, E.I.; Montorsi, M.; Mugoni, C.; Montorsi, M.; Siligardi, C. Improvement of the adhesion between TiO$_2$ nanofilm and glass substrate by roughness modifications. *Phys. Procedia* **2013**, *40*, 19–29. [CrossRef]

35. Shibasaki, M.; Warburg, N.; Eyerer, P. Upscaling effect and life cycle assessment. In Proceedings of the 13th CIRP International Conference on Life Cycle Engineering, Leuven, Belgium, 31 May–2 June 2006.

36. Gavankar, S.; Suh, S.; Keller, A.A. The role of scale and technology maturity in life cycle assessment of emerging technologies—A case study on carbon nanotubes. *J. Ind. Ecol.* **2015**, *19*, 51–60. [CrossRef]

37. Caduff, M.; Huijbregts, M.A.J.; Althaus, H.-J.; Koehler, A.; Hellweg, S. Wind power electricity: The bigger the turbine, the greener the electricity? *Environ. Sci. Technol.* **2012**, *46*, 4725–4733. [CrossRef] [PubMed]

38. Arvidsson, R.; Kushnir, D.; Molander, S.; Sandén, B.A. Energy and resource use assessment of graphene as a substitute for indium tin oxide in transparent electrodes. *J. Clean. Prod.* **2016**, *132*, 289–297. [CrossRef]

39. Li, Q.; McGinnis, S.; Sydnor, C.; Wong, A.; Renneckar, S. Nanocellulose life cycle assessment. *ACS Sustain. Chem. Eng.* **2013**, *1*, 919–928. [CrossRef]

40. Product Ecology Consultants (PRè). SimaPro database manual—Methods library. Available online: http://discounthardware.us/read-online/download-now/simapro-database-manual-methods-library.pdf (accessed on 4 January 2017).

41. Life Cycle Inventories, Ecoinvent Database v. 2.0. Available online: http://www.ecoinvent.ch (accessed on 12 December 2010).

42. Demeestere, K.; Dewulf, J.; De Witte, B.; Beeldens, A.; Van Langenhove, H. Heterogeneous photocatalytic removal of toluene from air on building materials enriched with TiO_2. *Build. Environ.* **2008**, *43*, 406–414. [CrossRef]

43. Klöpffer, W.; Curran, M.A.; Frankl, P.; Heijungs, R.; Köhler, A.; Olsen, S.I. Nanotechnology and Life Cycle Assessment. A Systems Approach to Nanotechnology and the Environment. In Proceedings of the Nanotechnology and Life Cycle Assessment Workshop, Washington, DC, USA, 2–3 October 2006.

44. *BS EN 149:2001+A1:2009 Respiratory Protective Devices. Filtering Half Masks to Protect against Particles. Requirements, Testing, Marking*; British Standards Institution (BSI): London, UK, 2011.

45. General Programme Instructions for Environmental Product Declarations EPD, the International EPD Cooperation, version 1.0. Available online: http://www.environdec.com/en/The-International-EPD-System/General-Programme-Instructions/ (accessed on 9 January 2017).

46. Jolliet, O.; Margni, M.; Charles, R.; Humbert, S.; Payet, J.; Rebitzer, G.; Rosenbaum, R. IMPACT2002+: A new life cycle impact assessment methodology. *Int. J. LCA* **2003**, *8*, 324–330. [CrossRef]

47. USEtox™ (2016), User Manual. Available online: http://www.usetox.org/support/tutorials-manuals (accessed on 23 March 2016).

48. Baldi, G.; Bitossi, M.; Barzanti, A. Method for the Preparation of Aqueous Dispersions of TiO_2 in the Form of Nanoparticles, and Dispersions Obtainable with This Method. U.S. Patent 20080317959A1, 25 December 2008.

49. Fujishima, A.; Rao, T.N.; Tryk, D.A. Titanium dioxide photocatalysis. *J. Photoch. Photobio. C* **2000**, *1*, 1–21. [CrossRef]

50. IGEL (Initiative for Global Environmental Leadership), Making the Transition to a Low-Carbon Economy. Available online: http://knowledge.wharton.upenn.edu/special-report/making-the-transition-to-a-low-carbon-economy/ (accessed on 20 December 2016).

12

Theoretical Studies of the Adsorption and Migration Behavior of Boron Atoms on Hydrogen-Terminated Diamond (001) Surface

Xuejie Liu *, Congjie Kang, Haimao Qiao, Yuan Ren, Xin Tan and Shiyang Sun

School of Mechanical Engineering, Inner Mongolia University of Science & Technology, Baotou,
Inner Mongolia 014010, China; happykcj@163.com (C.K.); 15540153167@163.com (H.Q.);
renyuan_bt@126.com (Y.R.); heart_tan@126.com (X.T.); sunshy@imust.cn (S.S.)
* Correspondence: xuejieliu2000@yahoo.com

Academic Editors: Quanshun Luo and Yongzhen Zhang

Abstract: The adsorption and migration activation energies of boron atoms on a hydrogen-terminated diamond (001) surface were calculated using first principles methods based on density functional theory. The values were then used to investigate the behavior of boron atoms in the deposition process of B-doped diamond film. On the fully hydrogen-terminated surface, the adsorption energy of a boron atom is relatively low and the maximum value is 1.387 eV. However, on the hydrogen-terminated surface with one open radical site or two open radical sites, the adsorption energy of a boron atom increases to 4.37 eV, and even up to 5.94 eV, thereby forming a stable configuration. When a boron atom deposits nearby a radical site, it can abstract a hydrogen atom from a surface carbon atom, and then form a BH radical and create a new radical site. This study showed that the number and distribution of open radical sites, namely, the adsorption of hydrogen atoms and the abstraction of surface hydrogen atoms, can influence the adsorption and migration of boron atoms on hydrogen-terminated diamond surfaces.

Keywords: B-doped diamond films; adsorption energy; migration behavior; first principles methods

---

1. Introduction

Boron-doped diamond (BDD) films deposited by chemical vapor deposition (CVD) have attracted increasing research attentions owing to their outstanding properties, including mechanical and electrochemical stability, as well as a wide potential window, low and stable background current, good biocompatibility, high corrosion resistance, and high efficiency in electrochemical oxidation processes. BDD films have been used in a number of electrochemical applications, such as water treatment, electro-synthesis, electro-catalysis, electro-analytical measurement, electrochemical energy technology, and biosensing [1–6], and BDD depositions have also been employed to prepare smooth coatings for machining carbon fiber reinforced plastics [7]. BDD film-coated rectangular-hole-shaped drawing dies have been discovered to not only display much better adhesion and wear-resistant properties, but also have higher initial surface smoothness before polishing and relatively lower film hardness, making it liable to be polished to the appropriate surface smoothness [8]. The performances of BDD films depend heavily on the films' microstructure, which mainly relies on the formation mechanism of BDD films in the CVD process. Therefore, the surface chemical process involved in the CVD diamond and BDD film growth has been extensively investigated [9–15]. With the method used to investigate the surface migration in diamond growth [10–13] and the insertion of boron into methane [16], Ashfold's group studied boron doping during diamond CVD, including boron addition to the cluster surface, boron insertion into the diamond surface, boron loss processes from the diamond surface [14], B and BH insertion reactions, and BH migration on the diamond surface [15].

In the present research, the diamond (001) surface plane is considered, because the (001) plane is one of most frequently obtained surface planes in CVD-grown crystalline diamond. Further, the {100} diamond exhibits a cubic morphology, with smooth, flat crystallite surfaces, and low defect densities, ensuring that {100} growth has long been a focus of experimental and theoretical study [11,17,18].

The method employed in the present work is slightly different from those of the previous studies [10–16]. First, a relatively large slab model $4 \times 4 \times (8 + 1 + 8)$ was employed as a hydrogen-terminated diamond (001) surface model. Second, the adsorption energies of an atom at certain highly symmetrical positions on the diamond (001) surface were calculated. Third, the potential energy surface (PES) was displayed using the calculated adsorption energy values. Fourth, the migration path was determined by analyzing the PES. Fifth, several image atoms were set along the migration path. Sixth, the minimum energies along the migration path were calculated using the nudged elastic band (NEB) method [19], and the saddle point energy or the energy barrier was then determined.

According to the gas-phase chemistry and composition in microwave plasma activated $B_2H_6/Ar/H_2$ mixtures [20,21], boron atoms are the most abundant gas-phase BH_x species adjacent to the growing diamond surface. Plasma field research [22] indicated that a great amount of hydrogen atoms exist on the substrate surface in a hydrogen-rich diamond deposition. As a result, the adsorption of hydrogen atoms and the abstraction of surface hydrogen atoms play important roles in the diamond growth process [10,23]. Therefore, the present study focused on the adsorption and migration behavior of boron atoms on a hydrogen-terminated diamond (001) surface. The investigation focused on boron atoms on a fully hydrogen-terminated surface and boron atoms on hydrogen-terminated surfaces with one and two open radical sites.

2. Calculation Details

Calculations were performed using a Vienna Ab-Initio Simulation Package (VASP) (VASP5.2, University Wien, Wien, Austria) code [24–26], which was based on density functional theory. In specific calculations, a plane-wave basis was set and periodic boundary conditions were used to determine the Kohn–Sham ground state. The projector-augmented wave (PAW) method [27,28] was used to describe electronic and ionic interactions. Local density was described with generalized gradient approximation (GGA) based on the Perdew-Burke-Ernzerhof (PBE) formulation [29,30]. Electron and ion relaxation convergence precisions were 10^{-4} and 10^{-3} eV, respectively. The Brillouin zone was sampled using the Monkhorst-Pack k-point grid [31] during a self-consistent calculation to identify the electronic ground state. A $5 \times 5 \times 1$ k-point mesh was utilized for slab calculations. The parameter settings were optimized for the calculations. Moreover, spin-polarized calculations were performed to optimize the structure and configuration. The NEB method in VASP [19] was used to calculate the minimum migration energy of atoms on the hydrogen-terminated diamond (001) surface.

Prior to calculating atom migration, a lattice constant of 0.3568 nm was determined by adjusting the calculation parameters to optimize the diamond crystalline structure, in which the plane-wave cut-off energy was 350 eV. The calculated lattice constant was close to the experimental value of 0.3567 nm. A $4 \times 4 \times (8 + 1 + 8)$ slab composed of eight carbon layers (with 16 carbon atoms per plane), one hydrogen layer, and eight vacuum layers was used to model the hydrogen-terminated diamond (001) film, as shown in Figure 1. The height of the vacuum layer was approximately 0.8020 nm, and a periodically arranged interference was prevented. The hydrogen-terminated diamond (001) film model was selected because of the abutment of hydrogen atoms on the surface in the diamond film synthesis [22]. During relaxation, the three bottom layers of carbon atoms were fixed, whereas the others moved freely. After relaxation, the hydrogen-terminated diamond (001) surface underwent reconstruction. The length of the dimer bond was 0.162 nm, the bond length between the surface carbon atom and the hydrogen atom was 0.102 nm, and the total energy of the configuration was -1164.233 eV.

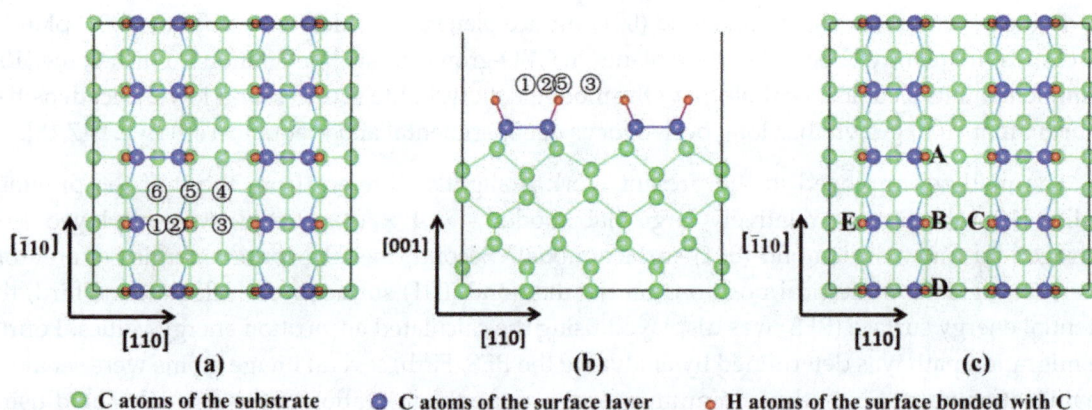

● C atoms of the substrate   ● C atoms of the surface layer   ● H atoms of the surface bonded with C

Figure 1. Six highly symmetrical positions and the open radical sites on the reconstructed hydrogen-terminated diamond (001) slab: (**a**) main view; (**b**) top view; (**c**) The open radical sites: when Hydrogen B was abstracted, the configuration represented the model of a hydrogen-terminated diamond (001) slab with one open radical site, denoted as 1ORS; when Hydrogens A and B were abstracted, the configuration stood for the model with two open radical sites along the dimer row, denoted as 2ORS-R; when Hydrogens B and C were abstracted, the configuration represented the model with two open radical sites along the dimer chain by the open-ring side, denoted as 2ORS-CO; when Hydrogens E and B were abstracted, the configuration represented the model with two open radical sites along the dimer chain by the close-ring side, denoted as 2ORS-CC.

Energies of a single carbon atom and a single boron atom were determined (Table 1) by fitting the calculated cohesive energy with the experimental value. The calculated lattice constant and cohesive energy were consistent with experimental values. Thus, our calculations are reasonable and reliable for studying the adsorption and migration behaviors in this system.

Table 1. Lattice constants and cohesive energies of C and B, and energies of a single C atom and a single B atom.

Items	C Atom [32]	B Atom
Energy of single atom (eV)	−1.479	−0.324
Lattice from calculation (nm)	0.3568	0.181
Lattice from experiment (nm) [33]	0.3567	0.177
Difference in the percentage of lattice (%)	0.028	2.26
Cohesive energy from calculation (eV·atom^{-1})	7.626	5.96
Cohesive energy from experiment (eV·atom^{-1}) [33]	7.37	5.81
Difference in the percentage of cohesive energy (%)	3.47	2.58

Adsorption and migration behaviors of a single boron atom on the hydrogen-terminated diamond (001) surface were investigated. Based on the special structure of the diamond, six highly symmetrical positions existed on the reconstruction surface of hydrogen-terminated diamond (001), as shown in Figure 1a,b: (1) on the bridge site of the dimer ring-closing bond (denoted as P1); (2) on top of the carbon atom in the carbon dimer (denoted as P2); (3) at the bridge site of the dimer ring-opening bond (denoted as P3); (4) between the carbon dimer row located on top of the atom in the third layer (denoted as P4); (5) between the carbon dimer on top of the atom in the second layer (denoted as P5); and (6) between the carbon dimer on top of the atom in the third layer (denoted as P6). A single boron atom was set at each of the six positions (0.2 nm higher than the surface layer) for calculating the adsorption energy of a boron atom. The orientations of [$\bar{1}10$] and [110] were regarded as the dimer row and the dimer chain, respectively [10]. The open radical sites on the hydrogen-terminated diamond surface are shown in Figure 1c. When Hydrogen B was abstracted, the configuration represented the model of a hydrogen-terminated diamond (001) slab with one open radical site (denoted as 1ORS); when

Hydrogen A and B were abstracted, the configuration stood for the model with two open radical sites along the dimer row (denoted as 2ORS-R); when Hydrogens B and C were abstracted, the configuration represented the model with two open radical sites along the dimer chain by the ring-opening side (denoted as 2ORS-CO); and when Hydrogens E and B were abstracted, the configuration represented the model with two open radical sites along the dimer chain by the ring-closing side (denoted as 2ORS-CC).

3. Results and Discussion

3.1. Adsorption of a Single Boron Atom on a Fully Hydrogen-Terminated Diamond (001) Surface

The total energies of configurations were calculated to obtain the adsorption energies of the boron atom at the six highly symmetrical positions. The adsorption energy is defined as

$$E_{ad} = -(E_{tot} - E_{slab} - E_B) \tag{1}$$

where E_{tot} is the total energy of configuration with adsorbed atoms, E_{slab} is the total energy of the relaxation hydrogen-terminated diamond (001) slab without adsorbed atoms, and E_B is the energy of a single boron atom. The total energy of the hydrogen-terminated diamond (001) slab E_{slab} was -1164.233 eV.

The total energies and adsorption energies of a single boron atom on the hydrogen-terminated diamond (001) surface are listed in Table 2. At P4 and P6, the boron atom desorbed. At P1, P2, and P3, the distance between the boron atom and the surface carbon atoms was approximately 0.289–0.354 nm, which is much longer than the covalent radius sum of the boron atom and the carbon atom (i.e., 0.165 nm). Thus, the boron atom could not bond with the surface carbon atoms. Moreover, the adsorption energies were relatively low (0.08–0.162 eV). At P5, both distances between the boron atom and two surface carbon atoms were 0.158 nm, and the distance between the boron atom and the carbon atom in the second layer was 0.145 nm. These three bond lengths are all close to the covalent radius sum of the boron atom and the carbon atom, i.e., 0.165 nm. Obviously, the boron atom bonded with these three carbon atoms; thus, the adsorption energy reached the maximum value of 1.378 eV.

Table 2. Total energies and adsorption energies of a single boron atom on a fully hydrogen-terminated diamond (001) surface.

Position	P1	P2	P3	P4	P5	P6
E_{tot} (eV)	−1164.719	−1164.640	−1164.717	−1166.029	−1165.935	−1172.714
E_{ad} (eV)	0.162	0.083	0.160	(desorption)	1.378	(desorption)

The adsorbed boron atom could migrate along the path from P5 to P1 and further to another P5. The migration energies along the path were calculated using the NEB method in the VASP code. The calculation results are listed in Table 3. The minimum energy data indicate that the migration barrier along the path is 1.216 eV. Therefore, the boron atom migration along the path is easy at normal CVD diamond film deposition temperature (700–900 °C).

Table 3. Minimum energies during the migration of a single boron atom on a fully hydrogen-terminated diamond (001) surface.

Position	P5	I	II	III	IV	P1
E_{tot} (eV)	−1165.935	−1165.543	−1164.879	−1164.893	−1164.919	−1164.719
E_{ad} (eV)	1.378	0.986	0.322	0.336	0.362	0.162

3.2. Adsorption of a Single Boron Atom on a Hydrogen-Terminated Diamond (001) Surface with One Open Radical Site

When Hydrogen B was abstracted Figure 1c, the configuration represented the model of a hydrogen-terminated diamond (001) slab with one open radical site (denoted 1ORS). The charge analysis shows that when a hydrogen atom bonded with the surface carbon atom (C_{NN}) was removed, the total magnetic moment of C_{NN} changed from 0.000 to 0.412 μ_B, implying that the C_{NN} had unpaired electrons and was more reactive. The total energy of the configuration of 1ORS E_{slab} is -1158.567 eV. The total energies of the configurations were calculated to study the adsorption of a boron atom at the six symmetrical positions. The relevant adsorption energies were obtained using Equation (1). The total energies and adsorption energies of a single boron atom on the hydrogen-terminated diamond (001) surface with one open radical site are listed in Table 4, and the PES of the adsorption energies is displayed in Figure 2.

Table 4. Total energies and adsorption energies of a single boron atom on a hydrogen-terminated diamond (001) surface with one open radical site.

B-Igration	P1	P2	P3	P4	P5	P6
E_{tot} (eV)	−1163.260	−1162.551	−1161.086	−1159.741	−1164.087	−1160.106
E_{ad} (eV)	4.370	3.661	2.196	0.851	5.197	1.216

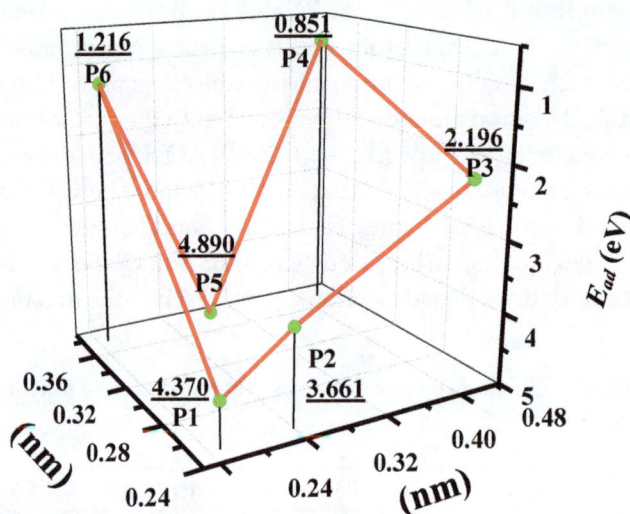

Figure 2. Potential energy surface (PES) of a single boron atom on a hydrogen-terminated diamond (001) surface with one open radical site.

Given that Hydrogen B was abstracted, the boron atom deposited at P2 could bond with the surface carbon atom; the bond length was approximately 0.154 nm, which is slightly less than their covalent radius sum of 0.165 nm. Thus, the adsorption energy reached the relatively large value of 3.66 eV. When the boron was deposited at P1, the boron atom bonded with the surface carbon dimer with bond lengths 0.152 and 0.142 nm, and the dimer bond was broken. The Bader charge analysis indicates that the charge of boron atom reduces from 2.823 to 1.555 electrons; the charge of two carbon atoms increase from 3.957 to 4.676 electrons and from 4.104 to 4.908 electrons. Thus, the adsorption energy increased to 4.37 eV, which was close to the result in previous study [15], which is 363 kJ/mol. When the boron atom was deposited at P5, the boron atom abstracted the hydrogen atom from a carbon atom nearby and bonded with two surface carbon atoms with bond lengths both at 0.159 nm. The boron atom bonded also with the carbon atom in the second layer with bond length of 0.185 nm. Therefore, the adsorption energy reached the maximum value of 5.197 eV. At P3, P4, and P6, the distance between the boron atom and the carbon atom with a dangling bond was relatively large, resulting in low adsorption energies.

The migration path can be determined by analyzing the PES. The boron atom could migrate from P1 to P5. At P5, the boron atom abstracted a hydrogen atom, and then the BH radical could migrate from P5 to P1. Along the migration path, the minimum energies were calculated with the NEB method. The calculation results are listed in Table 5 and the minimum energy path curves are presented in Figure 3a,b. The curve in Figure 3a shows that the boron atom migration from P1 to P5 is difficult, because of a high energy barrier of 2.36 eV. Once the boron atom overcomes the energy barrier, it can abstract the hydrogen atom from the nearby carbon atom and form a BH radical. The curve in Figure 3b shows that the migration of the BH radical is easy because of a low energy barrier of 1.53 eV. However, the BH radical tends to slide back to P5 and forms a stable configuration. The energy barrier for the BH radical to slide back to P5 is 0.017 eV.

Table 5. Minimum energies during the migration of a single boron atom or the BH on the hydrogen-terminated diamond (001) surface with one open radical site.

B-Migration	P1	I	II	III	IV	P5	E_a (eV)
E_{tot} (eV)	−1163.260	−1163.103	−1162.575	−1162.217	−1161.725	−1164.087	2.362
ΔE (eV)	0.827	0.984	1.512	1.870	2.362	0	
BH-Migration	**P5**	**I**	**II**	**III**	**IV**	**P1**	E_a (eV)
E_{tot} (eV)	−1164.087	−1163.855	−1163.240	−1162.626	−1162.558	−1162.575	1.529
ΔE (eV)	0	0.232	0.847	1.461	1.529	1.512	

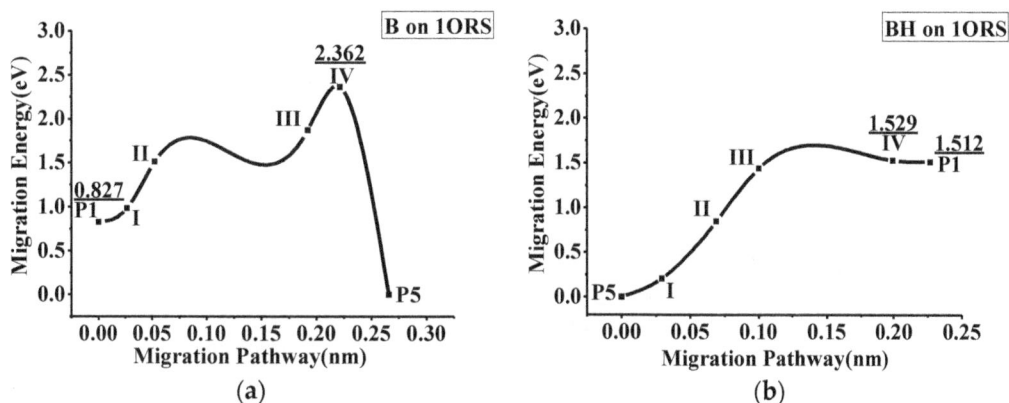

Figure 3. Migration minimum energy path curves of the boron atom and the BH radical on the 1ORS: (**a**) the boron atom migration from P5 to P1 and (**b**) the BH radical migration from P5 to P1.

3.3. Adsorption and Migration Behaviors of a Boron Atom on the Hydrogen-Terminated Diamond (001) Surface with Two Open Radical Sites

As shown in Figure 1c, when Atoms A and B were abstracted, the configuration represented the model of a hydrogen-terminated diamond (001) slab with two open radical sites along the dimer row (denoted as 2ORS-R). The total energy of the configuration of 2ORS-R E_{slab} is −1152.921eV. The total energies of the configurations of a single boron atom on 2ORS-R were calculated to study the adsorption of a boron atom at the six symmetrical positions. The relevant adsorption energies were obtained using Equation (1). The total energies and adsorption energies of a single boron atom on the hydrogen-terminated diamond (001) surface with two open radical site are listed in Table 6 and the PES of the adsorption energies is displayed in Figure 4a–c.

Table 6. Total energies and adsorption energies of a single boron atom on the hydrogen-terminated diamond (001) surface with two open radical sites.

2ORS-R	P1	P2	P3	P4	P5	P6
E_{tot} (eV)	−1157.594	−1156.973	−1158.131	−1155.112	−1159.184	−1155.899
E_{ad} (eV)	4.346	3.725	4.883	1.864	5.936	2.651
2ORS-CO	**P1**	**P2**	**P3**	**P4**	**P5**	**P6**
E_{tot} (eV)	−1157.692	−1156.867	−1158.499	−1154.609	−1158.435	−1154.403
E_{ad} (eV)	4.489	3.664	5.296	1.406	5.232	1.200
2ORS-CC	**P1**	**P2**	**P3**	**P4**	**P5**	**P6**
E_{tot} (eV)	−1159.006	−1157.629	−1158.042	−1154.527	−1158.514	−1155.008
E_{ad} (eV)	4.886	3.509	3.922	0.407	4.396	0.888

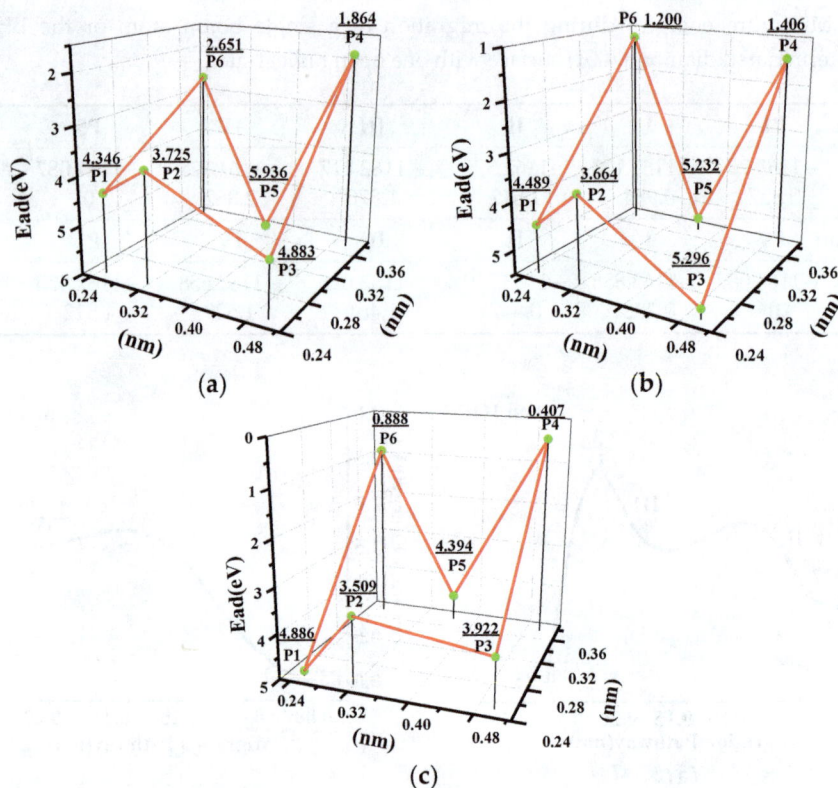

Figure 4. PES of a single boron atom on the hydrogen-terminated diamond (001) surface with two open radical sites: (**a**) on the 2ORS-R; (**b**) on the 2ORS-CO; and (**c**) on the 2ORS-CC.

Given that Atoms A and B were abstracted (Figure 1c), the boron atom deposited at P5 could bond with two carbon atoms on the surface, both with bond lengths of 0.152 nm, which is less than their covalent radius sum of 0.165 nm. In addition, the boron atom bonded with the carbon atom in the second layer, with a bond length of 0.169 nm. Therefore, the adsorption energy reached the maximum value of 5.936 eV. When deposited at P3, the boron atom abstracted a hydrogen atom from the nearby carbon atom and formed a BH radical. The boron atom bonded with two surface carbon atoms and both bond lengths were 0.171 nm. Thus, the adsorption energy reached 4.88 eV. Figure 4a represents the PES with three basins, which were used to determine the migration path. One path is for the boron atom to migrate from P1 to P5 and further to P1'; another path is for the boron atom to migrate from P5 to P3. At P3, the boron atom abstracted a hydrogen atom and formed a BH radical. After that, the BH radical can migrate from P3 to P5.

In the configuration of 2ORS-CO, hydrogen B and C were abstracted (Figure 1c). The boron atom deposited at P3 bonded with two surface carbon atoms, both with bond lengths of 0.158 nm. The adsorption energy reached the maximum value of 5.296 eV. The boron atom deposited at P5 could abstract a hydrogen atom from the nearby surface carbon atom and form a BH radical. The boron atom bonded also with two surface carbon atoms with bond lengths of 0.164 nm and 0.157 nm. In addition, the boron atom bonded with the carbon atom in the second layer, with bond length 0.183 nm. Thus, the adsorption energy was 5.23 eV. The migration path can be determined by analyzing the PES in Figure 4b. One path is for the boron atom to migrate from P3 to P1 and further to P3′; another path is for the BH radical to migrate from P5 to P3.

When Hydrogen B and E were abstracted (Figure 1c), the boron atom deposited at P1 could bond with the surface carbon dimer, both with bond lengths of 0.1478 nm, which is less than their covalent radius sum of 0.165 nm; in addition, the dimer bond was broken. Therefore, the adsorption energy reached the maximum value of 4.886 eV. The boron atom deposited at P3 abstracted a hydrogen atom from a carbon atom nearby and formed a BH radical. The boron atom bonded also with two surface carbon atoms with bond lengths of 0.171 nm and 0.172 nm. Thus, the adsorption energy reached 3.922 eV. Similar to the case of 2ORS-CO, the boron atom at P5 abstracted a hydrogen atom from the nearby surface carbon atom and formed a BH radical. The boron atom bonded with two surface carbon atoms with bond lengths of 0.160 nm and 0.158 nm. Moreover, the boron atom bonded with the carbon atom in the second layer with a bond length of 0.184 nm. Hence, the adsorption energy reaches 5.23 eV. The migration path can be determined by analyzing the PES in Figure 4c. One path is for the boron atom to migrate from P1 to P3; another path is for the BH radical to migrate from P3 to P1. The third path, similar to the case of 2ORS-CO, is for the BH radical from P5 to P1.

The migration energies along the path were calculated using the NEB method in the VASP code. The calculation results are listed in Table 7, and the minimum energy path curve is presented in Figure 5. The minimum energy path curve in Figure 5a shows a deep basin between two shallow basins. The boron atom on the configuration of 2ORS-R will slide to a shallow basin and overcomes an energy barrier of 0.54 eV from P1 to P5. However, the migration energy barrier of the boron atom from P5 to P1 is 1.71 eV. The energy barrier of the boron atom from P5 to P3 is 1.94 eV (Figure 5b). At P3, the boron atom abstracted a hydrogen atom and formed a BH radical. The migration of the BH radical from P3 to P5 is very easy because of the low energy barrier of 0.74 eV, as shown in Figure 5c.

The migration of a boron atom in the 2ORS-CO configuration along the path P1 to P3 (Figure 5d) is easy because of the low energy barrier of 0.825 eV. The return migration from P3 to P1 requires 1.632 eV to overcome the energy barrier. The migration of a BH radical in the 2ORS-CO configuration along the path from P5 to P3 is easy, given the low energy barrier of 0.765 eV, as shown in Figure 5e.

The migration of a boron atom in the 2ORS-CC configuration along the path P1 to P3 is relatively difficult because the energy barrier is high, i.e., 2.557 eV (Figure 5f). At P3, the boron atom abstracted a hydrogen atom and formed a BH radical. The migration of the BH radical from P3 to P1 (Figure 5g) needs to overcome an energy barrier of 1.882 eV. The migration of a BH radical in the 2ORS-CC configuration along the path from P5 to P1 is similar to that in the 2ORS-CO configuration.

The analyses of the PES and the migration of a boron atom in 2ORS-R, 2ORS-CO, and 2ORS-CC configurations indicate that the adsorption of a boron atom at two open radical sites forms a stable configuration and the adsorption energies are from 3.922 to 5.936 eV. All PES for the 2ORS-R, 2ORS-CO, and 2ORS-CC configurations represent the multiple-basin shapes. Moreover, the boron atom nearby the radical site could abstract a hydrogen atom from the surface carbon atom, form a BH radical, and create a new radical site.

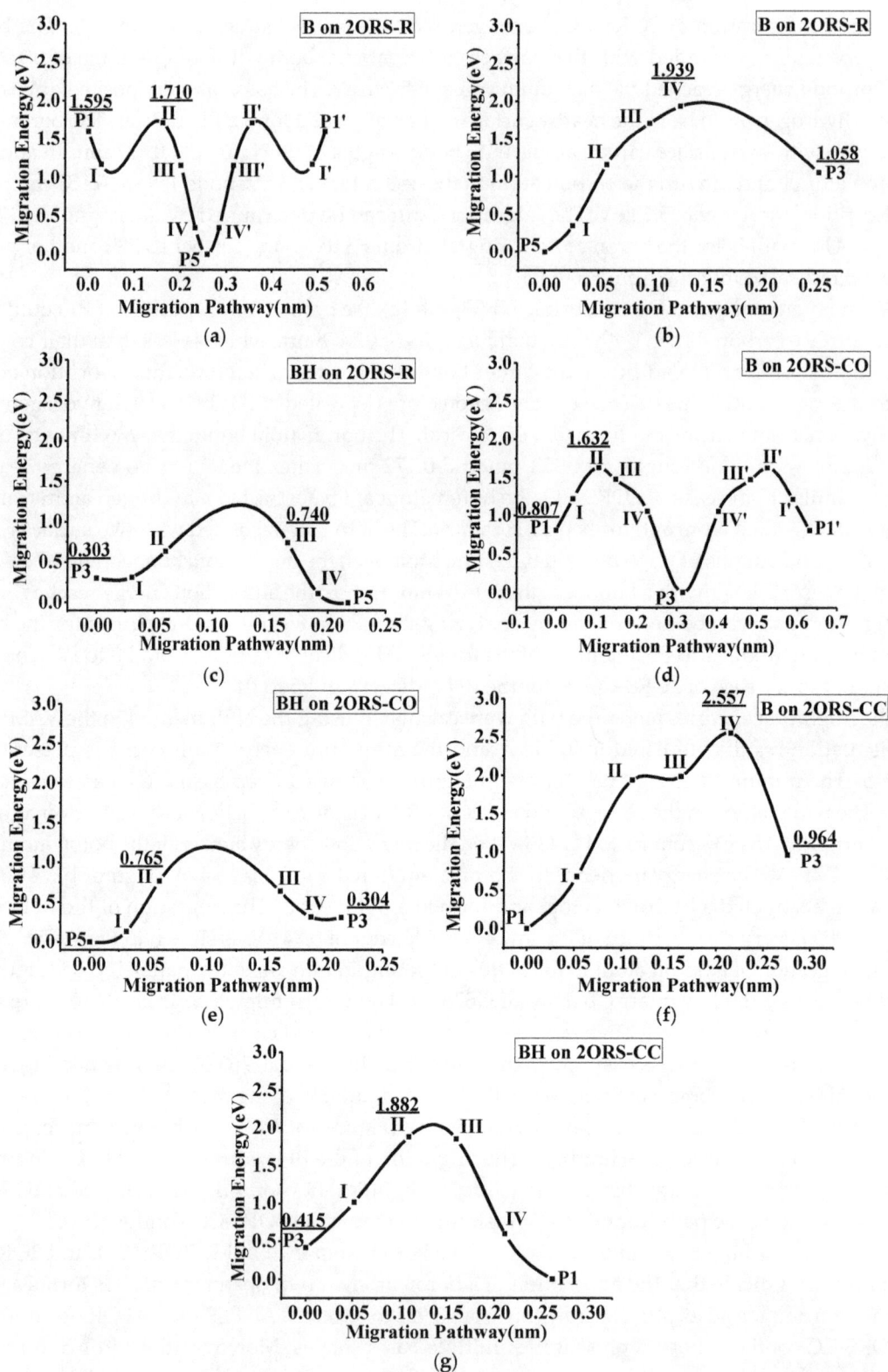

Figure 5. Minimum energy path curves for the boron (B) atom or the BH radical on the hydrogen-terminated diamond (001) surface with two open radical sites: (**a**) B migration from P1 to P5 on 2ORS-R; (**b**) B migration from P5 to P3 on 2ORS-R; (**c**) BH migration from P3 to P5 on 2ORS-R; (**d**) B migration from P1 to P3 on 2ORS-CO; (**e**) BH migration from P5 to P3 on 2ORS-CO; (**f**) B migration from P1 to P3 on 2ORS-CC; (**g**) BH migration from P3 to P1on 2ORS-CC.

Table 7. Minimum energies during the migration of a single B atom on the hydrogen-terminated diamond (001) surface with two open radical sites.

2ORS-R	P1	I	II	III	IV	P5	E_a (eV)
E_{tot} (eV)	−1157.594	−1158.018	−1157.479	−1158.032	−1158.841	−1159.189	1.710
ΔE (eV)	1.595	1.171	1.710	1.157	0.348	0	
2ORS-R	**P5**	**I**	**II**	**III**	**IV**	**P3**	E_a (eV)
E_{tot} (eV)	−1159.189	−1158.841	−1158.032	−1157.479	−1157.259	−1158.131	1.939
ΔE (eV)	0	0.348	1.157	1.71	1.939	1.058	
2ORS-R (B H)	**P3**	**I**	**II**	**III**	**IV**	**P5**	E_a (eV)
E_{tot} (eV)	−1158.131	−1158.121	−1157.800	−1157.694	−1158.321	−1158.434	0.740
ΔE (eV)	0.303	0.313	0.634	0.740	0.113	0	
2ORS-CO	**P1**	**I**	**II**	**III**	**IV**	**P3**	E_a (eV)
E_{tot} (eV)	−1157.692	−1157.280	−1156.867	−1157.022	−1157.441	−1158.499	1.632
ΔE (eV)	0.807	1.219	1.632	1.477	1.058	0	
2ORS-CO(BH)	**P5**	**I**	**II**	**III**	**IV**	**P3**	E_a (eV)
E_{tot} (eV)	−1158.435	−1158.322	−1157.670	−1157.799	−1158.121	−1158.131	0.765
ΔE (eV)	0	0.131	0.765	0.636	0.314	0.304	
2ORS-CC	**P1**	**I**	**II**	**III**	**IV**	**P3**	E_a (eV)
E_{tot} (eV)	−1159.006	−1158.329	−1157.069	−1157.020	−1156.449	−1158.042	2.557
ΔE (eV)	0	0.667	1.937	1.986	2.557	0.964	
2ORS-CC(BH)	**P3**	**I**	**II**	**III**	**IV**	**P1**	E_a (eV)
E_{tot} (eV)	−1158.042	−1157.438	−1156.575	−1156.604	−1157.860	−1158.457	1.882
ΔE (eV)	0.415	1.019	1.882	1.853	0.597	0	

4. Conclusions

In this study, the adsorption and migration activation energies of a single boron atom on a full hydrogen-terminated diamond (001) surface and on the hydrogen-terminated diamond (001) surface with several open radical sites were systematically investigated using first principles methods. The following conclusions can be drawn from the results:

1. On the fully hydrogen-terminated diamond surface, the adsorption energies of a boron atom are small and the maximum value is 1.387 eV. The migration barrier of the boron atom is 1.216 eV. Thus, boron atom migration on the surface is easy at the CVD diamond deposition temperature (700–900 °C).

2. On the hydrogen-terminated diamond surface with one open radical site or two open radical sites, the adsorption energies of a boron atom increase to 4.37 eV, even up to 5.94 eV, thereby forming a stable configuration. All PES on the hydrogen-terminated diamond surface with two open radical sites represent the multiple-basin shapes. Thus, a boron atom can migrate from a basin to another basin.

3. The boron atom by the radical site can abstract a hydrogen atom, form a BH radical, and create a new radical site.

4. The number and distribution of open radical sites, namely, the adsorption of hydrogen atoms and the abstraction of surface hydrogen atoms, can influence the adsorption and migration of boron atoms on the hydrogen-terminated diamond surface.

Acknowledgments: The authors acknowledge the financial support provided by the National Natural Science Foundation of China (Grant Nos. 50845065 and 51562031), the Natural Science Foundation of Inner Mongolia Autonomous Region (Grant Nos. 2014MS0516, 2015MS0550, and 2015MS0554), and the Science and Technology Foundation of Baotou (Grant No.2013J2001-1).

Author Contributions: Xuejie Liu conceived and designed the researches; Congjie Kang and Haimao Qiao performed the calculations; Yuan Ren, Xin Tan, and Shiyang Sun analyzed the data and participated in the discussions; Xuejie Liu and Congjie Kang wrote the paper.

Conflicts of Interest: The authors declare no conflict of interest.

References

1. Kraft, A. Doped diamond: A compact review on a new, versatile electrode material. *Int. J. Electrochem. Sci.* **2007**, *2*, 355–385.
2. Luong, J.H.T.; Male, K.B.; Glennon, J.D. Boron-doped diamond electrode: Synthesis, characterization, functionalization and analytical applications. *Analyst* **2009**, *134*, 1965–1979. [CrossRef] [PubMed]
3. Kalish, R. Diamond as a unique high-tech electronic material: Difficulties and prospects. *J. Phys. D Appl. Phys.* **2007**, *40*, 6467–6478. [CrossRef]
4. Waldvogel, S.R.; Elsler, B. Electrochemical synthesis on boron-doped diamond. *Electrochimca Acta* **2012**, *82*, 434–443. [CrossRef]
5. Svitkova, J.; Ignat, T.; Svorc, L. Chemical modification of boron-doped diamond electrodes for applications to biosensors and biosensing. *Crit. Rev. Anal. Chem.* **2016**, *46*, 248–256. [CrossRef] [PubMed]
6. Li, L.; Li, H.D.; Lü, X.; Cheng, S.; Wang, Q.; Ren, S.; Liu, J.; Zou, G. Dependence of reaction pressure on deposition and properties of boron-doped freestanding diamond films. *Appl. Surf. Sci.* **2010**, *256*, 1764–1768. [CrossRef]
7. Zhang, J.G.; Wang, X.C.; Shen, B.; Sun, F.H. Effect of boron and silicon doping on improving the cutting performance of CVD diamond coated cutting tools in machining CFRP. *Int. J. Refract. Met. Hard Mater.* **2013**, *41*, 285–292. [CrossRef]
8. Wang, X.C.; Lin, Z.C.; Zhang, T.; Shen, B.; Sun, F. Fabrication and application of boron-doped diamond coated rectangular-hole shaped drawing dies. *Int. J. Refract. Met. Hard Mater.* **2013**, *41*, 422–431. [CrossRef]
9. Butler, J.E.; Mankelevich, Y.A.; Cheesman, A.; Ma, J.; Ashfold, M.N.R. Understanding the chemical vapor deposition of diamond: recent progress. *J. Phys. Condens. Matter* **2009**, *21*, 36–42. [CrossRef] [PubMed]
10. Frenklach, M.; Skokov, S. Surface migration in diamond growth. *J. Phys. Chem. B* **1997**, *101*, 3025–3036. [CrossRef]
11. Cheesman, A.; Harvey, J.N.; Ashfold, M.N.R. Studies of carbon incorporation on the diamond [100] surface during chemical vapor deposition using density functional theory. *J. Phys. Chem. A* **2008**, *112*, 11436–11448. [CrossRef] [PubMed]
12. Richley, J.C.; Harvey, J.N.; Ashfold, M.N.R. On the role of carbon radical insertion reactions in the growth of diamond by chemical, vapor deposition methods. *J. Phys. Chem. A* **2009**, *113*, 11416–11422. [CrossRef] [PubMed]
13. Richley, J.C.; Harvey, J.N.; Ashfold, M.N.R. CH_2 group migration between the H-terminated 2×1 reconstructed {100} and {111} surfaces of diamond. *J. Phys. Chem. C* **2012**, *116*, 7810–7816. [CrossRef]
14. Cheesman, A.; Harvey, J.N.; Ashfold, M.N.R. Computational studies of elementary steps relating to boron doping during diamond chemical vapor deposition. *Phys. Chem. Chem. Phys.* **2005**, *7*, 1121–1126. [CrossRef] [PubMed]
15. Richley, J.C.; Harvey, J.N.; Ashfold, M.N.R. Boron incorporation at a diamond surface: A QM/MM study of insertion and migration pathways during chemical vapor deposition. *J. Phys. Chem. C* **2012**, *116*, 18300–18307. [CrossRef]
16. Fang, W.H.; Peyerimhoff, S.D. Theoretical studies on mechanisms of the insertion of boron into methane and its consequent reactions. *Mol. Phys.* **1998**, *93*, 329–339. [CrossRef]
17. Das, D.; Singh, R.N. A review of nucleation, growth and low temperature synthesis of diamond thin films. *Int. Mater. Rev.* **2007**, *52*, 29–64. [CrossRef]
18. Manelli, O.; Corni, S.; Righi, M.C. Water adsorption on native and hydrogenated diamond (001) surfaces. *J. Phys. Chem. C* **2010**, *114*, 7045–7053. [CrossRef]
19. Henkelman, G.; Jóhannesson, G.; Jónsson, H. Methods for finding saddle pointsand minimum energy paths. In *Theoretical Methods in Condensed Phase Chemistry*; Schwartz, S.D., Ed.; Kluwer Academic Publishers: New York, NY, USA; Boston, MA, USA, 2002; Volume 5, pp. 269–302.

20. Ma, J.; Richley, J.C.; Davies, D.R.W.; Ashfold, M.N.R.; Mankelevich, Y.A. Spectroscopic and modeling investigations of the gas phase chemistry and composition in microwave plasma activated $B_2H_6/CH_4/Ar/H_2$ mixtures. *J. Phys. Chem. A* **2010**, *114*, 10076–10089. [CrossRef] [PubMed]

21. Mankelevich, Y.A.; Ashfold, M.N.R.; Comerford, D.W.; Ma, J.; Richley, J.C. Boron doping: B/H/C/O gas-phase chemistry; H atom density dependences on pressure and wire temperature; puzzles regarding the gas-surface mechanism. *Thin Solid Films* **2011**, *519*, 4421–4425. [CrossRef]

22. Asmussen, J.; Grotjohn, T.A.; Schuelke, T. Advances in plasma synthesis of UNCD films. In *Ultrananocrystalline Diamond: Synthesis, Properties and Applications*, 2nd ed.; Shenderova, O.A., Gruen, M.D., Eds.; Oxford: New York, NY, USA, 2012; pp. 53–83.

23. Butler, J.E.; Woodin, R.L. Thin-film diamond growth mechanisms. In *Thin Film Diamond*; Lettington, A.H., Steeds, J., Eds.; Springer Science + Business Media Dordrecht: London, UK, 1994; pp. 209–224.

24. Kresse, G.; Hafner, J. Ab initio molecular dynamics for liquid metals. *Phys. Rev. B* **1993**, *47*, 558–561. [CrossRef]

25. Kresse, G.; Furthmüller, J. Efficient iterative scheme for ab initio total energy calculation using a plane-wave basis set. *Phys. Rev. B* **1996**, *54*, 11169–11186. [CrossRef]

26. Kresse, G.; Furthmuller, J. Efficiency of ab initio total energy calculation for metals and semiconductors using a plane-wave basis set. *Comput. Mater. Sci.* **1996**, *6*, 15–50. [CrossRef]

27. Blochl, P. Projector augmented-wave method. *Phys. Rev. B* **1994**, *50*, 17953–17979. [CrossRef]

28. Kresse, G.; Joubert, D. From ultrasoft pseudopotentials to the projector augmented-wave method. *Phys. Rev. B* **1999**, *59*, 1758–1775. [CrossRef]

29. Perdew, J.P.; Burke, K.; Ernzerhof, M. Generalized gradient approximation made simple. *Phys. Rev. Lett.* **1996**, *77*, 3865–3868. [CrossRef] [PubMed]

30. Perdew, J.P.; Ruzsinszky, A.; Csonka, G.I.; Vydrov, O.A.; Scuseria, G.E.; Constantin, L.A.; Zhou, X.; Burke, K. Restoring the density-gradient expansion for exchange in solids and surfaces. *Phys. Rev. Lett.* **2007**, *100*, 136406. [CrossRef] [PubMed]

31. Monkhorst, H.; Pack, J.D. Special points for Brillouin-zone integrations. *Phys. Rev. B* **1976**, *13*, 5188–5192. [CrossRef]

32. Liu, X.J.; Zhang, S.H.; Jiang, Y.J.; Ren, Y. Interface structure of nanodiamond composite films: First-principles studies. *J. Alloys Compd.* **2014**, *599*, 183–187. [CrossRef]

33. Kittel, C. *Introduction to Solid State Physics*, 8th ed.; John Wiley & Sons Inc.: New York, NY, USA, 2005; p. 59.

13

Improving the Wear Resistance of Moulds for the Injection of Glass Fibre–Reinforced Plastics Using PVD Coatings: A Comparative Study

Francisco Silva [1,*], **Rui Martinho** [1], **Maria Andrade** [1], **António Baptista** [2] **and Ricardo Alexandre** [3]

[1]   ISEP—School of Engineering, Polytechnic of Porto, 4200-072 Porto, Portugal; rpm@isep.ipp.pt (R.M.); mfga@isep.ipp.pt (M.A.)

[2]   INEGI—Instituto de Ciência e Inovação em Engenharia Mecânica e Engenharia Industrial, 4200-465 Porto, Portugal; amb@fe.up.pt

[3]   TEandM—Tecnologia, Engenharia e Materiais; 3045-508 Taveiro, Portugal; ricardo@teandm.pt

*   Correspondence: fgs@isep.ipp.pt

Academic Editor: Alessandro Lavacchi

Abstract: It is well known that injection of glass fibre–reinforced plastics (GFRP) causes abrasive wear in moulds' cavities and runners. Physical vapour deposition (PVD) coatings are intensively used to improve the wear resistance of different tools, also being one of the most promising ways to increase the moulds' lifespan, mainly when used with plastics strongly reinforced with glass fibres. This work compares four different thin, hard coatings obtained using the PVD magnetron sputtering process: TiAlN, TiAlSiN, CrN/TiAlCrSiN and CrN/CrCN/DLC. The first two are monolayer coatings while the last ones are nanostructured and consist of multilayer systems. In order to carry out the corresponding tribological characterization, two different approaches were selected: A laboratorial method, using micro-abrasion wear tests based on a ball-cratering configuration, and an industrial mode, analysing the wear resistance of the coated samples when inserted in a plastic injection mould. As expected, the wear phenomena are not equivalent and the results between micro-abrasion and industrial tests are not similar due to the different means used to promote the abrasion. The best wear resistance performance in the laboratorial wear tests was attained by the TiAlN monolayer coating while the best performance in the industrial wear tests was obtained by the CrN/TiAlCrSiN nanostructured multilayer coating.

Keywords: PVD coating; sputtering; wear behaviour; TiAlN; TiAlSiN; diamond-like carbon (DLC); injection moulds; abrasion; glass-fibre reinforced plastics (GFRP)

1. Introduction

Injection moulding is probably the most widely used plastics manufacturing technology. Nowadays, increasing requirements demanded by designers have driven the use of short glass fibers as reinforcements in many parts, namely for the automotive industry. Abrasion is a typical phenomenon in the plastic injection moulding process and it becomes even more important when plastics are reinforced with glass fibres [1–5]. Otherwise, maintenance operations related to moulds represent the costs and inactivity periods of the mould which affect the competitiveness of the process and are strongly connected to the concept of the life cycle cost [6].

New developments in the polymer industry led to the use of new polymer formulations and blends that are more and more aggressive to steel mould cavities due to gases generated into the mould, causing serious corrosion problems, which bring new challenges to the coatings research and industry [7]. This problem is magnified by the temperature effect [8], and it is even more important when we are talking about lighting reflector systems and other automotive parts where transparency and/or surface quality are critical requirements for the customer and no scratches or other scars are allowed [7]. The Fresnel lens, manufactured by injection moulding, also requires a high quality level, preventing the occurrence of any kind of superficial defects in the mould cavities [9]. In these cases, usually no reinforcements are applied, and the corrosion and the high pressures exerted into the mould are the main problems to overcome.

However, despite some other important issues related to mould cavity degradation, such as adhesion and fatigue, abrasion plays an important role, severely affecting the surface quality of the mould and, subsequently, the quality and brightness of the injected part's surface. Indeed, the natural motion of the fibre tips during the injection flow induces small scratches both in the screw and barrels of the injection machine as well as in the mould cavity, leading to bright degradation of the mould cavity surface. Tribological problems associated with the plastics moulding process can often be solved by an appropriate surface modification. Several techniques can be efficient to increase the wear behaviour of the mould cavity surface or other mechanical tools, such as ion implantation, physical vapour deposition (PVD), chemical vapour deposition (CVD), and standard diffusion processes, namely plasma nitriding [10–13].

Some authors even developed test models in order to simulate the tribological mechanisms involved inside the injection machine barrel, but without complete success, because only some of the main wear mechanisms were effectively tested [14]. However, there are other subjects that deserve our attention, namely the heat insulation and/or heat transfer phenomena that cannot be neglected with the coating application, constituting a new concern in the coating selection process [15]. Effectively, the coating must help to improve the wear resistance but it must also help to maintain or even decrease the injection cycle time, leading to a productivity improvement by keeping or improving the part cooling process after injection.

In order to overcome these problems, some authors have investigated a diversified number of surface treatments and tribological coatings [1–3,5,15–20]. In fact, recent developments in coatings technology allow a multifaceted approach to each problem, using mono- or multilayered coatings, single composition or graded layers, conventional or nanostructured films. More recently, some authors have tested selected films by submitting them to real injection conditions [1–3,21], obtaining results that allow us to think that coatings should be applied to moulds in a competitive way. The wear improvement of some steel mould inserts was also tested using a polycrystalline diamond [22] provided with an interlayer in order to avoid the diffusion of C into the Fe (substrate).

In this work, two dissimilar approaches were conducted using different coating concepts: two monolayer films (TiAlN and TiAlSiN) and two multilayer nanostructured coatings (CrN/TiAlCrSiN and CrN/CrCN/DLC). The main goal of this work was to characterise and compare the wear behaviour of these coatings under real conditions of plastic injection moulding, using glass fibre–reinforced polypropylene. This work also intended to find a competitive way to increase the life cycle of the moulds, minimizing maintenance operations and costs.

2. Materials and Methods

2.1. Substrates Material and Preparation

Plastic injection moulds manufacturers typically use AISI P20 tool steel (Bohler Uddeholm, Vienna, Austria) as main material for mould cavities, which is usually supplied in the hardened state. Thus, this material was used as substrate for this work. AISI P20 steel was subject to mass spectroscopy analysis and the chemical composition (wt %) obtained was the following one: 0.35% C, 0.29% Si, 1.95% Cr, 1.39% Mn, 0.19% Mo, 1.00% Ni and 0.01% S. The hardness of the steel was assessed using a universal hardness tester EMCO (model M4U, EMCO-TEST Prüfmaschinen GmbH, Kellau, Kuchl, Austria) carrying out five tests and taking the average value of 380 HBW 2.5/187.5/5. In order to carry out the wear tests, two different kinds of samples were used: laboratorial samples with a quadrangular shape of 25 mm × 25 mm and 2 mm thickness, and industrial samples with a specific shape allowing its insertion on proper cavities machined in the mould, allowing the contact of the samples with the plastic flow during each injection cycle. The cavities dug into the mould have a slight conical shape, allowing a better adjustment of the samples in the cavity. The shape of the industrial samples can be observed in Figure 1. These samples were placed in the plastic feed channel of the mould, letting its surface coincident with the remaining surface of the feed channel. Surface of both kind of samples was milled in a Computer Numeric Control machining centre and further ground until an average surface roughness of 0.06 µm has been retrieved, which is enough to observe wear phenomena and is similar to surface roughness usually used in this kind of injection tools.

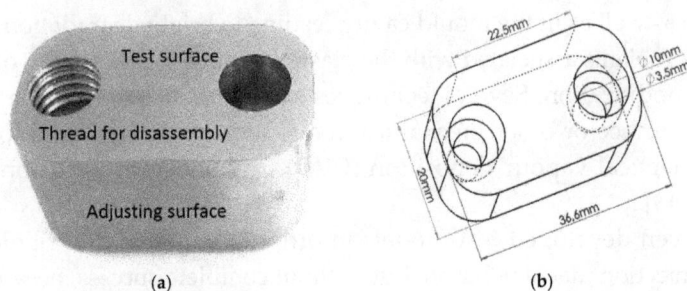

Figure 1. Industrial sample shape and corresponding dimensions: (**a**) view of the inserts used in the moulds and (**b**) corresponding dimensions.

2.2. Coatings Deposition

Coatings were produced using a CemeCon CC800/9 PVD unbalanced magnetron sputtering reactor (CemeCon, A.G., Wurselen, Germany). The following deposition parameters were similar for every coatings here referred: Gas pressure 500 mPa, Temperature 500 °C, Target power density 16 A·cm^{-2}, Bias in the range of −120 V to −50 V, and deposition time 4 h. However, in order to produce each coating, different targets and gases were used, as pointed in Table 1, which summarises these conditions.

Table 1. Targets and gases used in the coatings deposition process.

Coating	Targets	Gases
TiAlN	TiAl (×4)	N
TiAlSiN	TiAlSi (×4)	N
CrN/TiAlCrSiN	TiAlSi (×2) + Cr (×2)	N
CrN/CrCN/DLC	Cr (×2) + Graphite (×2)	$Ar^+ + N + C_2H_2$

In order to ensure the correct exposure of the samples to the targets and the best homogeneity of the films, sample's holder was animated with a circular motion of 1 rotation per minute.

In the particular case of the CrN/CrCN/DLC multi-layered coating, the CrN bottom and intermediate layers were generated from two Cr targets using N and Ar^+ feed gas, when for the last one C_2H_2 was added to N and Ar^+. For the top layer, the Cr targets were hidden behind the shutters and two graphite targets were exposed.

2.3. Coatings Characterization

After deposition, both film thickness and morphology were assessed by SEM (scanning electron microscopy, FEI, Hillsboro, OR, USA) using a FEI Quanta 400FEG scanning electron microscope provided with an EDAX Genesis X-ray spectroscope (EDS—energy dispersive spectroscopy, EDAX, Mahwah, NJ, USA). For the thickness measurement, a metallographic preparation was done; samples were partially cut in the reverse side of the coating and then they are submerged in liquid nitrogen during 20 min in order to promote a brittle behavior. At that time, samples were taken and mechanically broken, minimizing by this way plastic deformations close to the cutting edge. Then, they were embedded in resin with cross section turned to the working face. Further, the assembled set was ground with F1200 sandpaper until main grooves disappear and then, it was polished with diamond grit solutions of 3 and 1 μm during 10 min each one. Surface morphology and roughness were also evaluated by Atomic Force Microscopy (AFM) (VEECO Instruments, Ltd., Woodbury, NY, USA) using a VEECO Multimode atomic force microscope system (7 nm tip radius) provided with the NanoScope 6.13 software (version 6.13, Bruker, New York, NY, USA). In these analyses, two different areas were considered: 10 μm × 10 μm and 50 μm × 50 μm, comparing the results. A CAMECA Electron Probe Micro-Analysis (EPMA) system (SX-50 model, CAMECA Instruments, Inc., Madison, WI, USA) was also used equipped with wavelength dispersive spectroscopy (WDS) system (version 2.0, CAMECA Instruments, Inc., Madison, WI, USA), allowing confirm the coatings chemical composition. Micro-hardness was also quantified using Fischerscope® H100 equipment (Fischerscope, Windsor, CT, USA), provided with Vickers indenter. A normal load of 50 mN was selected keeping it constant during 30 s, avoiding by this way creep phenomena. This equipment produces a values chart that allows obtaining "load–depth" curves, which let to compute hardness and Young's modulus values.

2.4. Adhesion Analysis

The adhesion between each coating and substrate was assessed by scratch test and Rockwell indentation. Scratch tests were carried out using a CSM REVETEST equipment (CSEM, Neuchatel, Switzerland) following the procedures recommended by the BS EN 1071-3 (2005) standard [23]. The normal load was increasing from 0 to 80 N using a growing up rate of 100 N·min^{-1} and the indenter sliding speed used was 10 mm·min^{-1}. Regarding the surface texture effect due to grinding process, two orthogonal measurements were considered allowing to understand the texture effect on the adhesion failure mechanisms. In each orthogonal direction three tests were made in order to maximise the results accuracy. Then, grooves were carefully observed by optical microscopy, correlating the distance of the groove and film detachment phenomena detected on the scratch with the applied load in each point. The grooves observation lets identifying when cohesive and adhesive failures occur, allowing to determine the corresponding load. Another technique was also used permitting to confirm qualitative results. Following the VDI 3198:1991 standard [24], Rockwell indentations were done at 294 N (30 kgf) and 980 N (100 kgf) normal loads, using an EMCO M4U 025 Universal Hardness Tester (EMCO-TEST Prüfmaschinen GmbH, Kellau, Kuchl, Austria). Indentation borders were observed in order to identify the presence of cracks and its pattern.

2.5. Micro-Abrasion Tests

In order to evaluate the wear behaviour of the three coatings and obtain comparable results with other types of coatings, micro-abrasion test configuration is an accurate option, being perfectly suitable for thin films wear characterization. Thus, a PLINT TE 66 tribometer (Figure 2) using ball-cratering configuration was used. In this test, a ball of AISI 52100 steel with 25 mm diameter was used as counter body. This ball, initially in polished state, was etched using a 10% NITAL solution during 20 s, increasing its surface roughness and allowing better abrasive particles motion. These tests use abrasive slurry composed by 35.4 g of SiC F1200 powder (following FEPA 42-2:2006 standard [25]), with average particle size of 4.3 μm, standard deviation 1.4 μm, homogeneously distributed in 100 mL distilled water. This slurry is continuously stirred by a magnetic issue into a glass container, during the wear tests. The sample surface in test is pressed against the ball by a vertical bar sustained by a pivot system, rotating on it. The ball rotation speed was 80 rpm, corresponding to 0.105 m·s^{-1} and a normal load of 0.25 N was applied in all tests. Three different test lengths were agreed: 200, 500 and 700 cycles, corresponding to sliding distances of 15.71, 39.27 and 54.98 m, respectively. Attending each test conditions set, five different tests were carried out trying to increase the accuracy of the results. After the tests, all wear scars were observed by optical microscopy, using an OLYMPUS BX51M microscope provided with 12.5 megapixel OLYMPUS digital camera and AnalySIS DOCU image software (version 5.0, Olympus, Tokyo, Japan). Some results were later explored by SEM, using the above-mentioned equipment.

Figure 2. PLINT TE66 equipment (Phoenix Technology, London, UK) based on ball-cratering wear test configuration.

2.6. Industrial Wear Tests

As previously referred, the main objective of this work is to study the wear resistance of the coatings when applied in injection mould cavities used to produce glass fibre–reinforced plastic parts for automotive industry. Thus, an industrial mould often used for the production of radiator plastic fans was selected and three cavities symmetrically distributed and crossing the runners were machined into the mould allowing that samples surface fits well on the cavity due to the progressive section of the sample and cavity and allow that the sample exposed face is in the same plane of the mould surface in this area. The cavities were located centred on each plastic runner as depicted in Figure 3. During the injection cycle, glass fibre–reinforced plastic flows by the main feed channel (sprue) at the centre, being then split in three different ways. Arrows included in this figure intend to show the three directions of the plastic flow towards the secondary plastic feed channels (runners). These arrows were drawn over the sample cavities specially produced on the mould. Inserts were located in a turbulent area, in order to maximise the abrasive effect of the glass fibres, due to a previous quick flow direction shift. The composite used in the process is polypropylene reinforced with 30 wt % glass fibres and 9×10^4 injection cycles were performed allowing study the wear behaviour of the coating. It is well-known that, in these conditions, uncoated AISI P20 tool steel presents eye-visible severe wear scars. In these conditions, mould needs a complicate and expensive maintenance process, involving production breaks, accurate production plans and smart stock management or delivery delays. For this purpose, a KRAUSS MAFFEI injection machine (Krauss-Maffei Wegmann GmbH & Co. KG, Munich, Germany) was used presenting 5000 kN clamp force and inner initial pressure of 140 bar on the mould. The injection speed used was 50 m·s^{-1} and injection temperature was about 250 °C.

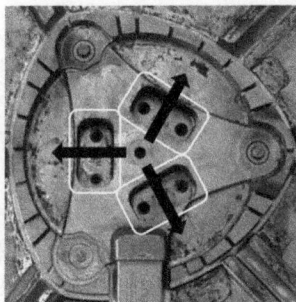

Figure 3. Industrial wear tests setup, showing the cavities location in the mould (white boxes) and the plastic flow (black arrows).

3. Results and Discussion

The chemical composition of the films was evaluated using electron-probe microanalysis. The results can be seen in Table 2, showing that the compositions are in line with the targets and gases used in the deposition process and fulfil the initial expectations. The coatings' thickness was assessed by SEM, leading to the values listed in Table 3, in the range of 3.5 to 4.5 µm, as desired, having been controlled by both the deposition rate and time. The morphology of the coatings was analysed by SEM as well, denoting that the films tend to follow the substrate topography, keeping the thickness approximately constant along the surface. The coatings' surface was relatively smooth, as shown in Figure 4, as expected when the films are produced using the magnetron sputtering technique. The nanostructured coating shown in Figure 4d corresponded to different CrCN phases, one of them more carbon-rich than the other, which was obtained providing C_2H_2 periodically and controlled over time. However, the TiAlSiN film presented some aggregates at the surface, which can originate from tribological problems, both through the generation of preferred sites for crack nucleation and problems related to the release of these large particles in contact the counter-face. The roughness was evaluated by profilometry and yields to confirm that Ra values were in the range of 0.033 µm to 0.061 µm for all coatings. The R_a values obtained by the AFM evaluation were slightly different, varying in the range of 0.028 µm to 0.053 µm, this difference being mainly attributed to the lower area of analysis. The values obtained are in line with the requirements usually defined for inner mould surfaces for the plastic injection moulding of non-visible automotive parts, such as the case studied in this work. The micro-hardness was measured using different techniques due to the low thickness of the CrN/CrCN/DLC final layers, revealing that the hardest coating was the CrN/TiAlCrSiN coating and the softest was the CrN/CrCN/DLC coating. Taking into account the curves' load/depth of displacement of the indenter generated by the micro-hardness equipment, Young's modulus was also assessed, allowing us to compute the reduced Young's modulus, the values of which are shown in Table 3, which allowed us to compute the values of the H^3/E_r^2 ratio, a good indicator for the coatings' resistance to plastic deformation. In this case, the best and worst plastic deformation resistance was presented by CrN/TiAlCrSiN and CrN/CrCN/DLC coatings, respectively. Believing just in the hardness values obtained, the CrN/TiAlCrSiN coating would present the best wear resistance due to it having highest value in the range of coatings tested in this work.

Adhesion tests were carried out using two different techniques: scratch tests and indentation tests, as described in the Experimental Section. The scratches produced during the tests were carefully analysed by optical microscopy and allowed us to observe some typical phenomena usually occurring in these tests, namely conformal cracking (TiAlSiN) plus lateral chipping (TiAlN), coating delamination in the scratch channel and lateral cracking (CrN/TiAlCrSiN). In general, the critical loads that caused the coatings' failure in these tests were below the expectations. These values varied between 20 and 25 N, considering the two orthogonal directions selected. Films with similar compositions usually present critical loads for adhesion or cohesion failure at about 70 or even 75 N. Thus, the indentation technique was also used in order to cross the results and to guarantee that the coatings' adhesion was enough to continue to the wear tests. Except the CrN/CrCN/DLC coating, all coatings were subject to indentation tests using 30 and 100 kgf as described in the Experimental Section and subsequently observed by optical microscopy, allowing us to observe that indentation borders presented mainly a cracking pattern of small radial cracks in all the films observed. Despite some reports that CrN-based coatings present delamination in the borders of the indentation [26], no delaminations were observed in the samples used in this work. Thus, all the films were considered suitable for wear tests.

Table 2. Chemical characterization of the coatings (performed by EPMA).

Coating	N (at.%)	O (at.%)	Al (at.%)	Si (at.%)	Ti (at.%)	Cr (at.%)
TiAlN	49.77 ± 1.09	1.17 ± 0.17	26.91 ± 0.67	–	22.15 ± 0.47	–
TiAlSiN	46.24 ± 0.49	1.33 ± 0.17	10.56 ± 0.14	4.56 ± 0.04	37.31 ± 0.48	–
CrN/TiAlCrSiN	44.92 ± 1.91	–	2.55 ± 0.19	1.26 ± 0.08	1.290 ± 0.11	49.97 ± 1.60
CrN/CrCN/DLC			Not evaluated			

Table 3. General characterization of the coatings (performed by profilometry and AFM).

Coating	Thickness (μm)	Roughness (R_a, μm)	Roughness (R_t, μm)
Uncoated	–	0.060 ± 0.005/0.051 ± 0.002	0.365 ± 0.018/0.312 ± 0.009
TiAlN	3.5 ± 0.2	0.058 ± 0.006/0.049 ± 0.003	0.517 ± 0.029/0.488 ± 0.021
TiAlSiN	4.5 ± 0.3	0.061 ± 0.004/0.053 ± 0.003	0.635 ± 0.021/0.598 ± 0.023
CrN/TiAlCrSiN	4.3 ± 0.4	0.033 ± 0.003/0.028 ± 0.001	0.414 ± 0.011/0.375 ± 0.013
CrN/CrCN/DLC	3.6 ± 0.2	0.054 ± 0.005/0.045 ± 0.002	0.385 ± 0.010/0.351 ± 0.008

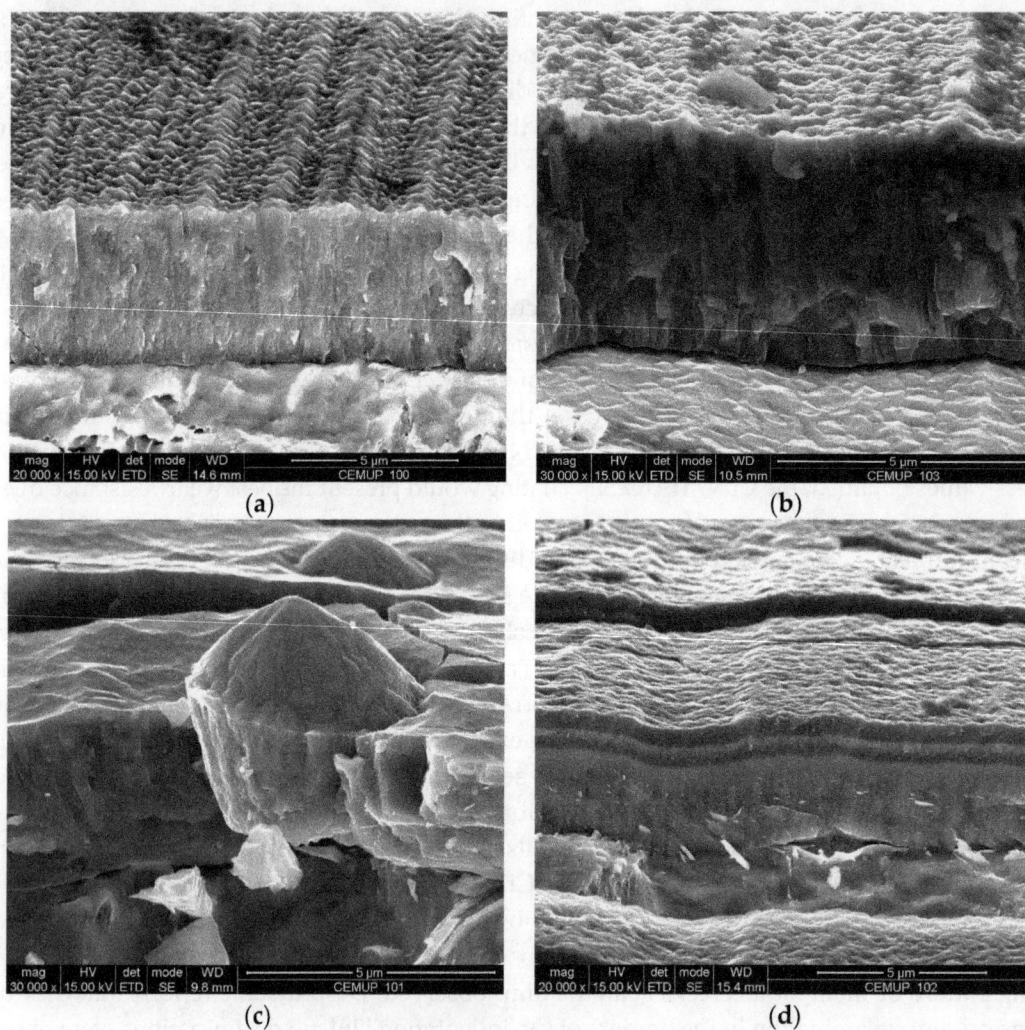

Figure 4. Morphological characterization of the following coatings: (a) TiAlN; (b) TiAlSiN; (c) CrN/TiAlCrSiN and (d) CrN/CrCN/DLC assessed by SEM.

The micro-abrasion tests were carried out following the conditions already mentioned in the Experimental Section. Due to the low thickness of the films, after some ball revolutions in the micro-abrasion tests, the coating was consumed by abrasion and just the film border prevented greater wear [27]. Thus, in these cases, it was necessary to measure the inner and outer diameter of the crater, which corresponded to the substrate and coating-resistant area. In order to analyse the coatings' behaviour, just the removed material corresponding to the coating was considered for calculations (c—coating). The results showed that the wear coefficient (k_c) was the smallest for the TiAlN coating while it was the greatest for the CrN/TiAlCrSiN coating, corresponding to the better wear resistance (k^{-1}) of the TiAlN coating, which also presented a relatively low reduced Young's modulus. The results summary of these tests can be seen in Table 4. Figure 5 depicts the usual shape of each crater with the typical border corresponding to the film's resistant thickness, whereas the inner area corresponds to the substrate that was in contact with the ball and the abrasive slurry after coating perforation. Figure 5b shows the typical way out border of the slurry, displaying how the abrasive particles scratched the coating into the crater, reducing this effect as the ball left the contact area of the sample. As shown in Table 4, regarding just the micro-abrasion tests, the best wear behaviour was shown by the CrN/TiAlCrSiN coating, allowing a useful lifespan 65.5 times greater than the uncoated substrate. However, these tests should be considered comparative, using the wear resistance already obtained and computed in similar conditions when testing other coatings. The wear phenomena expected in the plastic injection moulding process are quite different, just because the difference in the hardness between materials and the density of abrasive particles present in the contact are also quite different.

Table 4. Hardness and micro-abrasion wear resistance of the coatings used in this work.

Coating	Thickness (μm)	Hardness (Gpa)	E (GPa)	E_r (GPa)	H^3/E^2 (Gpa)	k_c (mm^3/N·m)	k^{-1} (N·m/mm^3)
TiAlN	3.5 ± 0.2	22.7 ± 1.2	392	304	0.076	6.51E-5	15.37
TiAlSiN	4.5 ± 0.3	21.8 ± 1.7	491	282	0.043	24.6E-5	4.068
CrN/TiAlCrSiN	4.3 ± 0.4	30.9 ± 2.1	422	325	0.166	37.3E-5	2.683
CrN/CrCN/DLC	3.6 ± 0.2	19.2 ± 1.1	340	288	0.061	31.8E-5	3.141

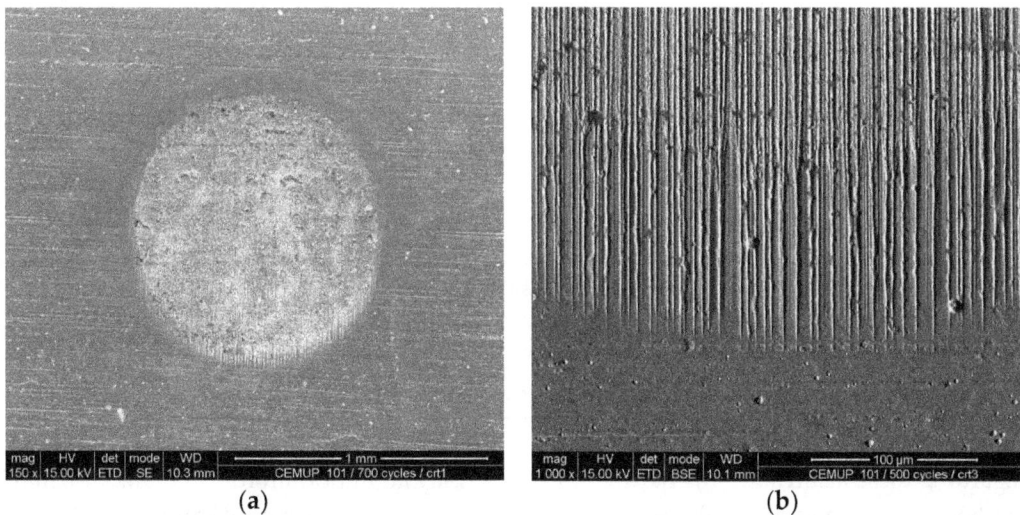

(a)   (b)

Figure 5. (a) Crater pattern and (b) border detail corresponding to wear test of the CrN/TiAlCrSiN coating after 700 wear cycles in the TE66 PLINT Tribometer under the test conditions mentioned above.

Assuming that industrial tests could lead to significantly different results, the industrial tests were performed by inserting three different samples into cavities previously machined in the mould and adjusting them in order to keep their surface on the same plane of the remaining inner surface of the mould. Thus, the reinforced plastic could flow through the feed channels, tangentially abrading the mould and sample surface during its motion along the feed channels, with some turbulence caused by the directional flow shift of the reinforced plastic close to the sample's position. Each sample was provided with a different coating, allowing us to test three different coatings in each setup. The injection conditions and reinforced plastic properties had already been previously announced.

After 9×10^4 injection cycles, the lowest volume of the coating removed by abrasion and adhesion was achieved by the hardest multilayer coating, CrN/TiAlCrSiN. The results in this test were obtained by analysing the material removed from the surface due to the glass fibres' abrasion during the injection cycle runs, as can be seen in Figure 6. The surface of the coatings was analysed by SEM and wear scars or detachments were identified by computed with homemade software (PAQI) which allows us to determine the percentage of the coating area removed. Complementary EDS tests were also conducted in order to verify the absence of coating in those areas.

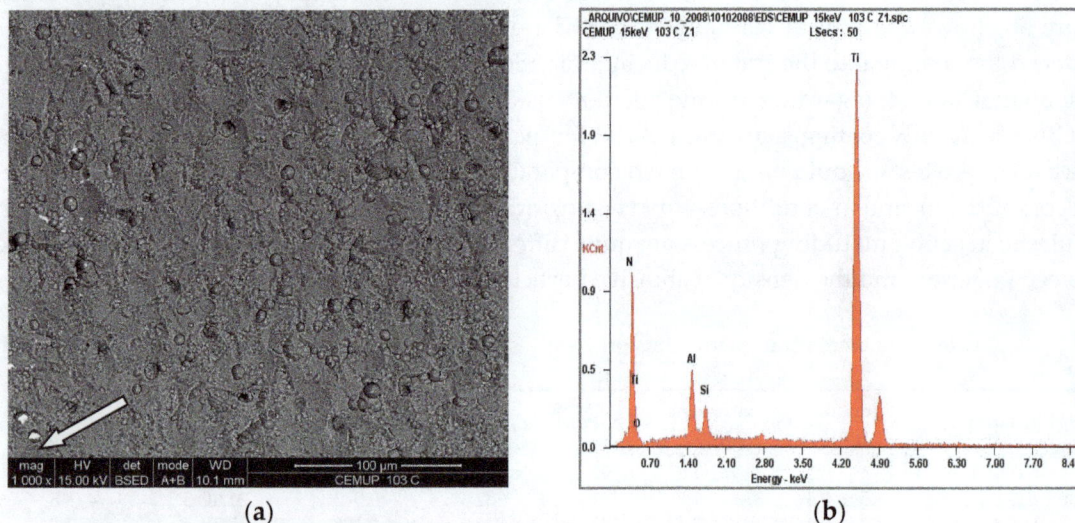

(a)   (b)

Figure 6. (a) Coating surface view after 90,000 injection cycles, denoting some punctual wear close to the left border of the image and (b) corresponding EDS analysis of the points indicated by the white arrow.

The thickness of the coatings was already analysed by profilometry, based on the surface of the samples not exposed to the plastic flow. The results can be seen in Table 5.

Table 5. Industrial wear test results summary.

Coating	Hardness (Gpa)	Removed Height (μm)	Removed Coating Area (%)	Life vs. Steel
TiAlN	22.7 ± 1.2	0.035 ± 0.003	0.125 ± 0.005	30.8×
TiAlSiN	21.8 ± 1.7	0.049 ± 0.006	0.134 ± 0.003	25.0×
CrN/TiAlCrSiN	30.9 ± 2.1	0.014 ± 0.003	0.013 ± 0.001	65.5×
CrN/CrCN/DLC	19.2 ± 1.1	0.017 ± 0.004	0.032 ± 0.002	58.2×

Regarding these last results, the hardness seems to play an important role in the wear behavior of the multilayered coating when submitted to the injection of reinforced polypropylene. On the other hand, TiAlN presented the best wear behavior when submitted to micro-abrasion tests, even presenting a lower hardness than the CrN/TiAlCrSiN coating and also a relatively low H3/E2 ratio, showing low resistance to plastic deformation. Thus, this work allowed us to observe that micro-abrasion tests cannot be used to predict the coating lifespan when applied to mould cavities for the reinforced plastic injection moulding process.

4. Conclusions

Four different coatings were deposited, characterised and tested using two different methods in order to verify their ability to be used in mould cavities for the reinforced plastic injection moulding process. After all analyses and tests, some conclusions can be drawn, as follows:

- The morphology of the coatings is adequate for use in mould cavities of the reinforced plastic injection moulding process, allowing roughnesses (R_a) in the range of 0.033 and 0.061 μm (results obtained by profilometry).

- The coatings' adhesion, measured by scratch-test method, was low, allowing us to observe the failure mechanisms of the films in the scratches. However, indentation tests revealed that no indentation border detachments were observed, showing that the coatings' adhesion was enough to ensure the best wear resistance of the films.

- Micro-abrasion ball-cratering tests allowed us to determine that, under these wear conditions, the most wear-resistant coating in the range considered in this work was the TiAlN monolayered coating. However, as previously mentioned, these test conditions cannot be compared with the wear mechanisms usually performed in the reinforced plastic injection process due to different means used to promote the abrasion, i.e., SiC abrasive particle in the case of the lab tests and glass fibres in the case of industrial tests. The different hardness presented by these means endorses different responses of the coated surfaces.

- Industrial tests revealed that the best coating under the plastic injection conditions was the CrN/TiAlCrSiN coating, allowing a considerable lifespan benefit that was 65.5 times greater than that of the uncoated substrate.

Thus, considering these results, the industry can adopt the CrN/TiAlCrSiN coating as the best choice in terms of the wear caused by the glass fibre tips due to their random motion during the injection process.

Acknowledgments: The authors gratefully acknowledge the financial support of INEGI—Instituto de Ciência e Inovação em Engenharia Mecânica e Engenharia Industrial as well as the permission to use their facilities. The strong cooperation of the TEandM—Tecnologia, Engenharia e Materiais, S.A. and Plastaze—Plásticos de Azeméis, S.A. (SIMOLDES Group) companies is also deeply acknowledged.

Author Contributions: Francisco Silva conceived and designed the experiments and wrote the paper; Rui Martinho performed the experiments, collected the data and made the critical analyses; Maria de Fátima Andrade prepared all samples; António Baptista analysed the data, provided critical analyses and supervised all works; Ricardo Alexandre provided the coatings and helped in critical analyses.

Conflicts of Interest: The authors declare no conflict of interest.

References

1. Martinho, R.P.; Silva, F.J.G.; Alexandre, R.J.D.; Baptista, A.P.M. TiB$_2$ Nanostructured Coating for GFRP Injection Moulds. *J. Nanosci. Nanotechnol.* **2011**, *11*, 5374–5382. [CrossRef]

2. Silva, F.J.G.; Martinho, R.P.; Alexandre, R.J.D.; Baptista, A.P.M. Wear Resistance of TiAlSiN Thin Coatings. *J. Nanosci. Nanotechnol.* **2012**, *12*, 9094–9101. [CrossRef] [PubMed]

3. Silva, F.J.G.; Casais, R.C.B.; Martinho, R.P.; Baptista, A.P.M. Mechanical and Tribological Characterization of TiB$_2$ Thin Films. *J. Nanosci. Nanotechnol.* **2012**, *12*, 9187–9194. [CrossRef] [PubMed]

4. Silva, F.J.G.; Martinho, R.P.; Baptista, A.P.M. Characterization of laboratory and industrial CrN/CrCN/diamond-like carbon coatings. *Thin Solid Films* **2014**, *550*, 278–284. [CrossRef]

5. Bobzin, K.; Brögelmann, T.; Grundmeier, G.; de los Arcos, T.; Wiesing, M.; Kruppe, N.C. (Cr,Al)N/(Cr,Al)ON Oxy-nitride Coatings deposited by Hybrid dcMS/HPPMS for Plastics Processing Applications. *Surf. Coat. Technol.* **2016**, *308*, 394–403. [CrossRef]

6. Folgado, R.; Peças, P.; Henriques, E. Life cycle cost for technology selection: A Case study in the manufacturing of injection moulds. *Int. J. Prod. Econ.* **2010**, *128*, 368–378. [CrossRef]

7. Mitterer, C.; Holler, F.; Reitberger, D.; Badisch, E.; Stoiber, M.; Lugmair, C.; Nobauer, R.; Muller, T.; Kullmer, R. Industrial applications of PACVD hard coatings. *Surf. Coat. Technol.* **2003**, *163*, 716–722. [CrossRef]

8. Boey, P.; Ho, W.; Bull, S.J. The effect of temperature on the abrasive wear of coatings and hybrid surface treatments for injection-moulding machines. *Wear* **2005**, *258*, 149–156. [CrossRef]

9. Tosello, G.; Hansen, H.N.; Gasparin, S.; Albajez, J.A.; Esmoris, J.I. Surface wear of TiN coated nickel tool during the injection moulding of polymer micro Fresnel lenses. *CIRP Ann. Manuf. Technol.* **2012**, *61*, 535–538. [CrossRef]

10. Fox-Rabinovich, G.S.; Yamamoto, H.; Kovalev, A.I.; Veldhuis, S.C.; Ning, L.; Shuster, L.S.; Elfizy, A. Wear behavior of adaptive nano-multilayered TiAlCrN/NbN coatings under dry high performance machining conditions. *Surf. Coat. Technol.* **2008**, *202*, 2015–2022. [CrossRef]

11. Bouzakis, K.-D.; Pappa, M.; Skordaris, G.; Bouzakis, E.; Gerardis, S. Correlation between PVD coating strength properties and impact resistance at ambient and elevated temperatures. *Surf. Coat. Technol.* **2010**, *205*, 1481–1485. [CrossRef]

12. Dosbaeva, G.K.; Veldhuis, S.C.; Yamamoto, K.; Wilkinson, D.S.; Beake, B.D.; Jenkins, N.; Elfizye, A.; Fox-Rabinovich, G.S. Oxide scales formation in nano-crystalline TiAlCrSiYN PVD coatings at elevated temperature. *Int. J. Refract. Met. Hard Mater.* **2010**, *28*, 133–141. [CrossRef]

13. Bouzakis, K.-D.; Klocke, F.; Skordaris, G.; Bouzakis, E.; Gerardis, S.; Katirtzoglou, G.; Makrimallakis, S. Influence of dry micro-blasting grain quality on wear behaviour of TiAlN coated tools. *Wear* **2011**, *271*, 783–791. [CrossRef]

14. Bull, S.J.; Zhou, Q. A simulation test for wear in injection moulding machines. *Wear* **2001**, *249*, 372–378. [CrossRef]

15. Chen, S.-C.; Chang, Y.; Chang, Y.-P.; Chen, Y.-C.; Tseng, C.-Y. Effect of cavity surface coating on mold temperature variation and the quality of injection molded parts. *Int. Commun. Heat Mass Transfer* **2009**, *36*, 1030–1035. [CrossRef]

16. Bull, S.J.; Davidson, R.I.; Fisher, E.H.; McCabe, A.R.; Jones, A.M. A simulation test for the selection of coatings and surface treatments for plastics injection molding machines. *Surf. Coat. Technol.* **2000**, *130*, 257–265. [CrossRef]

17. Van Stappen, M.; Vandierendonck, K.; Mol, C.; Beeckman, E.; De Clercq, E. Practice vs. laboratory tests for plastic injection moulding. *Surf. Coat. Technol.* **2001**, *142–144*, 143–145. [CrossRef]

18. Cunha, L.; Andritschky, M.; Pischow, K.; Wang, Z.; Zarychta, A.; Miranda, A.S.; Cunha, A.M. Performance of chromium nitride and titanium nitride coatings during plastic injection moulding. *Surf. Coat. Technol.* **2002**, *153*, 160–165. [CrossRef]

19. Santos, C.S.C.; Neto, V. Nanostructured coatings in micromoulding injection—A case study. *Mater. Today Proc.* **2015**, *1*, 414–422. [CrossRef]

20. Ozturk, O.; Onmus, O.; Williamson, D.L. Microstructural, mechanical, and corrosion characterization of nitrogen-implanted plastic injection mould steel. *Surf. Coat. Technol.* **2005**, *196*, 333–340. [CrossRef]

21. Stock, H.-R.; Diesselberg, M.; Zoch, H.-W. Investigation of magnetron sputtered titanium–nickel–nitride thin films for use as mould coatings. *Surf. Coat. Technol.* **2008**, *203*, 717–720. [CrossRef]

22. Neto, V.F.; Vaz, R.; Oliveira, M.S.A.; Grácio, J. CVD diamond-coated steel inserts for thermoplastic mould tools—Characterization and preliminary performance evaluation. *J. Mater. Process. Technol.* **2009**, *209*, 1085–1091. [CrossRef]

23. *BS EN 1071-3 Advanced Technical Ceramics, Methods of Test for Ceramic Coatings-Determination of Adhesion and Other Mechanical Failure Modes by a Scratch Test*; British Standards Institution: London, UK, 2005.

24. *VDI 3198-Beschichten von Werkzeugen der Kaltmassivumformung CVD- und PVD-Verfahren*; Verein Deutscher Ingenieure: Dusseldorf, Germany, 1991.

25. *FEPA Standard 42-2-Grains of Fused Aluminium Oxide, Silicon Carbide and Other Abrasive Materials for Bonded Abrasives and for General Industrial Applications Microgrits F230 to F2000*; Federation of European Producers of Abrasives: Darmstadt, Germany, 2006.

26. Vidakis, N.; Antoniadis, A.; Bilalis, N. The VDI 3198 indentation test valuation of a reliable qualitative control for layered compounds. *J. Mater. Process. Technol.* **2003**, *143*, 481–485. [CrossRef]

27. Kusano, Y.; Van, A.K.; Hutchings, I.M. Methods of data analysis for the micro-scale abrasion test on coated substrates. *Surf. Coat. Technol.* **2004**, *183*, 312–327. [CrossRef]

14

Numerical and Experimental Investigation on the Spray Coating Process Using a Pneumatic Atomizer: Influences of Operating Conditions and Target Geometries

Qiaoyan Ye * and Karlheinz Pulli

Fraunhofer Institute for Manufacturing Engineering and Automation, Nobelstr. 12, Stuttgart 70569, Germany; karlheinz.pulli@ipa.fraunhofer.de
* Correspondence: qiaoyan.ye@ipa.fraunhofer.de

Academic Editor: Robert B. Heimann

Abstract: This paper presents a numerical simulation of the spray painting process using a pneumatic atomizer with the help of a computational fluid dynamics code. The droplet characteristics that are necessary for the droplet trajectory calculation were experimentally investigated using different shaping air flow rates. It was found that the droplet size distribution depends on both the atomizing and the shaping air flow rate. An injection model for creating the initial droplet conditions is necessary for the spray painting simulation. An approach for creating these initial conditions has been proposed, which takes different operating conditions into account and is suitable for practical applications of spray coating simulation using spray guns. Further, tests on complicated targets and complex alignments of the atomizer have been carried out to verify this numerical approach. The results confirm the applicability and reliability of the chosen method for the painting process.

Keywords: spray coating simulation; pneumatic atomizer; droplet size; film thickness distribution; particle injection model; 3D-target geometry

1. Introduction

Pneumatic atomizers are widely used in painting industries, especially in the automotive industry. For instance, they are used in the coating of the engine compartment and for spray coating using metallic paints, where a special appearance and high optical quality are required. Numerical simulation as an effective tool plays an important role for planning and optimizing in spray painting processes.

With the help of numerical simulations, it is currently possible to model the complete air flow field between the nozzle of an atomizer and the target [1]. Some experimental and numerical studies [2–8] of air and air-assisted spray guns have been carried out in recent years, focusing mainly on the characterization of droplet size distribution, on the effects of air flow and on the particle trajectory. However, modeling of the spray painting process using pneumatic atomizers—especially modeling the droplet phase—is still quite difficult: The mechanism of liquid atomization is quite complicated. There is no physical model that can be used to predict the atomization process, although currently more and more numerical studies [9–11] on air-assisted primary break-up processes have been performed by means of direct numerical (DNS) and large-eddy simulation (LES). Furthermore, for instance, it is difficult to perform measurements of the droplet size distribution close to the atomizer nozzle by using optical methods, because of the high velocities and the high number density of droplets.

Experimental and numerical investigations on the spray characteristics of a pneumatic atomizer have been carried out by Domnick et al. [2,3]. Basically, they measured details of the spray structure at a 100 mm distance below the spray nozzle by using Phase-Doppler Anemometry (PDA) (Dantec).

An elliptical spray region with a 220 mm long axis and an 80 mm short axis was obtained. Altogether, approximately 100 measuring points were taken in a quadrant of the spray region. These spray measurements were used to perform first numerical simulations of the coating process. However, PDA measurements are complicated and therefore time consuming when important atomizer parameters (e.g., atomizing and shaping air flow rate) are changed. This finally limits the practical relevance and applicability of the simulations. Another limitation is that PDA could not be applied to paints that have non-homogeneous optical properties.

In our previous study [1], the spray behavior of a coaxial jet-type pneumatic atomizer was numerically modeled by means of a commercial Computational Fluid Dynamics (CFD) code. The air flow field between the nozzle and the target was calculated applying the known air inlet conditions directly at the atomizer. The measured integral droplet size distribution obtained at 50 mm distance below the spray nozzle was used for trajectory simulations, assuming that it is also valid close to the nozzle. The initial droplet velocity close to the nozzle was fitted by using measured downstream droplet velocity and film thickness pattern on the target. This way, the two-way coupling process between droplets and air flow—which alters the spray cone shape—can also be taken into account. A similar approach [8] to creating initial conditions for droplet injection has been used in the CFD simulation of paint deposition using an air spray gun. There too, the droplets were injected quite close to the nozzle and the initial velocity was assumed to be that of the liquid jet at the moment of breakage. The flow field of the continuous phase was extensively analyzed and compared with PDA-measurements. Film thickness distributions on targets have been predicted. However, there was no validation between simulated and measured film thickness distribution.

In this paper, more detailed numerical investigations—namely, sensitivity studies considering different operating conditions and complicated target geometries—are presented. Measured and calculated film thicknesses on the targets are compared to each other. A good agreement between coating experiment and simulation has been found.

2. Experimental Investigation

2.1. Geometry of the Pneumatic Atomizer

The basic geometry of the pneumatic atomizer used in this study is shown in Figure 1. Atomization occurs by a coaxial jet arrangement, in which the central paint jet with diameter 1.4 mm (not shown in the figure) is surrounded by high-speed air, leaving the annular ring around the paint nozzle under supersonic conditions. Around the annular ring, there are eight small holes with 0.6 mm diameter, from which the high-speed air is supplied. Four of them are not coaxial, but are oriented at an angle of 45° to the spray axis. In addition, the so-called shaping air nozzles with 2 mm diameter on the two sides of the center jet are used to deform the spray cone for painting larger work pieces.

Figure 1. Inlet airflow in the nozzles of the pneumatic atomizer (Computational Fluid Dynamics (CFD) simulation of the atomizer).

2.2. Droplet Size Measurements

As already mentioned in the introduction, the CFD model is not able to simulate the atomization process, and thus requires the measured droplet size distribution as necessary prerequisites for the droplet trajectory calculation. In the current study, the atomization characteristics were measured at locations 50 mm below the liquid nozzle using a Spraytec Fraunhofer type particle sizer, which can measure in a 560 µm size range with a $f = 300$ mm receiving lens. The measuring volume of the laser beam has a diameter of 9 mm.

The major characteristics of the atomizer and the operating conditions investigated are summarized in Table 1. Owing to the shaping air from the atomizer, an elliptical spray cone is formed downstream the atomizer. Size distributions along the long axis of the spray were obtained by moving the atomizer along the x-direction, as shown in Figure 2. Figure 3 shows the individual droplet size distributions in the elliptical spray region; for instance, for the shaping air flow rate of 220 L/min. The integral distribution of the whole spray region was then calculated based on the individual droplet size distributions and the measured particle concentration in the measuring volume. Figure 4 shows integral droplet size distributions for each shaping air flow rate. It can be seen that a finer droplet size distribution is obtained by increasing the shaping air flow rate, which means that the atomization process is also affected by the shaping air flow. The experimental results provide important information for the particle injection model in the following numerical simulations.

Table 1. Atomizer characteristics.

Atomizing Air Flow Rate	Shaping Air Flow Rate	Liquid Flow Rate	Liquid Phase
260 L/min	150, 220, 360 L/min	300 mL/min	Alpine white paint

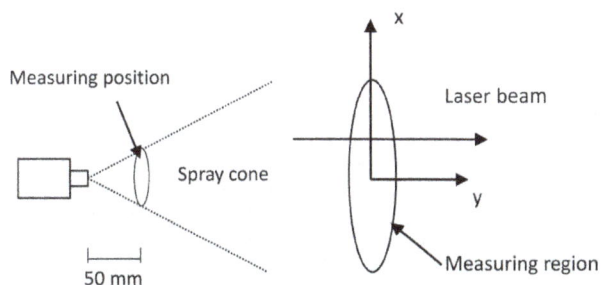

Figure 2. Setup of droplet size measurement.

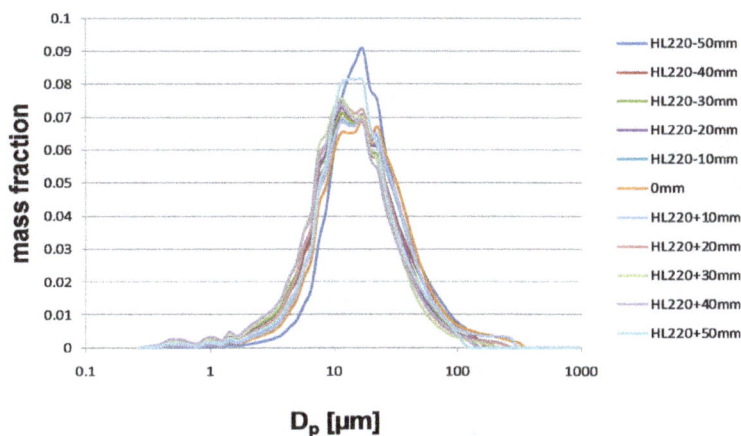

Figure 3. Individual droplet size distributions in the whole elliptical spray region for the shaping air flow rate of 220 L/min. 0 mm is the measuring position that is located at $x = 0$ in Figure 2, the spray cone center. ± 50 mm are located at the edges of the elliptical spray region ($x = \pm 50$ mm).

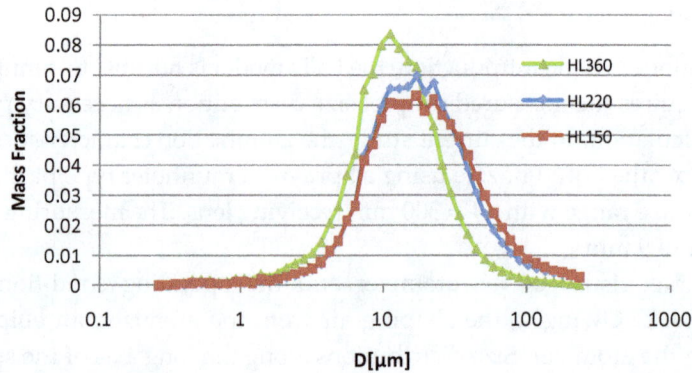

Figure 4. Measured integral droplet size distributions for three shaping air flow rates: 150, 220, 360 L/min.

2.3. Measurement of Paint Film Thickness on the Target Plate

Dynamic spray painting with a relative speed between gun and target of 0.15 m/s and a gun-to-target distance of 250 mm was performed. The target—a flat steel panel with a size of 200×800 mm^2—was positioned horizontally. The gun axis was perpendicular to the target surface so that the major axis of the spray pattern was formed along the 800 mm direction. After painting, the panel was put horizontally into an oven for baking. The dry film thickness on the panel was then measured by means of magneto-inductive method. Figure 5 shows the mean values of the measured film thickness distribution with uncertainty bars (standard deviation) for a typical shaping air flow rate of 360 L/min. Spray painting was later also carried out on a more complicated substrate, a 3D-target that consists of four panels. The atomizer was located at the center directly above panel C, moving along the channel. The corresponding film thickness distributions on different panels are shown in Figure 6. The measured film thickness distributions were applied to compare with the simulation results that will be shown in the next section.

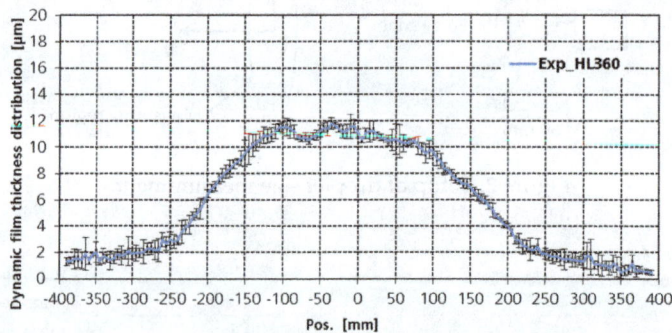

Figure 5. Measured dynamic film thickness distribution on a flat steel plate with the shaping air flow rate of 360 L/min.

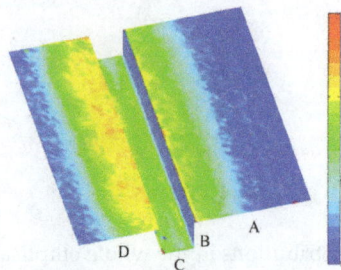

Figure 6. Measured dynamic film thickness (μm) on a 3D-target for the shaping air flow rate 360 L/min.

3. Computational Method

The commercial CFD code ANSYS-Fluent 12.0.16—based on the finite-volume approach—was used for the numerical simulations. The gas phase was modeled using the Eulerian conservation equations of mass, momentum, and energy. The 3-D compressible airflow was directly simulated from the nozzle using the coupled solver, as it was found to be more stable than the segregated solver for this specific airflow calculation. As inlet boundary conditions, air mass flow rate and stagnation temperature were used at the air nozzles. Turbulent transport was modeled using the Realizable k-ε model with scalable standard wall function. An unstructured mesh with 1.8 million cells was used to discretize the computational domain of $600 \times 2000 \times 2000$ mm^3, and mesh refinement was carried out.

Droplet trajectories were calculated using the Lagrangian tracking method by integration of the equation of motion,

$$\frac{\mathrm{d}u_\mathrm{p}}{\mathrm{d}t} = f_\mathrm{D}(u - u_\mathrm{p}) + F_\mathrm{G} \, , \quad \frac{\mathrm{d}x}{\mathrm{d}t} = u_\mathrm{p} \qquad (1)$$

in which the drag force $f_\mathrm{D}(u - u_\mathrm{p})$ and the gravity force F_G (force/unit particle mass) were taken into account. Other forces, such as "virtual mass" and Saffman's lift force may be neglected, since the density ratio between air and liquid is 1 to 1000, and the mean droplet diameter is about 40 μm in current painting processes. In the above equation, u_p is the particle velocity and u the instantaneous air velocity that was calculated by superimposing the local mean velocity and a fluctuating velocity component corresponding to the local turbulence level, using a stochastic tracking model. The effect of the turbulence dispersion on the droplet motion was thereby taken into account with an integral time scale constant of 0.3. In the current study, the interaction between droplets was neglected due to the low mass flow rate of the liquid. Computational particles—namely droplet parcels, representing a number of real droplets with the same properties—were used. The number of computational particles plays an important role in the Lagrangian tracking method. However, in order to save computation resources, 50,000 particles were used per trajectory calculation, which proved to be sufficiently accurate for current industry applications.

In order to carry out the particle trajectory calculation, initial conditions of droplet trajectory must be determined. A method similar to that in our previous study [1] was applied in the present study. Basically, the droplet injection data was applied in a circular region with a radius of 1 mm, below the liquid nozzle and above the cross-section of the shaping air flow jets, where the axial velocity dominates the air flow field. The integral distributions shown in Figure 4 were used as droplet size distributions. The droplets are uniformly distributed in the injection region. Nowadays there are more and more numerical studies available on the primary break-up of the liquid based on the Volume-of-Fluid method using different atomizers [9,11,12]. The velocity of shot liquid filaments based on the atomization study [12] using a coaxial jet atomizer was analyzed, which delivers useful information for the determination of the droplet initial velocity. In the present study, the axial and radial velocity of the droplets were fitted by using the information from the gas flow field and by matching the film thickness distribution on a flat plate. The droplet initial velocity was obtained from the local air velocity with a reduction factor in the order of 0.2–0.7, depending on the shaping air flow rate. The resulting injection data were used not only for spray coating on the flat plate, but also on a complicated three-dimensional work piece, as well as for the case of an inclined atomizer. The simulation results are discussed in the following section.

Static film thickness distribution—namely, the film growth rate (μm/s) on the target—can be obtained after particle trajectory calculation, assuming uniform distribution of droplet mass in a computational cell. Since the spray gun is static in the present numerical simulation, in order to compare with the measured dynamic film thickness distribution (as shown in Figure 5), integration of the simulated static film pattern has to be carried out [13], taking into consideration the robot velocity and the wet and dry density of the paint material.

4. Simulation Results

4.1. Spray Painting Calculation by Using a Flat Plate

The calculated velocity contours of the airflow field in the plane $z = 0$ are shown in Figure 7. A supersonic velocity (about 545 m/s) is located at the atomizing air nozzle exits. To provide a sensible resolution of the entire flow field, the velocity contours are depicted in the range of 0–50 m/s. Air velocity close to the liquid nozzle and above the cross-section of shaping air jets is about 360 m/s, where computational droplets are injected with the maximum axial velocity of 72 m/s. The corresponding reducing factor for the initial droplet axial velocity is about 0.2 for the case of the shaping air flow rate of 360 L/min. Due to the shaping air, the gas flow field is deformed, deviating from the standard symmetric free jet of coaxial jet atomizers. Here, a quite narrow elliptic flow region is formed with a narrow extension along z, which results in a narrow film pattern on the plate (the so-called static film thickness distribution or static film growth rate (μm/s), since the spray gun is static in the simulation). With decreasing shaping air flow rate, the spray approaches a round spray. The liquid paint is concentrated in the center of the plate. The dependency of the film thickness on the shaping air flow rate is shown in Figure 8, where for a movement of the atomizer along z direction, the so-called dynamic film thickness distributions along the x direction are derived by integrating the static film thickness in the z direction. The calculated dynamic dry film thickness profiles were compared with the experimental results. A good agreement between measured and predicted film thickness was obtained.

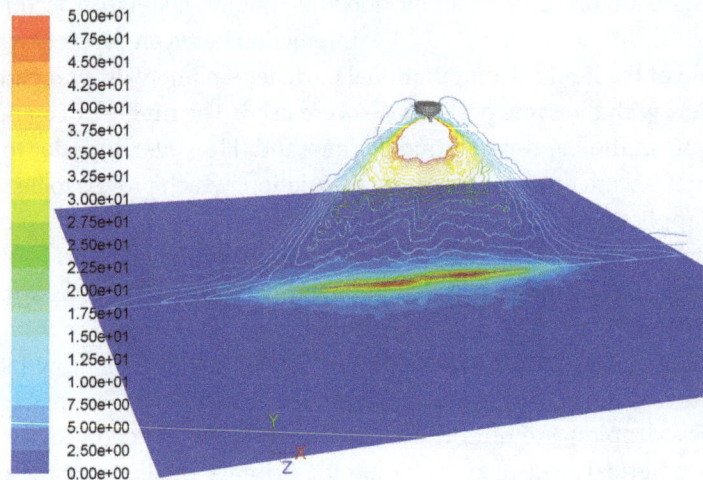

Figure 7. Calculated velocity contours colored by velocity magnitude (m/s) in the plane $z = 0$ (shaping air flow rate: 360 L/min). The static film thickness distribution on the plate is also overlaid.

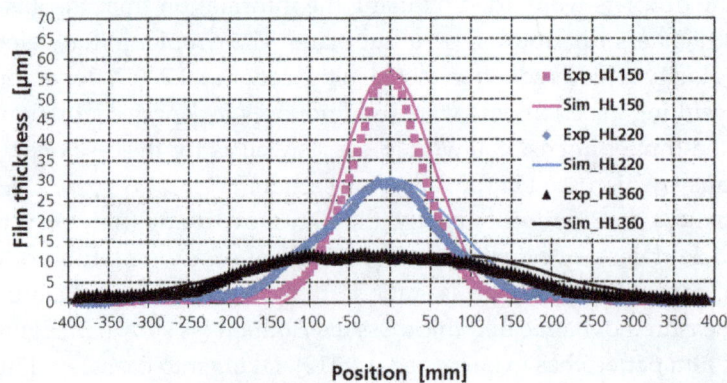

Figure 8. Comparison of measured and calculated dynamic film thickness distributions for three shaping air flow rates: 150, 220, 360 L/min.

4.2. Spray Painting Calculation for a Complicated Geometry of the Work Piece and an Inclined Atomizer

In order to verify the applicability of the numerical approach (especially the method for creating initial droplet conditions), additional tests were performed by purposefully increasing the complexity of the target geometry and the alignment of the atomizer. The target geometry considered herein is shown in Figure 9. Essentially, it is a stylized rear part of a car body with a water groove. The 3D target consists of four panels. The atomizer is located at the center directly above and (initially) perpendicular to panel C. The same droplet injection data that were used for the simulations with the flat plate were used here. As an example, velocity contours in a cross section through the center of the atomizer for two shaping air flow rates are shown in Figures 9 and 10, respectively. The liquid paint is distributed mainly on panels A, C, and D for the flat spray with the shaping air flow rate of 360 L/min. For the quasi round spray, almost all of the liquid paint is deposited on panel C. Detailed comparisons of the measured and the simulated film thickness distributions are depicted in Figure 11. Error bars in the measured film thickness distributions were derived based on a few film thickness profiles in the center region on the 3D-target. Higher standard deviations of the measured film thickness were observed at the panel edges or corners, especially for the round spray.

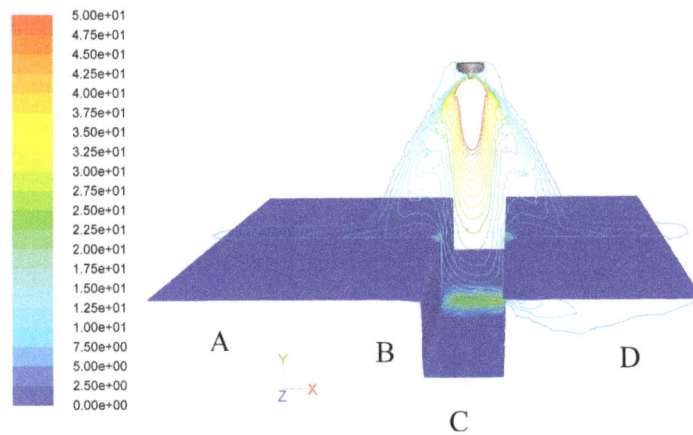

Figure 9. Calculated velocity contours colored by velocity magnitude (m/s) in the plane $z = 0$ (shaping air flow rate: 150 L/min). The static film thickness distribution on the plates is also overlaid.

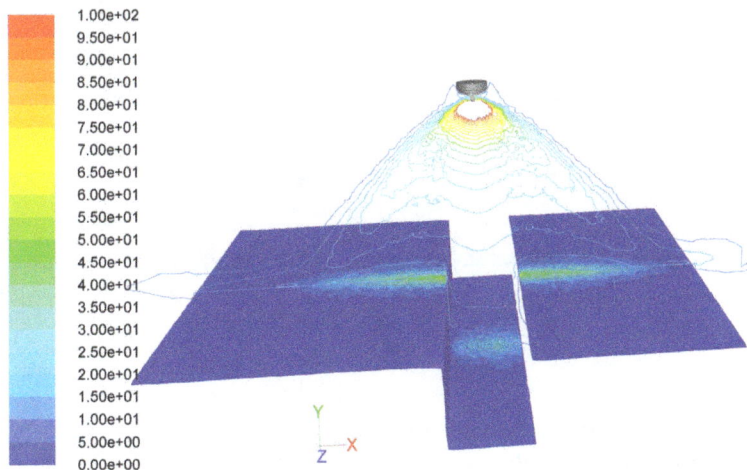

Figure 10. Calculated velocity contours colored by velocity magnitude (m/s) in the plane $z = 0$ (shaping air flow rate: 360 L/min). The static film thickness distribution on the plates is also overlaid.

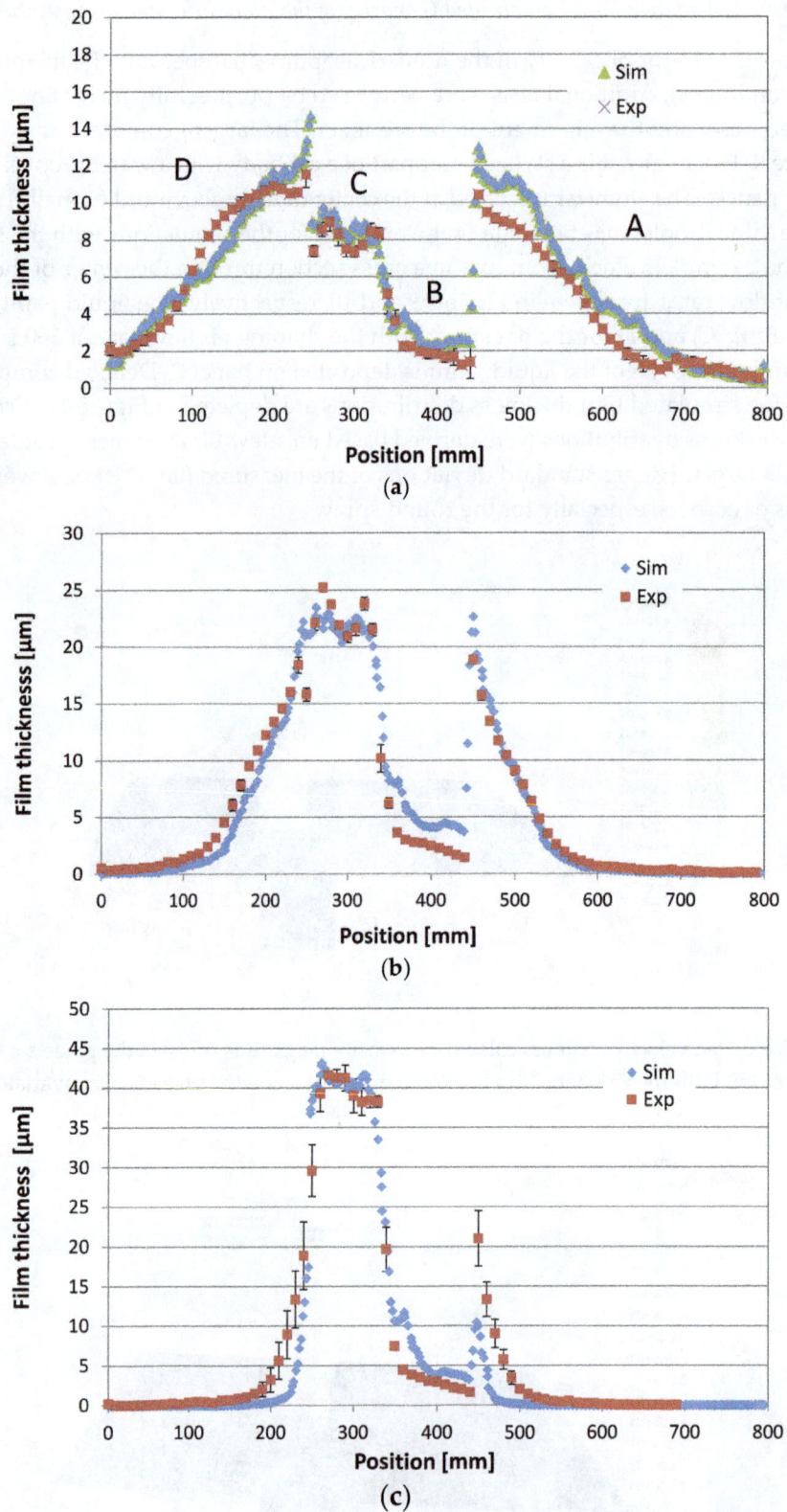

Figure 11. Comparison of the measured and calculated dynamic film thickness distributions: (a) shaping air flow rate 360 L/min; (b) 220 L/min; (c) 150 L/min.

Finally, the spray painting simulation with an inclination angle of 20° between atomizer axis and panel C was carried out. The velocity contours overlaid with the static film thickness distribution on the work piece and the comparison between simulation and experiment are shown in Figures 12 and 13. Clearly, a reasonably good agreement between the measurements and the simulations has been obtained both without inclined atomizer (Figure 11) and with inclined atomizer (Figure 13), except at positions very close to the edge. Based on the numerical simulation results, the numerical approaches used in this study can be put into use for industrial computations of spray painting using a pneumatic atomizer with complicated substrate geometries and arbitrary atomizer alignments.

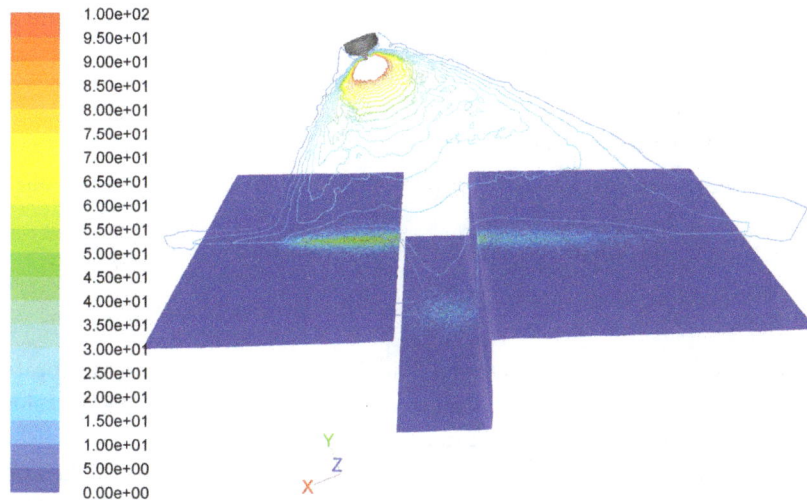

Figure 12. Velocity contours (m/s) at a cross section $z = 0$ overlaid with static film thickness on the target.

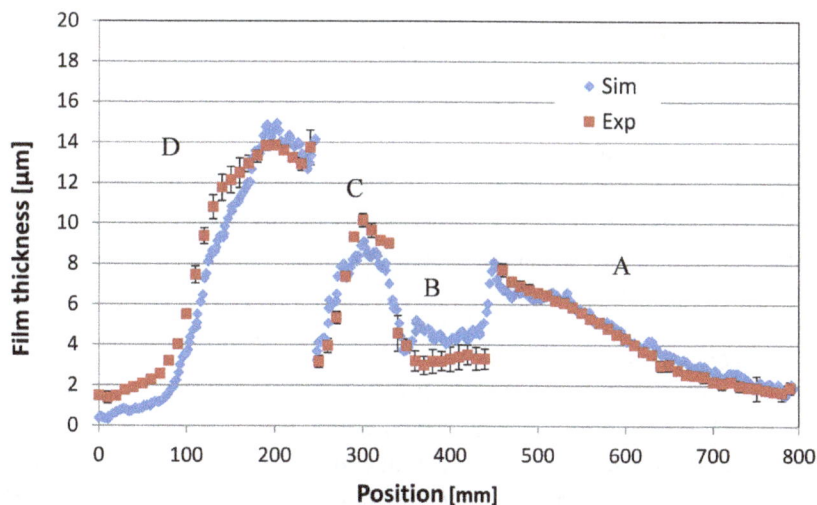

Figure 13. Comparison of the measured and simulated dynamic film thickness distributions (shaping air flow rate: 360 L/min).

5. Conclusions

In this paper, a numerical simulation of the spray coating process using a pneumatic atomizer has been presented. The compressible airflow was directly calculated from the nozzles.

Droplet size distributions were measured by applying a Spraytec Fraunhofer-type particle sizer. It was found that for a given paint material, the atomization process—and in particular, the droplet size distribution—depends not only on the atomizing air flow rate, but also on the shaping air flow rate. Based on the experimental results of film thickness distributions on a flat plate and air flow field around the atomizer nozzles, the injection model for calculating initial conditions for the droplet trajectory calculation was created, which takes into account the influence of the operating conditions on the droplet size distribution. The droplet phase calculations were then carried out by using the obtained injection data that were set very close to the nozzle. In this way, the spray painting simulation can be used for complex geometries of work pieces and arbitrary alignments of the atomizer, which is meaningful for practical applications.

Special attention was also paid to the verification of the current numerical approach by using a complex work piece geometry and an inclined atomizer alignment. The calculated dynamic film thickness distributions on the work piece were compared with measurements. A quite good agreement between the simulation and the experiment was obtained.

Acknowledgments: This study was partly funded by the BMW in Germany. We would like to thank our colleague Paustian for his assistance in the experiments.

Author Contributions: Qiaoyan Ye wrote the paper. Qiaoyan Ye and Karlheinz Pulli carried out the numerical simulations.

Conflicts of Interest: The authors declare no conflict of interest.

References

1. Ye, Q.; Domnick, J.; Khalifa, E. Effect of Inlet and Boundary Conditions on the Numerical Modelling of the Spray Coating Process Using a Pneumatic Atomizer. In Proceedings of the 10th Workshop on Two-Phase Flow Predictions, Merseburg, Germany, 9–12 April 2002.
2. Domnick, J.; Lindenthal, A.; Rüger, M.; Sommerfeld, M.; Svejda, P. *Zwischenbericht I zum AiF-Forschungsvorhaben 9076: Entwicklung und Anwendung der PDA-Messtechnik für die Bestimmung des Oversprays in der Lackiertechnik, Bericht des Lehrstuhls für Strömungsmechanik 393/I/93*; Universität Erlangen-Nürnberg: Erlangen, Germany, 1993. (In German)
3. Domnick, J.; Lindenthal, A.; Tropea, C.; Xu, T.-H. Application of Phase-Doppler anemometry in paint sprays. *At. Sprays* **1994**, *4*, 437–450.
4. Lindenthal, A. Verbesserung der Effizienz der pneumatischen Lackapplikation mit Hilfe von Phasen-Doppler-Anemometrie-Untersuchungen. Ph.D. Thesis, Universität Erlangen-Nürnberg, Erlangen, Germany, 1997. (In German)
5. Diwakar, R.; Fansler, T.D.; French, D.T.; Ghandi, J.B.; Dasch, C.J.; Heffelfinger, D.M. Liquid and vapor fuel distributions from an air-assist injector-an experimental and computational study. *SAE Int.* **1992**. [CrossRef]
6. Cheng, D.L.; Lee, C.F. Numerical studies of air-assisted sprays. *At. Sprays* **2002**, *12*, 463–500. [CrossRef]
7. Guézennec, N.; Poinsot, T. Large eddy simulation and experimental study of a controlled coaxial liquid-air jet. *AIAA J.* **2010**, *48*, 2596–2610. [CrossRef]
8. Fogliati, M.; Fontana, D.; Garbero, M.; Vanni, M.; Baldi, G. CFD simulation of paint deposition in an air spray process. *JCT Res.* **2006**, *3*, 117–125. [CrossRef]
9. Desjardins, O.; McCaslin, J.O.; Owkes, M.; Brady, P. Direct numerical and Large-Eddy simulation of primary atomization in complex geometries. *At. Sprays* **2013**, *23*, 1001–1048. [CrossRef]
10. Lebas, R.; Menard, T.; Beau, P.A.; Berlemont, A.; Demoulin, F.X. Numerical simulation of primary break-up and atomization: DNS and modelling study. *Int. J. Multiph. Flow* **2009**, *35*, 247–260. [CrossRef]
11. Hirt, C.W.; Nichols, B.D. Volume of fluid (VOF) Method for the dynamics of free boundaries. *J. Comput. Phys.* **1981**, *39*, 201–225. [CrossRef]
12. Shen, B.; Ye, Q.; Tiedje, O.; Westkämper, W. Primary Breakup of a Paint Liquid by a Coaxial High-Speed Gas Jet Used in Spray Painting Processes. In Proceedings of the 13th Triennial International Conference on Liquid Atomization and Spray Systems, Tainan, Taiwan, 23–27 August 2015.
13. Ye, Q.; Shen, B.; Tiedje, O.; Domnick, J. Investigations of spray painting processes using an airless spray gun. *J. Energy Power Eng.* **2013**, *7*, 74–81.

15

Internally Oxidized Ru–Zr Multilayer Coatings

Yung-I Chen *, Tso-Shen Lu and Zhi-Ting Zheng

Institute of Materials Engineering, National Taiwan Ocean University, 2 Pei-Ning Road, Keelung 20224, Taiwan; x76825@gmail.com (T.-S.L.); 10455001@ntou.edu.tw (Z.-T.Z.)
* Correspondence: yichen@mail.ntou.edu.tw

Academic Editors: Tony Hughes and Russel Varley

Abstract: In this study, equiatomic Ru–Zr coatings were deposited on Si wafers at 400 °C by using direct current magnetron cosputtering. The plasma focused on the circular track of the substrate holder and the substrate holder rotated at speeds within 1–30 rpm, resulting in cyclical gradient concentration in the growth direction. The nanoindentation hardness levels of the as-deposited Ru–Zr coatings increased as the stacking periods of the cyclical gradient concentration decreased. After the coatings were annealed in a 1% O_2–99% Ar atmosphere at 600 °C for 30 min, the internally oxidized coatings shifted their respective structures to a laminated structure, misaligned laminated structure, and nanocomposite, depending on their stacking periods. The effects of the stacking period of the cyclical gradient concentration on the mechanical properties and structural evolution of the annealed Ru–Zr coatings were investigated in this study.

Keywords: cyclical gradient concentration; internal oxidation; multilayer coating; nanocomposite coating

---

1. Introduction

Multilayer nitride coatings with nanoscale layer thickness have exhibited extremely high mechanical hardness due to dislocation blocking by layer interfaces and Hall–Petch strengthening [1]. By contrast, the hardness enhancement in the Y_2O_3/ZrO_2 superlattice has been limited because oxides are brittle materials that are deformed by fracture mechanisms [2]. Two metallic multilayer coatings deposited by cosputtering for immiscible systems, W–Cu [3,4] and Cu–Ta [5,6], have developed a phase-separated nanostructure. However, Ru/Al multilayers have been deposited to fabricate a B2-RuAl intermetallic compound through annealing at approximately 600 °C in a vacuum or Ar [7,8]. Oxide-dispersion-strengthened platinum materials [9] and Ag-oxide-based electric contact material [10] are conventional applications of internal oxidation [11]. Our previous studies [12–15] investigated the internal oxidation of Ru-based alloy multilayer coatings annealed at 600 °C in oxygen-containing atmospheres for the application of protective coatings on glass molding dies. The specific cosputtering processes, which were performed using a substrate holder rotating at a slow speed of one to seven revolutions per minute, have been examined in detail for fabricating Ru–Ta coatings [14]; the fabricated coatings had exposed substrates alternately to the sputter sources without shutter shielding, forming a multilayer structure with a cyclical gradient concentration period at a nanometer scale. An oxidized laminated structure formed because of the inward diffusion of oxygen during the annealing process; this structure comprised alternating oxygen-rich and oxygen-deficient sublayers stacked adjacent to the surface. The inward diffusion of oxygen at 600 °C was dominated by lattice diffusion in the active element-enriched regions [13,16,17]. Because the elements were stacked on the substrate with an alternating gradient concentration, the O atoms could easily diffuse through the paths in the transverse direction, thereby forming oxide sublayers. After the oxygen content in the oxide sublayers reached a saturation level, the grainboundary diffusion along the original columnar structure drove oxygen to the next period of the laminated structure. During an annealing process conducted at 600 °C in a 1% O_2–99% Ar atmosphere, internal oxidation occurred for Ti–Ru, Zr–Ru, Nb–Ru, Mo–Ru, Hf–Ru, Ta–Ru, and W–Ru coatings, which were prepared using a substrate holder rotating at one revolution per minute [15]; the mechanical properties of the annealed coatings depended on the characteristics of the oxide sublayers. The nanoindentation hardness of the annealed $Zr_{0.30}Ru_{0.70}$ coating exhibited a relatively high value of 18.4 GPa. The widths of the oxide sublayers were restricted by the Ru-dominant sublayers [16,17]; therefore, the internally oxidized coatings can be categorized as nonisostructural oxide/metal multilayers [1]. The substrate holder rotation speed in sputtering affects the stacking period of the laminated structure [14]; therefore, assessing the effect of the stacking period on the mechanical properties of the internally oxidized Ru–Zr coatings is imperative.

2. Materials and Methods

Ru–Zr coatings with a Cr interlayer were fabricated by using magnetron cosputtering onto silicon wafers. Pure metal targets of Ru (99.95%), Zr (99.9%), and Cr (99.95%) with diameters of 50.8 mm each were adopted as source materials for sputtering. The sputter guns were inclined to focus plasma on the circular track of the substrate holder, as described in detail in a previous study [13]. The target-to-substrate distance was maintained at 90 mm for all sputtering runs. The chamber was evacuated down to 2.7×10^{-4} Pa, followed by the inlet of argon gas as a plasma source. The substrate holder was heated to 400 °C and the Ar flow rate was controlled at 20 sccm; the resulting working pressure was 0.7 Pa. The substrate holder was rotated at 1 rpm for depositing the Cr interlayer. Then, Ru–Zr coatings with fixed DC sputtering powers (W_{Ru} = 100 W and W_{Zr} = 200 W) and various substrate holder rotation speeds were deposited on the Cr interlayer for 25 min. To investigate the internal oxidation phenomenon after performing heat treatments, the Ru–Zr coatings were further annealed at 600 °C in a 1% O_2–99% Ar atmosphere by introducing O_2–Ar mixed gas into a quartz tube furnace.

Chemical composition analysis was conducted by using energy dispersive spectrometry (EDS, Horiba, Kyoto, Japan) equipped with a scanning electron microscope (SEM, S3400N, Hitachi, Tokyo, Japan) on the surface. Surface morphology and thickness evaluation of the coatings were performed by using a field emission scanning electron microscope (FE-SEM, S4800, Hitachi, Tokyo, Japan) at a 15-kV accelerating voltage. A conventional X-ray diffractometer (XRD, X'Pert PRO MPD, PANalytical, Almelo, The Netherlands) with Cu Kα radiation was adopted to identify the phases of the coatings, using the grazing incidence technique with an incidence angle of 1°. The Cu Kα radiation was generated from a Cu anode operated at 45 KV and 40 mA. The nanostructure was examined by using a transmission electron microscope (TEM, JEM-2010F, JEOL, Tokyo, Japan) at a 200-kV accelerating voltage. The TEM samples were prepared by applying a focused ion beam system (FEI Nova 200, Hillsboro, OR, USA) operated at an accelerating voltage of 30 kV with a gallium ion source. A Pt layer was deposited to protect the free surface in the sample preparation. The chemical states of the constituent elements were examined by using an X-ray photoelectron spectroscope (XPS, PHI 1600, PHI, Kanagawa, Japan) with an Mg Kα X-ray beam (energy = 1253.6 eV and power = 250 W) operated at 15 kV. The XPS spectra of O 1s, Ru 3d, and Zr 3d core levels were recorded. Ar+ ion beam of 3 keV was used to sputter the coatings for depth profiling. The surface hardness and Young's modulus of Ru–Zr coatings were measured with a nanoindentation tester (TI-900 Triboindenter, Hysitron, Minneapolis, MN, USA). The nanoindenter was equipped with a Berkovich diamond-probe tip. The applied load was controlled to produce an indentation depth of 80 nm, which is 1/10 of the film thickness [18]. The loading, holding, and unloading times were 5 s each. The nanoindentation hardness and elastic modulus of each indent were calculated using the Oliver and Pharr method [19]. The standard deviations for nanoindentation data were calculated from 8 measurements made at different locations on one sample. The surface roughness values of the coatings, R_a [20], were evaluated by using an atomic force microscope (AFM, Dimension 3100 SPM, NanoScope IIIa, Veeco, Santa Barbara, CA, USA). The scanning area of each image was set at 5×5 μm^2 with a scanning rate of 1.0 Hz.

3. Results

3.1. As-Deposited Equiatomic Ru–Zr Coatings

Table 1 lists the chemical compositions of the as-deposited equiatomic Ru–Zr coatings prepared at various substrate holder rotation speeds of 1–30 rpm. The samples were denoted as $Ru_xZr_{1-x}(Ry)$, or Ry, where Ry indicated that the sample prepared using the substrate holder was rotated at y rpm. All the coatings exhibited similar atomic ratios $Ru/(Ru + Zr)$ within 0.46–0.50 after being examined using EDS on the surface, and a thickness of 870–920 nm after being evaluated using FE-SEM in the cross section. Oxygen content in the as-deposited coatings was 0.1–0.5 at.% because of weak oxidation caused by the residual oxygen in the vacuum chamber.

Table 1. Chemical compositions, thickness values, laminated period, mechanical properties, and surface roughness values of $Ru_xZr_{1-x}(Ry)$ coatings as-deposited and annealed at 600 °C in 1% O_2–99% Ar for 30 min.

Sample	Chemical Composition (at.%)			Atomic Ratio	Thickness (nm)		Period (nm)	H (GPa)	E (GPa)	Roughness (nm)
	Ru	Zr	O	Ru:Zr	Coating	Interlayer				
As-deposited										
$Ru_{0.50}Zr_{0.50}$(R1)	50.2	49.6	0.2	50.3:49.7	870	90	34.8	9.1 ± 0.2	128 ± 3	1.76 ± 0.04
$Ru_{0.49}Zr_{0.51}$(R3)	48.7	50.9	0.4	48.9:51.1	900	100	12.0	10.3 ± 0.3	142 ± 3	2.51 ± 0.02
$Ru_{0.48}Zr_{0.52}$(R5)	47.9	51.8	0.3	48.0:52.0	890	100	7.2	10.5 ± 0.6	137 ± 6	3.35 ± 0.06
$Ru_{0.47}Zr_{0.53}$(R10)	47.2	52.3	0.5	47.4:52.6	900	100	3.6	11.0 ± 0.4	161 ± 4	2.62 ± 0.05
$Ru_{0.46}Zr_{0.54}$(R15)	46.3	53.4	0.3	46.4:53.6	900	90	2.4	11.1 ± 0.5	177 ± 7	4.08 ± 0.04
$Ru_{0.47}Zr_{0.53}$(R20)	46.6	53.4	0.1	46.6:53.4	920	95	1.8	11.4 ± 0.6	171 ± 6	1.25 ± 0.01
$Ru_{0.46}Zr_{0.54}$(R30)	46.4	53.6	0.1	46.4:53.6	920	90	1.2	13.1 ± 0.5	172 ± 5	1.37 ± 0.01
Annealed										
$Ru_{0.50}Zr_{0.50}$(R1)	21.3	21.1	57.5	50.1:49.9	1380	110	55.2	15.5 ± 0.4	157 ± 10	5.33 ± 0.50
$Ru_{0.49}Zr_{0.51}$(R3)	21.2	21.2	59.2	48.0:52.0	1370	110	18.1	16.1 ± 0.8	158 ± 8	4.26 ± 0.10
$Ru_{0.48}Zr_{0.52}$(R5)	20.6	20.6	60.7	47.6:52.4	1390	110	11.1	17.2 ± 0.4	178 ± 9	7.02 ± 0.37
$Ru_{0.47}Zr_{0.53}$(R10)	20.3	20.3	61.5	47.3:52.7	1390	110	5.5	12.3 ± 2.1	164 ± 16	17.32 ± 0.53
$Ru_{0.46}Zr_{0.54}$(R15)	20.2	20.2	62.5	46.1:53.9	1390	110	3.7	16.4 ± 1.0	182 ± 6	7.05 ± 0.20
$Ru_{0.47}Zr_{0.53}$(R20)	20.8	20.8	61.3	46.3:53.7	1380	110	2.8	16.1 ± 0.8	160 ± 6	1.89 ± 0.00
$Ru_{0.46}Zr_{0.54}$(R30)	21.0	21.0	61.4	45.7:54.3	1390	110	1.9	17.9 ± 0.7	175 ± 6	5.90 ± 1.05

Figure 1 shows cross-sectional SEM images of the as-deposited Ru–Zr coatings, which exhibit a columnar structure. Laminated structures stacked along the growth direction were observed in the $Ru_{0.50}Zr_{0.50}$(R1) and $Ru_{0.49}Zr_{0.51}$(R3) coatings, for which the equilibrated laminated layer periods were 35 and 12 nm, respectively, as determined using the thickness recorded from the SEM observation divided by the number of laminated layers; in other words, the number of revolutions of the substrate holder. Each equilibrated laminated layer period formed as a result of cyclical gradient concentration deposition. The laminated structures of the Ru–Zr(Ry) coatings prepared at higher substrate holder rotation speeds such as $Ru_{0.47}Zr_{0.53}$(R10) and $Ru_{0.46}Zr_{0.54}$(R30) exhibited narrower equilibrated laminated layer periods that could not be evaluated through SEM images.

Figure 2 shows the XRD patterns of the as-deposited Ru–Zr(Ry) coatings. The $Ru_{0.50}Zr_{0.50}$(R1), $Ru_{0.49}Zr_{0.51}$(R3), $Ru_{0.48}Zr_{0.52}$(R5), and $Ru_{0.47}Zr_{0.53}$(R10) coatings exhibited reflections of hexagonal Ru [ICDD 06-0663], cubic RuZr [ICDD 18-1147], and hexagonal Zr [ICDD 05-0665] phases, implying that these coatings consisted of laminated sublayers. The equilibrated laminated layer periods for the R5 and R10 coatings were 7.2 and 3.6 nm, respectively. By contrast, XRD patterns of the as-deposited $Ru_{0.46}Zr_{0.54}$(R15), $Ru_{0.47}Zr_{0.53}$(R20), and $Ru_{0.46}Zr_{0.54}$(R30) coatings exhibited a RuZr phase dominant structure. The cubic RuZr phase exhibited XRD reflections of (110), (200), and (211), which are comparable with previous XRD results reported by Mahdouk et al. [21]. RuZr exhibited a B2 structure (CsCl type) [21–25]. Figure 3 depicts a cross-sectional TEM image of the as-deposited $Ru_{0.46}Zr_{0.54}$(R15) coating, which comprises a columnar structure without evident laminated sublayers; the diffraction pattern of the selected area shows a cubic RuZr phase. The equilibrated laminated layer periods for the as-deposited $Ru_{0.46}Zr_{0.54}$(R15), $Ru_{0.47}Zr_{0.53}$(R20), and $Ru_{0.46}Zr_{0.54}$(R30) coatings were 2.4, 1.8, and 1.2 nm, respectively, which were too thin to construct the laminated structure. Under such conditions, the equilibrated laminated layer periods were equal to a variation period of cyclical gradient concentration. Because the substrate temperature was sustained at 400 °C during cosputtering, the deposited atoms formed an intermetallic RuZr compound, as observed by the XRD patterns. In our previous study [26], B2-RuAl phase was observed for Ru–Al multilayer coatings prepared at 400 °C.

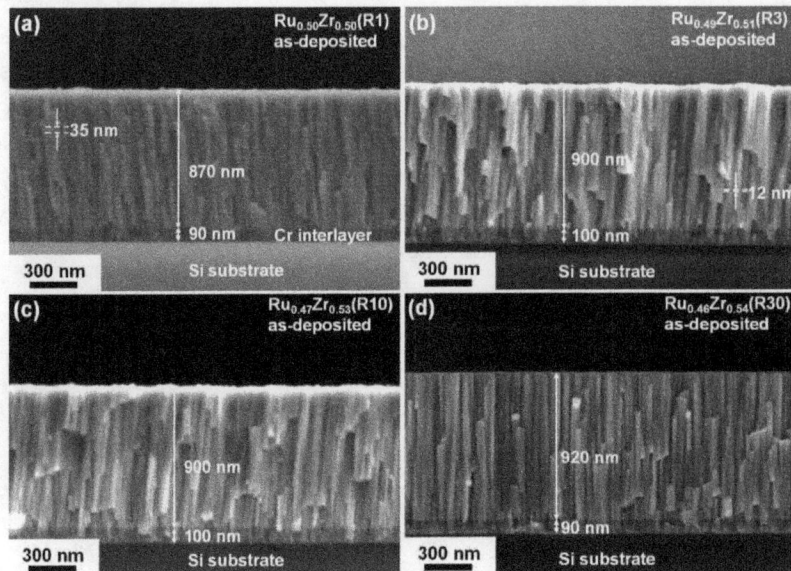

Figure 1. Cross-sectional SEM images of the as-deposited (**a**) $Ru_{0.50}Zr_{0.50}$(R1); (**b**) $Ru_{0.49}Zr_{0.51}$(R3); (**c**) $Ru_{0.47}Zr_{0.53}$(R10); and (**d**) $Ru_{0.46}Zr_{0.54}$(R30) coatings.

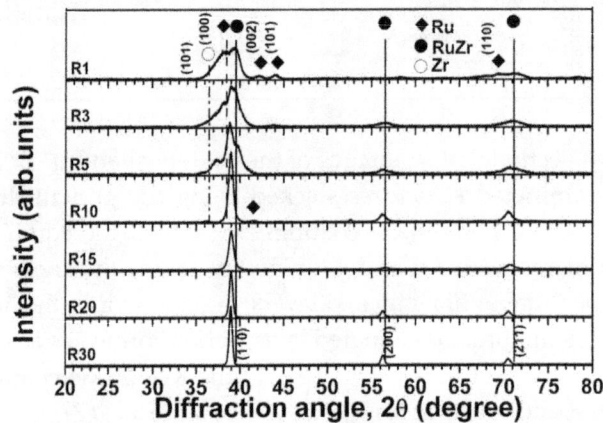

Figure 2. XRD patterns of the as-deposited Ru–Zr(Ry) coatings.

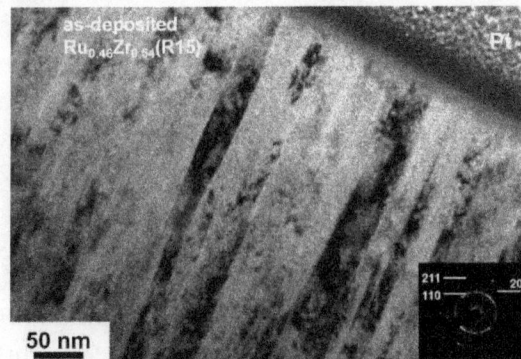

Figure 3. Cross-sectional TEM image and selected area diffraction pattern of the as-deposited $Ru_{0.46}Zr_{0.54}$(R15) coating.

3.2. Internally Oxidized Ru–Zr Coatings

Figure 4 shows the cross-sectional SEM image of the annealed $Ru_{0.50}Zr_{0.50}$(R1) coating, which exhibited a laminated structure with an equilibrated laminated layer period of 55 nm. However, the features of the other coatings could not be identified through SEM. Figure 5 presents the XRD patterns of the Ru–Zr coatings after annealing in 1% O_2–99% Ar at 600 °C for 30 min; all patterns exhibited monoclinic ZrO_2 [ICDD 32-1484], tetragonal ZrO_2 [ICDD 42-1164], and Ru phases. The Ru:Zr atomic ratios were maintained at levels similar to those of the as-deposited coatings (Table 1), implying that no volatile oxides were formed during annealing. The O content in the annealed coatings increased to within 58–62 at.%, indicating that extra O was trapped because the stoichiometric ratio of ZrO_2 was two, enabling partial Ru atoms to be oxidized.

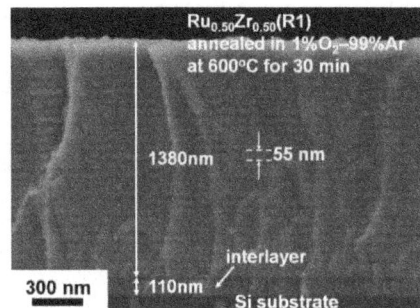

Figure 4. Cross-sectional SEM image of the $Ru_{0.50}Zr_{0.50}$(R1) coating after annealing in 1% O_2–99% Ar at 600 °C for 30 min.

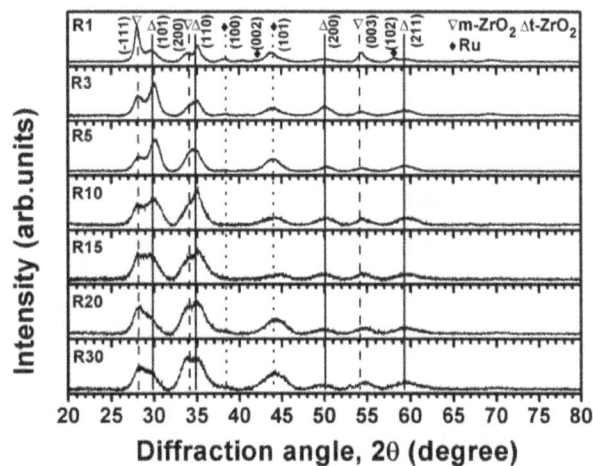

Figure 5. XRD patterns of the Ru–Zr(Ry) coatings after annealing in 1% O_2–99% Ar at 600 °C for 30 min.

Figure 6a–c illustrates the XPS spectra of O 1s, Zr 3d, and Ru 3d core levels, respectively, at various thickness levels of the annealed $Ru_{0.50}Zr_{0.50}$(R1) coating. The detected depth crossed six periods of the laminated layers. The O and Zr species were identified as O^{2-} and Zr^{4+}, whereas Ru was identified as Ru^0 except for the spectra near the surface region (depth < 13 nm), where the Ru^{x+} and Ru^{4+} signals were split. The binding energy value of Ru^0 $3d_{5/2}$ (279.96 ± 0.08 eV) was consistent with that of other coatings (279.69–280.16 eV) reported in the literature [13,16,17,27], whereas the binding energies of Ru^{x+} and Ru^{4+} $3d_{5/2}$ were 280.45 ± 0.11 and 282.57 ± 0.15 eV, respectively. Previous studies reported 281.4–282.2 eV [26,28–30] for the binding energy of Ru^{4+} $3d_{5/2}$. Ru of 17%–20% exhibited the Ru^{4+} state at a depth of 0–13 nm. Ru atoms remained in its metallic state beneath the near surface region. Figure 6d shows the intensity variations of O^{2-} 1s, Ru^0 $3d_{5/2}$, and Zr^{4+} $3d_{5/2}$ signals along the depth, which indicates that the variation trend of the

O^{2-} 1s profile coincides with that of the Zr^{4+} $3d_{5/2}$ profile and is in contrast to that of the Ru^0 $3d_{5/2}$ profile, implying that ZrO_2 is the dominant oxide. Therefore, the annealed $Ru_{0.50}Zr_{0.50}(R1)$ coating comprised alternating oxygen-rich and oxygen-deficient layers stacked along the O-diffusion direction. The binding energy value of Zr^{4+} $3d_{5/2}$ deviated within 182.05–183.35 eV (Figure 6e). Moreover, this range decreased to 182.71–183.35 eV after the data in the first laminated period had been excluded. Previous studies have reported 182.75 [31], 182.8 [32], and 182.9 eV [33] for the binding energy value of Zr^{4+} $3d_{5/2}$. The binding energy value of O^{2-} 1s demonstrated a variation pattern similar to that of the binding energy value of Zr^{4+} $3d_{5/2}$ (Figure 6e). The charging effect of analyzing insulators [34] caused substantial deviation in binding energy. The binding energy difference $\Delta = (O^{2-} \, 1s - Zr^{4+} \, 3d_{5/2})$ was 347.92 ± 0.05 eV at the analyzed depth of 19.5–318.5 nm. This difference was highly consistent with the reported difference of 348.01 and 348.2 eV, calculated using 530.76 and 182.75 eV [31] or 531.1 and 182.9 eV [33] for O^{2-} 1s and Zr^{4+} $3d_{5/2}$, respectively. The periodic changes of nonoxidized metallic Ru suggested the influence of oxygen in the Zr-deficient sublayers.

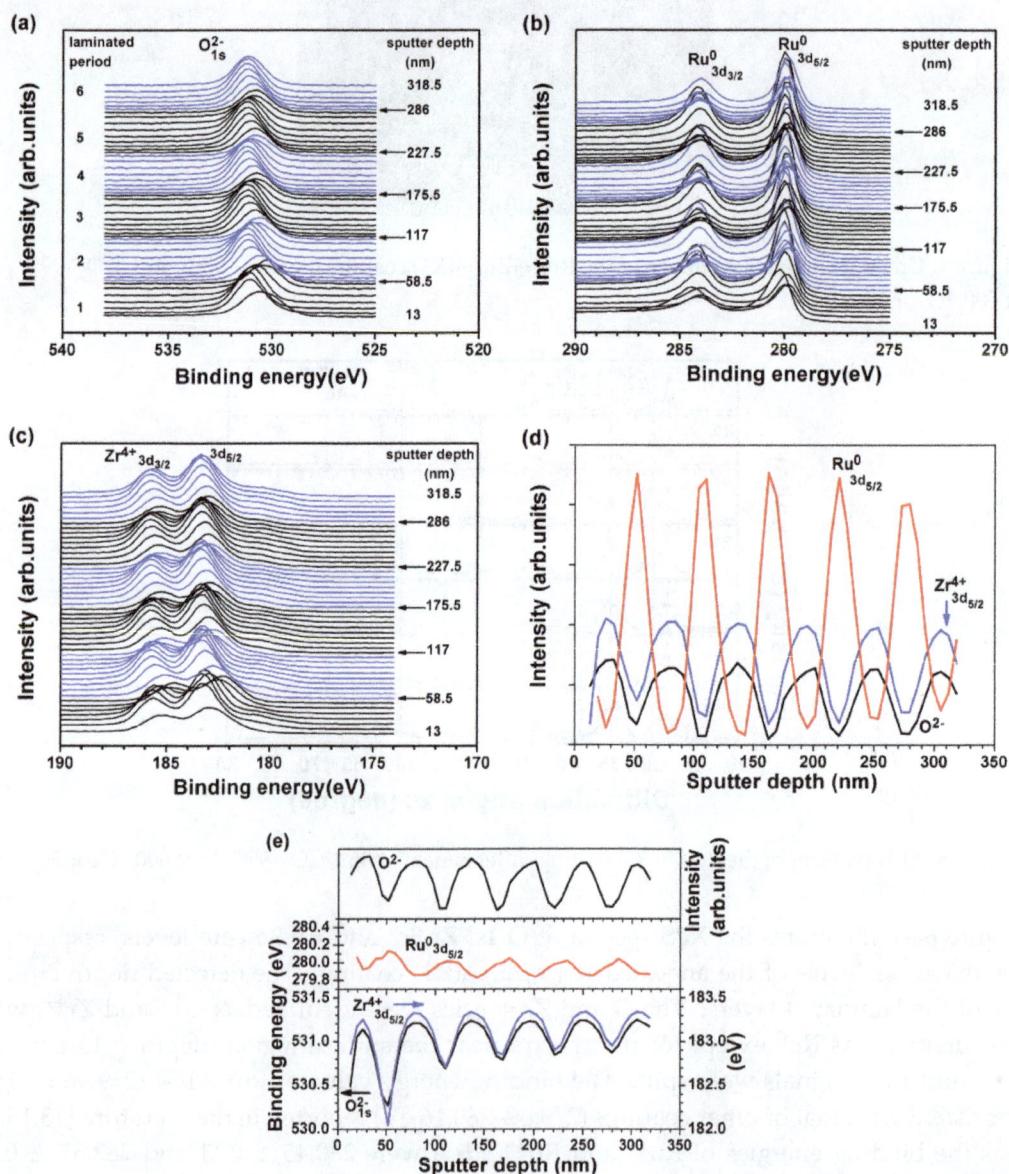

Figure 6. XPS spectra of (**a**) O 1s, (**b**) Ru 3d, and (**c**) Zr 3d core levels of the $Ru_{0.50}Zr_{0.50}(R1)$ coating after annealing in 1% O_2–99% Ar at 600 °C for 30 min; variation patterns of (**d**) intensity and (**e**) binding energy of O^{2-}1s, $Zr^{4+}3d_{5/2}$, and Ru^0 $3d_{5/2}$.

Figure 7a,b shows the cross-sectional TEM images of the annealed $Ru_{0.48}Zr_{0.52}$(R5) coating; the laminated structure was evident. Figure 7c shows a high-resolution TEM image of the near-surface region of the annealed coating. The lattice fringes of particular areas indicated that the annealed $Ru_{0.48}Zr_{0.52}$(R5) coating comprised ZrO_2- and Ru-dominant sublayers, which linked together across the original columnar boundaries such that the annealed $Ru_{0.48}Zr_{0.52}$(R5) coatings were laminated and the columnar boundaries were unresolved. Figure 8a depicts the cross-sectional TEM image of the annealed $Ru_{0.47}Zr_{0.53}$(R10) coating. The laminated sublayers were curved, because of which the stacks of sublayers among neighboring columnar structures were disconnected. Figure 8b shows the high-resolution TEM image of the middle region of the annealed $Ru_{0.47}Zr_{0.53}$(R10) coating. The Ru-dominant sublayers were two-nanometers thick only, and disconnected regions of the sublayers among neighboring columnar structures were observed. The fast variation of cyclical gradient concentration for the R10 coatings prepared with a quick substrate holder rotation speed resulted in the formation of grooved sublayers. For the coatings with thicker Ru sublayers, R1, R3, and R5, the sublayers became flat. The misaligned connections were more evident in the near-surface region (Figure 8c).

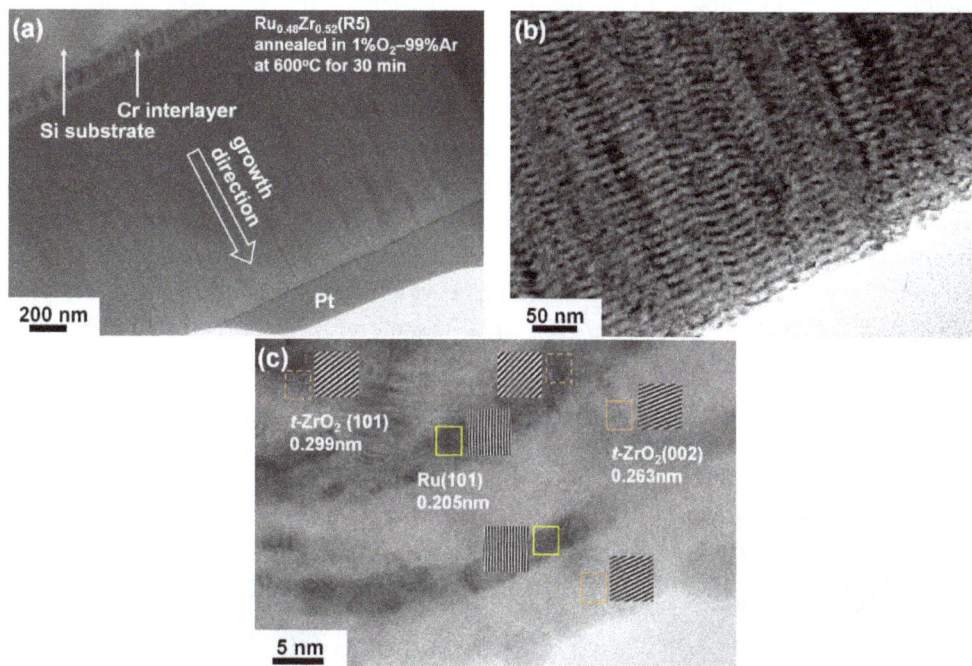

Figure 7. (a,b) cross-sectional TEM images of the oxidized $Ru_{0.48}Zr_{0.52}$(R5) coating; (c) high-resolution image of the near surface region of the annealed $Ru_{0.48}Zr_{0.52}$(R5) coating.

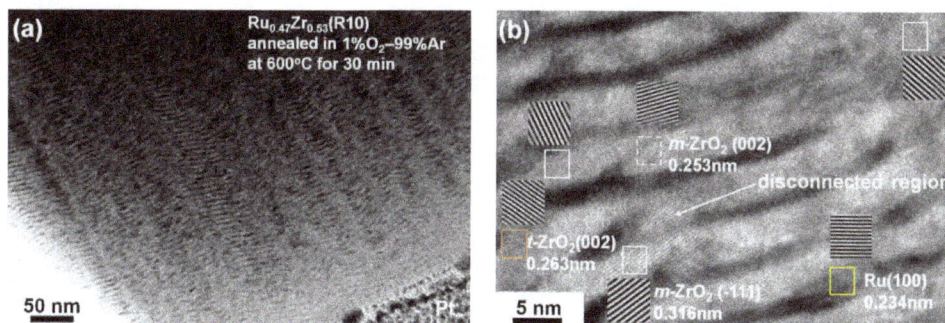

Figure 8. Cont.

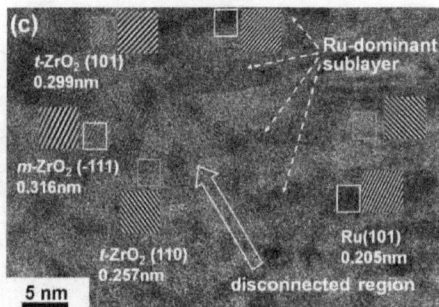

Figure 8. (**a**) Cross-sectional TEM image of the oxidized Ru$_{0.47}$Zr$_{0.53}$(R10) coating; high-resolution images of the (**b**) middle region and (**c**) near-surface region of the annealed coating.

Figure 9a shows a cross-sectional TEM image of the annealed Ru$_{0.46}$Zr$_{0.54}$(R30) coating, in which the original columnar boundaries are evident, but no laminated structures were observed. A high-resolution TEM image (Figure 9b) revealed nanocrystalline grains of ZrO$_2$ and Ru, each approximately five nanometers in diameter, implying that a nanocomposite structure had been constructed. Furthermore, Ru grains, the dark regions in the image, tended to concentrate along the columnar boundaries. Figure 10a–c illustrates the XPS spectra of the annealed Ru$_{0.46}$Zr$_{0.54}$(R30) coating. The XPS spectra of Ru 3d core levels indicated the presence of minor Ru^{4+} (3d$_{5/2}$: 282.11 eV) in addition to Ru^0 (3d$_{5/2}$: 280.19 $\pm$ 0.07 eV) at the near-surface region (Figure 10b), which was attributed to the incorporation of Ru into the ZrO$_2$ grains because RuO$_2$ and ZrO$_2$ possessed a similar tetragonal structure. Figure 10d shows that the intensities of O^{2-} 1s, Ru^0 3d$_{5/2}$, and Zr^{4+} 3d$_{5/2}$ signals were constant along the depth due to the limit of XPS analyses. Similar binding energy trends were observed for O^{2-} and Zr^{4+} (Figure 10e). The binding energy difference $\Delta = $ (O^{2-} 1s $-$ Zr^{4+} 3d$_{5/2}$) was 348.00 $\pm$ 0.02 eV at the analyzed depth (5.7–96.9 nm). Therefore, Zr reacted with O during annealing, and the annealed coating exhibited a nanocomposite comprising ZrO$_2$ and Ru grains.

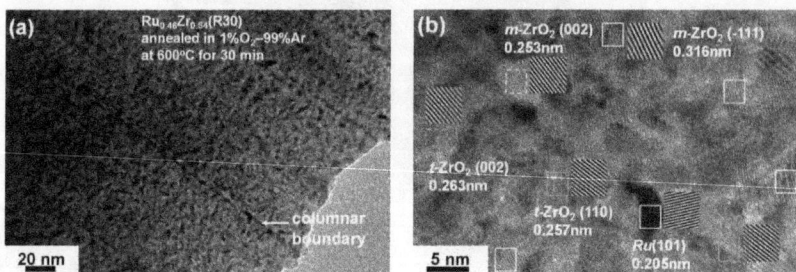

Figure 9. (**a**) Cross-sectional TEM image and (**b**) high-resolution image of the oxidized Ru$_{0.46}$Zr$_{0.54}$(R30) coating.

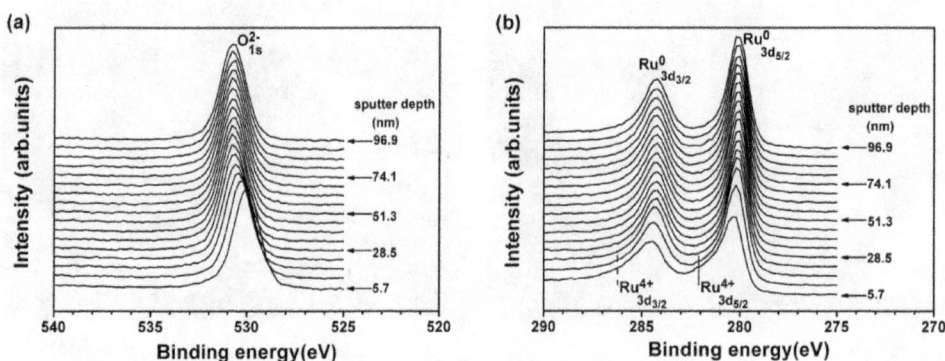

Figure 10. *Cont.*

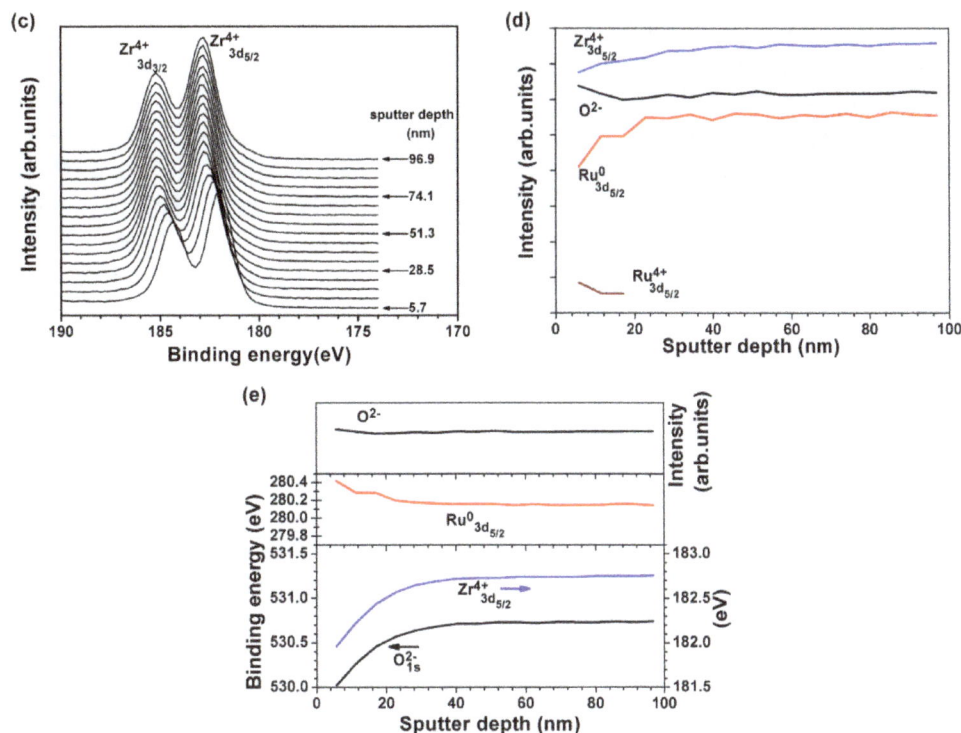

Figure 10. XPS spectra of (**a**) O 1s, (**b**) Ru 3d, and (**c**) Zr 3d core levels of the $Ru_{0.46}Zr_{0.54}$(R30) coating after annealing in 1% O_2–99% Ar at 600 °C for 30 min; variation patterns of (**d**) intensity and (**e**) binding energy of O^{2-} 1s, $Zr^{4+}3d_{5/2}$, and Ru^0 $3d_{5/2}$.

3.3. Mechanical Properties of Internally Oxidized Ru–Zr Coatings

Figure 11 depicts the nanoindentation hardness variations of the as-deposited and internally oxidized Ru–Zr coatings prepared at various substrate holder rotation speeds through sputtering. The hardness of the as-deposited coatings increased from 9.1 to 13.1 GPa with the substrate holder rotation speed and decreasing equilibrated laminated layer period. This hardness increase was attributed to the decrease in crystalline size and structural variation. The nanoindentation hardness of all Ru–Zr coatings increased after annealing in 1% O_2–99% Ar at 600 °C for 30 min. The hardness variation curve of the internally oxidized Ru–Zr coatings exhibited three divisions representing a laminated structure, a disconnected laminated structure, and a nanocomposite region. The hardness increased from 9.1, 10.3, and 10.5 to 15.5, 16.1, and 17.2 GPa for the annealed $Ru_{0.50}Zr_{0.50}$(R1), $Ru_{0.49}Zr_{0.51}$(R3), and $Ru_{0.48}Zr_{0.52}$(R5) coatings, respectively, which exhibited equilibrated laminated layer periods of 55, 18, and 11 nm, respectively. This result indicates that the hardness of the internally oxidized Ru–Zr coatings, which exhibited crystalline phases identical to those identified through XRD analysis and appropriately maintained their multilayer structures, was affected by the layer period. These internally oxidized Ru–Zr multilayer coatings were categorized as nonisostructural oxide/metal multilayers [1]. Dislocation could not be moved across oxide/metal interfaces because oxides are brittle materials that deform through fracture mechanisms, limiting the hardness enhancement [2]; therefore, the hardness of oxide/metal multilayers approached that of the oxide ZrO_2. Gan et al. reported a nanoindentation hardness of 18 GPa for monoclinic ZrO_2 thin films [35]. By contrast, the hardness of the annealed $Ru_{0.47}Zr_{0.53}$(R10) coatings with an equilibrated laminated layer period of 5.6 nm exhibited a relatively low level of 12.3 GPa. Although the internally oxidized $Ru_{0.47}Zr_{0.53}$(R10) coatings were laminated in each columnar structure, the same sublayers among neighboring columnar structures were misaligned and disconnected, which reduced the hardness. The internally oxidized $Ru_{0.46}Zr_{0.54}$(R15), $Ru_{0.47}Zr_{0.53}$(R20), and $Ru_{0.46}Zr_{0.54}$(R30) coatings exhibited high hardness within 16.1–17.9 GPa and were categorized as nanocrystalline composites consisting of hard ZrO_2 grains and soft Ru grains. Figure 12 shows the variation in Young's moduli of the as-deposited and internally oxidized Ru–Zr coatings. The Young's moduli increased from 130 to 140 GPa for R1, R3, and R5 coatings, to 160 GPa for R10 coatings and 170–180 GPa for R15, R20, and R30 coatings. Because the internally oxidized Ru–Zr coatings exhibited similar phases, ZrO_2 and Ru, the differences in Young's moduli among the annealed coatings were limited (i.e., 160–180 GPa). The surface roughness values of the Ru–Zr coatings are shown in Table 1. When evaluating the mechanical properties of coatings, previous studies [36–38] have reported that coatings with higher surface roughness exhibit larger standard deviation values or lower mean values. The effect of surface roughness on the mechanical properties of as-deposited Ru–Zr coatings was unclear. By contrast, the mechanical properties of the annealed coatings revealed larger deviations and higher surface roughness values than did those of the as-deposited coatings.

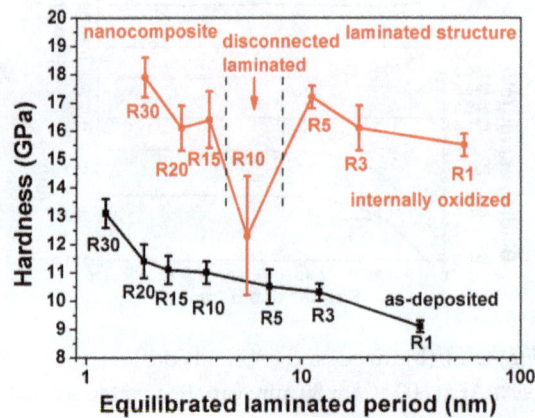

Figure 11.   Nanoindentation hardness values of the as-deposited and internally oxidized Ru–Zr coatings.

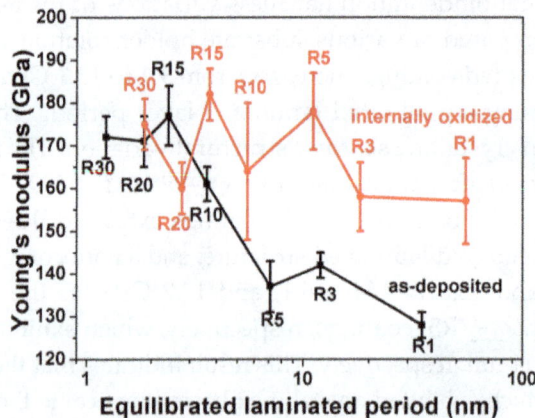

Figure 12. Young's modulus values of the as-deposited and internally oxidized Ru–Zr coatings.

4. Conclusions

Rotation speeds of the substrate holder during sputtering affected the crystalline structure and mechanical properties of Ru–Zr coatings both in the as-deposited and internally oxidized states. Because Ru–Zr coatings were fabricated using a cyclical gradient concentration stacked constitution, the coatings prepared at low rotation speeds (1–10 rpm) exhibited a laminated structure in addition to a columnar structure. The as-deposited Ru–Zr coatings exhibited nanoindentation hardness of 9.1–13.1 GPa, and the coatings prepared at higher substrate holder rotation speeds exhibited higher hardness. After annealing in a 1% O_2–99% Ar atmosphere at 600 °C for 30 min accompanied by the conduction of internal oxidation, the coatings prepared at a substrate holder rotation speed of one to five revolutions per minute maintained a laminated structure; this structure comprised alternately stacked Ru-dominant and ZrO$_2$-dominant sublayers whose nanoindentation hardness increased to 15.5–17.2 GPa because of the formation of ZrO$_2$ phase and the maintenance of sublayer interfaces. By contrast, the annealed coatings prepared at a rotation speed of 10 rpm maintained a similar laminated structure; however, the stacks of sublayers among neighboring columnar structures were misaligned and disconnected, resulting in relatively low nanoindentation hardness of 12.3 GPa. The annealed coatings prepared at a substrate holder rotation speed of 15–30 rpm exhibited nanocomposite coatings comprising Ru and ZrO$_2$ grains within evident columnar boundaries and a high nanoindentation hardness of 16.1–17.9 GPa.

Acknowledgments: The financial support of this work from the Ministry of Science and Technology, Taiwan, under Contract No. 102-2221-E-019-007-MY3 is appreciated.

Author Contributions: Yung-I Chen designed the experiments and wrote the paper; Tso-Shen Lu performed the experiments; Zhi-Ting Zheng analyzed the XPS data.

Conflicts of Interest: The authors declare no conflict of interest. The founding sponsors had no role in the design of the study; in the collection, analyses, or interpretation of data; in the writing of the manuscript, and in the decision to publish the results.

References

1. Yashar, P.C.; Sproul, W.D. Nanometer scale multilayered hard coatings. *Vacuum* **1999**, *55*, 179–190. [CrossRef]
2. Yashar, P.C.; Barnett, S.A.; Hultman, L.; Sproul, W.D. Deposition and mechanical properties of polycrystalline Y$_2$O$_3$/ZrO$_2$ superlattices. *J. Mater. Res.* **1999**, *14*, 3614–3622. [CrossRef]
3. Vüllers, F.T.N.; Spolenak, R. From solid solutions to fully phase separated interpenetrating networks in sputter deposited "immiscible" W–Cu thin films. *Acta Mater.* **2015**, *99*, 213–227. [CrossRef]
4. Beainou, R.E.; Martin, N.; Potin, V.; Pedrosa, P.; Yazdi, M.A.P.; Billard, A. Correlation between structure and electrical resistivity of W–Cu thin films prepared by GLAD co-sputtering. *Surf. Coat. Technol.* **2017**, *313*, 1–7. [CrossRef]
5. Müller, C.M.; Sologubenko, A.S.; Gerstl, S.S.A.; Süess, M.J.; Courty, D.; Spolenak, R. Nanoscale Cu/Ta multilayer deposition by co-sputtering on a rotating substrate. Empirical model and experiment. *Surf. Coat. Technol.* **2016**, *302*, 284–292. [CrossRef]
6. Müller, C.M.; Spolenak, R. An in situ X-ray diffraction study of phase separation in Cu–Ta alloy thin films. *Thin Solid Films* **2016**, *598*, 276–288. [CrossRef]
7. Zotov, N.; Woll, K.; Mücklich, F. Phase formation of B2-RuAl during annealing of Ru/Al multilayers. *Intermetallics* **2010**, *18*, 1507–1516. [CrossRef]
8. Guitar, M.A.; Aboulfadl, H.; Pauly, C.; Leibenguth, P.; Migot, S.; Mücklich, F. Production of single-phase intermetallic films from Ru-Al multilayers. *Surf. Coat. Technol.* **2014**, *244*, 210–216. [CrossRef]
9. Manhardt, H.; Lupton, D.F.; Kock, W. Gold-Free Platinum Material Dispersion-Strengthened by Small, Finely Dispersed Particles of Base Metal Oxide. U.S. Patent 6,663,728, 16 December 2003.
10. Nakamura, T.; Sakaguchi, O.; Kusamori, H.; Matsuzawa, O.; Takahashi, M.; Yamamoto, T. Method for Preparing Ag-ZnO Electric Contact Material and Electric Contact Material Produced Thereby. U.S. Patent 6,432,157, 13 August 2002.
11. Khanna, A.S. *Introduction to High Temperature Oxidation and Corrosion*; ASM International: Materials Park, OH, USA, 2002.
12. Chen, Y.I.; Chang, L.C.; Huang, R.T.; Tsai, B.N.; Kuo, Y.C. Internal Oxidation of Mo–Ru Coatings. *Thin Solid Films* **2010**, *518*, 3819–3824. [CrossRef]
13. Chen, Y.I.; Tsai, B.N. Annealing and oxidation study of Ta–Ru hard coatings. *Surf. Coat. Technol.* **2010**, *205*, 1362–1367. [CrossRef]
14. Chen, Y.I. Laminated structure in internally oxidized Ru–Ta coatings. *Thin Solid Films* **2012**, *524*, 205–210. [CrossRef]
15. Chen, Y.I.; Chu, H.N.; Chang, L.C.; Lee, J.W. Internal oxidation and mechanical properties of Ru based alloy coatings. *J. Vac. Sci. Technol. A* **2014**, *32*, 02B101. [CrossRef]

16. Chen, Y.I.; Lu, T.S. Internal oxidation of laminated ternary Ru–Ta–Zr coatings. *Appl. Surf. Sci.* **2015**, *353*, 245–253. [CrossRef]

17. Chen, Y.I.; Chu, H.N.; Kai, W. Internal oxidation of laminated Nb–Ru coatings. *Appl. Surf. Sci.* **2016**, *389*, 477–483. [CrossRef]

18. Saha, R.; Nix, W.D. Effects of the substrate on the determination of thin film mechanical properties by nanoindentation. *Acta Mater.* **2002**, *50*, 23–38. [CrossRef]

19. Oliver, W.C.; Pharr, G.M. An improved technique for determining hardness and elastic modulus using load and displacement sensing indentation experiments. *J. Mater. Res.* **1992**, *7*, 1564–1583. [CrossRef]

20. Bennett, J.M. *Rough Surfaces*, 2nd ed.; Imperial College Press: London, UK, 1999.

21. Mahdouk, K.; Elaissaoui, K.; Charles, J.; Bouirden, L.; Gachon, J.C. Calorimetric study and optimization of the ruthenium-zirconium phase diagram. *Intermetallics* **1997**, *5*, 111–116. [CrossRef]

22. Wang, F.E. Equiatomic binary compounds of Zr with transition elements Ru, Rh, and Pd. *J. Appl. Phys.* **1967**, *38*, 822–824. [CrossRef]

23. David, N.; Benlaharche, T.; Fiorani, J.M.; Vilasi, M. Thermodynamic modeling of Ru–Zr and Hf–Ru systems. *Intermetallics* **2007**, *15*, 1632–1637. [CrossRef]

24. Arikan, N.; Bayhan, Ü. *Ab initio* calculation of structural, electronic and phonon properties of ZrRu and ZrZn in B2 phase. *Phys. B* **2011**, *406*, 3234–3237. [CrossRef]

25. Xing, W.; Chen, X.Q.; Li, D.; Li, Y.; Fu, C.L.; Meschel, S.V.; Ding, X. First-principles studies of structural stabilities and enthalpies of formation of refractory intermetallics: TM and TM3 (T = Ti, Zr, Hf; M = Ru, Rh, Pd, Os, Ir, Pt). *Intermetallics* **2012**, *28*, 16–24. [CrossRef]

26. Chen, Y.I.; Zheng, Z.T.; Kai, W.; Huang, Y.R. Oxidation behavior of Ru–Al multilayer coatings. *Appl. Surf. Sci.* **2017**, *406*, 1–7. [CrossRef]

27. Chen, Y.I.; Chang, L.C.; Lee, J.W.; Lin, C.H. Annealing and oxidation study of Mo–Ru hard coatings on tungsten carbide. *Thin Solid Films* **2009**, *518*, 194–200. [CrossRef]

28. Rochefort, D.; Dabo, P.; Guay, D.; Sherwood, P.M.A. XPS investigations of thermally prepared RuO_2 electrodes in reductive conditions. *Electrochim. Acta* **2003**, *48*, 4245–4252. [CrossRef]

29. Chen, Y.I.; Chen, S.M.; Chang, L.C.; Chu, H.N. X-ray photoelectron spectroscopy and transmission electron microscopy study of internally oxidized Nb–Ru coatings. *Thin Solid Films* **2013**, *544*, 491–495. [CrossRef]

30. Cox, P.A.; Goodenough, J.B.; Tavener, P.J.; Telles, D.; Egdell, R.G. The electronic structure of $Bi_{2-x}GdxRu_2O_7$ and RuO_2: A study by electron spectroscopy. *J. Solid State Chem.* **1986**, *62*, 360–370. [CrossRef]

31. Morant, C.; Sanz, J.M.; Galán, L.; Soriano, L.; Rueda, F. An XPS study of the interaction of oxygen with zirconium. *Surf. Sci.* **1989**, *218*, 331–345. [CrossRef]

32. Maurice, V.; Salmeron, M.; Somorjai, G.A. The epitaxial growth of zirconium oxide thin films on Pt (111) single crystal surfaces. *Surf. Sci.* **1990**, *237*, 116–126. [CrossRef]

33. Wang, Y.M.; Li, Y.S.; Wong, P.C.; Mitchell, K.A.R. XPS studies of the stability and reactivity of thin films of oxidized zirconium. *Appl. Surf. Sci.* **1993**, *72*, 237–244. [CrossRef]

34. Ramana, C.V.; Atuchin, V.V.; Kesler, V.G.; Kochubey, V.A.; Pokrovsky, L.D.; Shutthanandan, V.; Becker, U.; Ewing, R.C. Growth and surface characterization of sputter-deposited molybdenum oxide thin films. *Appl. Surf. Sci.* **2007**, *253*, 5368–5374. [CrossRef]

35. Gan, Z.; Yu, G.; Zhao, Z.; Tan, C.M.; Tay, B.K. Mechanical properties of zirconia thin films deposited by filtered cathodic vacuum arc. *J. Am. Ceram. Soc.* **2005**, *88*, 2227–2229. [CrossRef]

36. Qasmi, M.; Delobelle, P. Influence of the average roughness R_{ms} on the precision of the Young's modulus and hardness determination using nanoindentation technique with a Berkovich indenter. *Surf. Coat. Technol.* **2007**, *201*, 1191–1199. [CrossRef]

37. Walter, C.; Antretter, T.; Daniel, R.; Mitterer, C. Finite element simulation of the effect of substrate roughness on nanoindentation of thin films with spherical indenters. *Surf. Coat. Technol.* **2007**, *202*, 1103–1107. [CrossRef]

38. Kim, J.Y.; Kang, S.K.; Lee, J.J.; Jang, J.; Lee, Y.H.; Kwon, D. Influence of surface-roughness on indentation size effect. *Acta Mater.* **2007**, *55*, 3555–3562. [CrossRef]

16

Combustion Synthesis during Flame Spraying ("CAFSY") for the Production of Catalysts on Substrates

Galina Xanthopoulou *, Amalia Marinou, Konstantinos Karanasios and George Vekinis

Institute of Nanoscience and Nanotechnology, National Center for Scientific Research"Demokritos",
Aghia Paraskevi, Athens15310, Greece; a.marinou@inn.demokritos.gr (A.M.); konkaranasios@yahoo.gr (K.K.);
g.vekinis@inn.demokritos.gr (G.V.)
* Correspondence: g.xanthopoulou@inn.demokritos.gr

Academic Editor: Eric Loth

Abstract: Combustion-assisted flame spraying ("CAFSY") has been used to produce catalytically active nickel aluminide coatings on ceramic substrates. Their catalytic activity was studied in CO_2 (dry) reforming of methane, which is particularly significant for environmental protection as well as production of synthesis gas ($CO + H_2$). By varying the CAFSY processing parameters, it is possible to obtain a range of Ni–Al alloys with various ratios of catalytically active phases on the substrate. The influence of the number of coating layers and the type of substrate on the final catalyst composition and on the catalytic activity of the CAFSY coatings was studied and is presented here. The morphology and microstructure of the composite coatings were determined by scanning electron microscopy (SEM) with energy-dispersive X-ray spectroscopy (EDX) elemental analysis, X-ray diffraction (XRD), and Brunauer–Emmett–Teller (BET) specific area analysis. Catalytic tests for dry reforming of methane were carried out using crushed pellets from the coatings at temperatures of 750–900 °C, and gas chromatography showed that methane conversion approached 88% whereas that of carbon dioxide reached 100%. The H_2/CO ratio in the synthesis gas produced by the reaction varied from about 0.7 to over 1.2, depending on the catalyst and substrate type and testing temperature.

Keywords: thermal spray coating; SHS; catalysts on substrates; dry reforming of methane; syngas

---

1. Introduction

Combustion-assisted flame spraying ("CAFSY") combines conventional flame spraying and powder combustion synthesis into a single step [1,2]. Parameters including the temperature of the flame, the temperature of the substrate, the powder ratio and any pre-treatment of the metal powders, the gas speed and feed rate, and size and condition of starting powders can all be optimized for any particular coating needed. This new method offers a number of advantages over spraying of ready alloyed powders to produce protective coatings on substrates, the main ones being cost and versatility.

Ni–Al alloys, made by slow diffusion or by SHS (self-propagating high-temperature synthesis), have been found to display good catalytic properties for a number of chemical reactions. There are many reports on the use of bulk SHS catalysts' activity in different processes [3–12]. Because of the high synthesis temperatures and fast cooling rates achieved, combustion synthesis (CS) and especially SHS produce materials that are characterised by heavily distorted atomic structures, with a very large proportion of catalytically active centres. Many CS catalysts display very enhanced activity compared to the same materials produced by other means, even if they have much lower specific surface area [13]. In this work, we report on the use of the CAFSY method for the production of catalytically active Ni–Al alloys sprayed on various substrates for dry (CO_2) reforming of methane. This process is environmentally important, because the mitigation and utilization of greenhouse gases, CO_2 and CH_4, are among the most significant global challenges. Dry reforming of methane has therefore received much attention from both an environmental and an industrial perspective, as the reaction can convert these greenhouse gases into industrially valuable synthesis gas with a wide range of H_2/CO ratios.

In a recent report [2], SHS catalysts based on Co–Al–O were shown to exhibit very high catalytic activity for dry reforming of methane with minimal coke production. At a temperature of 900 °C and atmospheric pressure, CH_4 conversion was 100%, CO_2 conversion was about 83%, and the H_2/CO ratio in the product was about 1.2. Furthermore, at 800 °C, the CH_4 conversion was also 100% but the H_2/CO ratio reached close to 2. Such highly promising results, in conjunction with extremely low coke deposition and a high performance maintained over many hours [14], indicates that the SHS method offers significant benefits.

During the past decades, Ni-based catalysts for dry reforming of methane have been extensively studied [15–23]. Previously, we reported [24] that Ni-based catalysts exhibit very high catalytic activity for dry reforming of methane with minimal coke production.

The CAFSY Method for Spraying Catalytically Active Coatings

Reactive processing of intermetallic materials has been reported in the past [25]. Reaction synthesis (RS) and SHS have been applied for low-energy formation of various binary alloys using a variety of configurations [26–28]. In these studies, the exothermic chemical energy from the reaction between elemental powders is used to synthesise the alloys, thereby reducing the amount of external heating required. CAFSY is attractive for production catalysts because (1) it is a simple operation (requiring minimal operator training and can be used with hand-held sprayers); (2) it gives high spray rates and deposit efficiencies; (3) it works also with low-cost base metal powders of sizes ranging from 5 to 300 μm; (4) it is possible to change, at any time, the composition of the initial charge, and by adjusting the flame spray conditions, different properties of the coating can be achieved; and, finally; (5) different intermetallic phases can be deposited in one step.

Our previous work [1,2] showed that the Ni–Al composite coatings applied by CAFSY were principally composed of three different intermetallic phases (i.e., NiAl, Ni_2Al_3, $NiAl_3$) with small splats of Al_2O_3 and $NiAl_2O_4$ phases as well as traces of unreacted nickel and aluminium. In a recent paper [29], we confirmed the catalytic properties of those CAFSY coatings, and the objective of the present paper is to elaborate further on the catalytic properties of Ni–Al coated by CAFSY on spinel and refractory substrates for dry reforming of methane.

2. Materials and Methods

2.1. Materials

The Ni and Al powders were obtained from Sulzer Metco 56C-NS (spheroidal/precipitated powder, particle size 45–75 μm), and Aluminium Powder Company Ltd (gas fragmentation, 100/D-10, particle size 45–90 μm), respectively. Substrates used for CAFSY were made of highly macroporous $MgAl_2O_4$ spinel (with some MgO, coded Sp-NiXX-Y, where XX is the amount of nickel in the initial powder mixture and Y is the number of gun passes, i.e., coating layers) prepared by SHS from an initial mixture of Al + Mg + $Mg(NO_3)_2$ + Al_2O_3. Square plates of dimensions 5 cm × 5 cm × 0.8 cm were produced using uniaxial compaction under a pressure of 10 MPa and SHS-fired at a preheating temperature of 900 °C in air. For comparison, substrates made from a porous Si–Al–O refractory brick (BNZ Materials Inc., Littleton, CO, USA, coded RB-NiXX-Y, dimensions 5 cm × 5 cm × 0.8 cm) and from a woven glass-fibre refractory mat (BNZ Materials Inc., Littleton, CO, USA, coded RM-NiXX-Y, dimensions 5 cm × 5 cm × 0.3 cm) were also used. The porosity of all the above substrates was about 50%–70%, which enabled crushing (after coating) to small pieces, which were used for the catalytic tests.

2.2. CAFSY Process

As discussed previously [2,30], CAFSY is based on the exothermic reactions between powders during thermal spraying, as shown schematically in Figure 1. The CAFSY parameters used in this work were described previously [2] and are as follows: ratio of oxygen to acetylene = 1.56 (acetylene pressure: 32 psi, oxygen: 50 psi), mechanical ball milling of initial mixture (Ni + Al) for 10 min, thermal spray distance of 4 cm, substrate temperature of 600 °C, and powder feeding rate of 25 g/min.

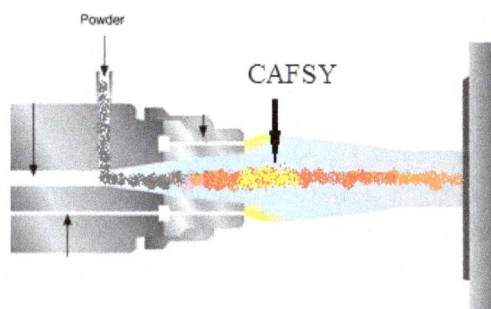

Figure 1. Scheme of combustion-assisted flame spraying (CAFSY) process.

Flame spraying was carried out by a Sulzer's Metco Thermospray Gun (5P-II) (Oerlikon Metco, Winterthur, Switzerland). During spraying, the Ni + Al powder mixtures are ignited by the combustion flame and react to produce nickel aluminide compounds, initially in the flame, which are then deposited on the surface and inside the pores of the substrate where the reactions are completed when the substrate temperature is suitable. The CAFSY process has been reported in detail previously [30], where it was shown conclusively that combustion synthesis occurs both in-flight as well as on the substrate. To help with the initiation of the combustion reactions in the flame, the Ni and Al particles are partially agglomerated after 10 min of mechanical treatment. The number of gun passes over the substrate corresponds to the number of coating layers produced on the substrate, which was also found to correlate with the concentration of the catalytic phases in the coatings on the substrate (Table 1). In this work, the following amounts of nickel were used in the Ni + Al mixtures: 42.1%, 59.3%, 65.1%, and 86.8% by weight, the remainder being aluminium (Table 1). The influence of the number of coating layers on the catalyst composition and the influence of the type of substrate on the catalytic activity of the CAFSY coatings are represented below.

Table 1. Various CAFSY tests and corresponding concentration of catalyst on various substrates.

Specimen	Ni in Powder Mixture, wt.%	Number of Layers (Gun Passes)	Concentration of Catalyst on the Substrate *, %
Sp-Ni65-2	65.1	2	4.70
Sp-Ni42-2	42.1	2	8.60
Sp-Ni59-2	59.3	2	8.50
Sp-Ni65-1	65.1	1	4.00
Sp-Ni87-2	86.8	2	13.00
Sp-Ni65-5	65.1	5	32.90
Sp-Ni65-8	65.1	8	70.60
Sp-Ni65-2	65.1	2	13.90
RB-Ni65-1	65.1	1	4.00
RM-Ni65-1	65.1	1	4.00

* Concentration of catalysts was determined by measuring the weight of the substrates before and after spraying, which gives the weight of the coating applied.

2.3. Catalyst Characterisation

The morphology and microstructure of the composite coatings obtained were examined by scanning electron spectroscopy (SEM) with energy-dispersive X-ray spectroscopy (EDX) elemental analysis and X-ray diffraction (XRD). The development of the various intermetallic phases was monitored by calculating the peak ratio of intensities of particular XRD peaks. The peaks we used were as follows. For aluminium metal, *hkl*: 111; for NiAl, *hkl*: 220; for Ni_3Al, *hkl*: 311; for $NiAl_3$, *hkl*: 112; Ni_2Al_3, *hkl*: 110. The peaks used were selected so that they were not clashing with any other peaks in the XRD spectra.

2.4. Conditions for Catalytic Dry Reforming of Methane

The catalytic activity of the synthesised catalysts was determined for dry (CO_2) reforming of methane using coarsely crushed coated substrates with an average pellet size of 3 mm, guided by usual characteristics for industrial catalysts. The tests were carried out in a fixed-bed free-flow quartz reactor without any pre-reduction. All of the tests were conducted at atmospheric pressure in a flow of a gas, which was a mixture of CO_2–CH_4–N_2 (ratio 1:1:1), at a flow rate of 30 mL/min. The total flow rate of reactants was set at 860 h^{-1} and the catalytic reaction was carried out at 750, 800, 850, and 900 °C. The products were analysed in a gas chromatograph using a 10 m steel column filled with polymer Highset-d. The outlet gases (O_2, N_2, CO, CO_2, CH_4) were separated at 40 °C by using a thermal conductivity detector and He as carrier gas.

3. Results andDiscussion

3.1. Catalyst Characterisation

The phase composition of the coated catalysts produced by CAFSY on $MgAl_2O_4$ substrates was determined by XRD analysis, and the results are shown in Figure 2 and Table 2.

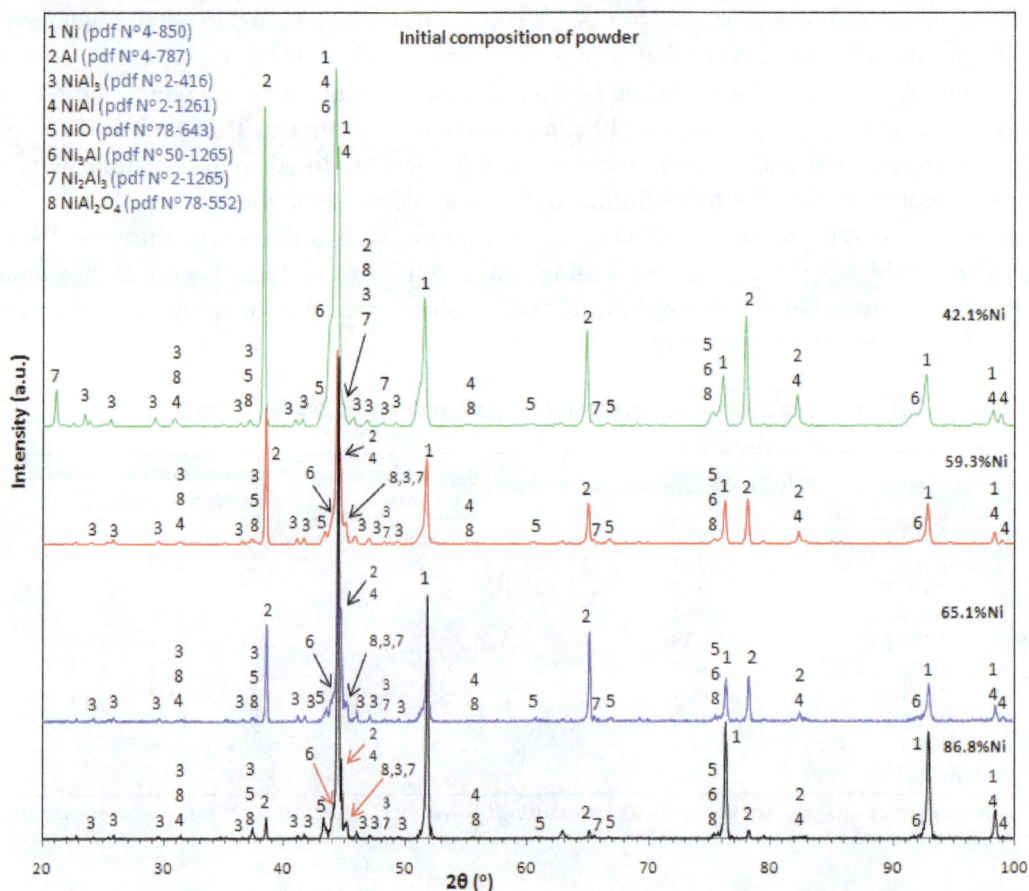

Figure 2. X-ray diffraction (XRD) of CAFSY coatings for different Ni + Al starting mixtures after two gun passes.

Table 2. Initial powder mixture composition and the main catalytically active alloys in the coatings produced by CAFSY on various substrates using only 2 gun passes.

Specimen	Initial Mixture Composition	Catalyst Composition
Sp-Ni42-2	42.1%Ni + 57.9%Al	Ni_2Al_3, NiAl, $NiAl_3$
Sp-Ni59-2	59.3%Ni + 40.7%Al	Ni_2Al_3, NiAl
Sp-Ni65-2	65.1%Ni + 34.9%Al	NiAl, Ni_2Al_3
Sp-Ni87-2	86.8%Ni + 13.2%Al	Ni_2Al_3, NiAl, Ni_3Al
RB-Ni65-2	65.1%Ni + 34.9%Al	NiAl, Ni_2Al_3, $NiAl_3$, Ni_3Al
RM-Ni65-2	65.1%Ni + 34.9%Al	NiAl, Ni_2Al_3, $NiAl_3$, Ni_3Al

In all samples, some unreacted Ni and Al (except Sp-Ni87-2) are present. This is due to the very limited contact time between particles of Al and Ni during flame spraying (of the order of a few milliseconds). Traces of NiO in all catalysts can be explained by partial oxidation of Ni. There are also traces of $NiAl_2O_4$ in all samples. Main products of nickel aluminides correspond to all stoichiometric compositions of Ni: 42.1 wt.% ($NiAl_3$), 59.3 wt.% (Ni_2Al_3), 65.1 wt.% (NiAl), and 86.8 wt.% (Ni_3Al). Also, in the case of 42.1% Ni in the starting powder mixture, an increased presence of Ni_2Al_3 in the coating was observed. At Sp-Ni87-2, the presence of Ni_2Al_3, NiAl, and Ni_3Al in products denotes the enrichment of the NiAl phase with Ni up to Ni_3Al. In the case of Sp-Ni65-2, only the NiAl phase is evident. This result supports the premise that $NiAl_3$, Ni_2Al_3, and Ni_3Al formed from enrichment of the NiAl phase. Figure 2 shows that the maximum content of NiAl and Ni_2Al_3 phases corresponds to their stoichiometric ratio, respectively (the development of the various phases calculated the peak ratio of intensities of the particular XRD peaks: NiAl, *hkl*: 220, Ni_2Al_3, *hkl*: 110).

Figure 3 shows the influence of Ni concentration in the powder mixture on the parameters of the crystal lattice of NiAl and Ni_2Al_3. The increase of Ni in the powder spraying mixture enriches constantly the NiAl and Ni_2Al_3 phases (up to 65.1% Ni for NiAl and 59.3% Ni for Ni_2Al_3 phase). Increasing the amount of Ni (Figure 3b) increases also the size of the crystal lattice spacing from 3.46 to 3.49 Å and from 3.44 to 3.53 Å for NiAl and Ni_2Al_3 phases, respectively. This probably occurs because the Ni^{2+} ionic radius is 0.69 Å and that of Al^{3+} is 0.51 Å.

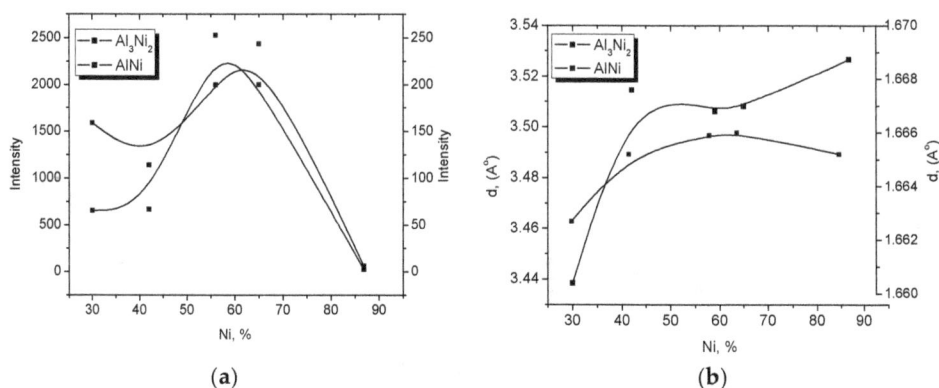

Figure 3. Influence of Ni concentration in the powder mixture (coating after two gun passes) on the yield of NiAl and Ni_2Al_3 (a) and on their crystal lattice parameters (b).

The formation of catalytically active coatings (Figure 2 and Table 2) can be explained by combustion synthesis reactions during CAFSY by the following reactions [31]:

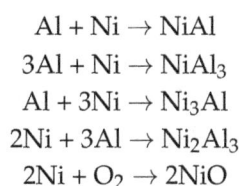

$$Al + Ni \rightarrow NiAl$$
$$3Al + Ni \rightarrow NiAl_3$$
$$Al + 3Ni \rightarrow Ni_3Al$$
$$2Ni + 3Al \rightarrow Ni_2Al_3$$
$$2Ni + O_2 \rightarrow 2NiO$$

$$4Al + 3O_2 \rightarrow 2Al_2O_3$$
$$Al_2O_3 + NiO \rightarrow NiAl_2O_4$$
$$NiAl + nNi \rightarrow Ni_xAl_y$$

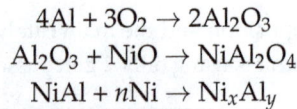

Nickel aluminides can be formed by the reaction of base metal powders (Ni + Al) followed by enrichment of the NiAl phase either during the flight or at the substrate: $NiAl + nNi \rightarrow Ni_xAl$. As the aluminium melts in the flame (melting point 660 °C) it reacts with the heated nickel (melting point 1440 °C) and NiAl (melting point 1911 °C) forms exothermically. The heated aluminium and nickel further reacts with NiAl, forming $NiAl_3$ and Ni_3Al, respectively. The simultaneous presence of oxides NiO and Al_2O_3, which form by partial oxidation of the base metals in the combustion flame, are also able to react, forming $NiAl_2O_4$.

The specific surface area of such catalysts is low—about 0.1–0.35 m^2/g (measured by the BET method (Gold APP Instrument Corporation, Beijing, China) on the coatings)—since the spinel substrates used were also made by SHS, which also have low specific surface area due to the large pores present in such substrates, caused by the melting of the coatings during spraying as evidenced by SEM examinations. A coating made by CAFSY from a powder mixture of 59.3%Ni + 40.7%Al on $MgAl_2O_4$ spinel studied by SEM/EDS is shown in Figure 4. The EDX spectra are in agreement with the XRD results shown in Figure 2.

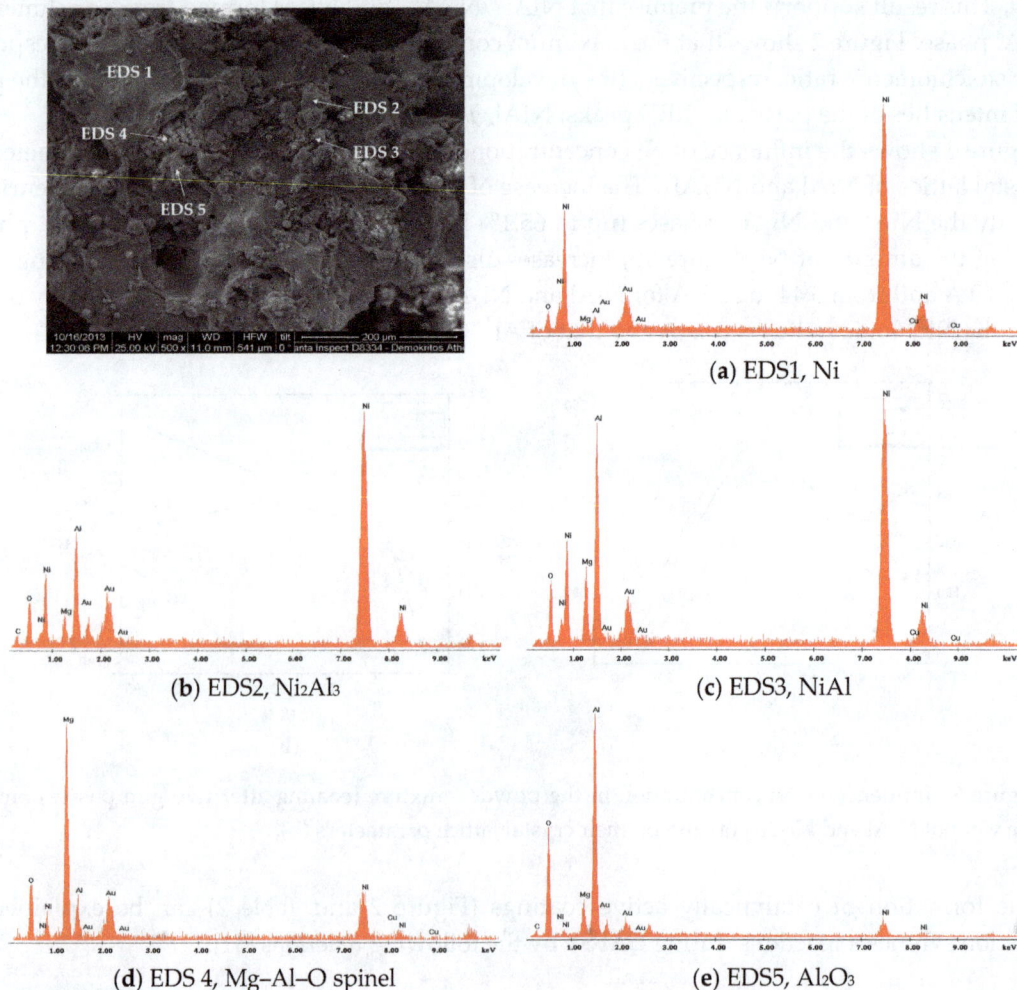

(a) EDS1, Ni

(b) EDS2, Ni₂Al₃

(c) EDS3, NiAl

(d) EDS 4, Mg–Al–O spinel

(e) EDS5, Al₂O₃

Figure 4. Scanning electron microscopy (SEM)/energy-dispersive X-ray spectroscopy (EDX) analysis of catalytically active coatings Sp-Ni65-2 (initial powder mixture of 65.1% Ni + 34.9% Al with two gun passes on $MgAl_2O_4$ substrate. The catalytic coating contains about 13.9% nickel aluminide). (a) EDS 1; (b) EDS 2; (c) EDS 3; (d) EDS 4; (e) EDS 5.

Increasing the number of gun passes (i.e., layers formed) during thermal spraying (compare SEM photo in Figure 4 (two passes) with SEM photo in Figure 5a (eight passes) increases the coating thickness by approximately 40 μm/pass and also heats up the coating, giving additional time to complete the reactions between NiAl alloys and Al, which gives rise to the $NiAl_3$ phase. A large number of gun passes also results in increased oxidation of nickel and aluminium, producing NiO and Al_2O_3, which react together to form spinel via the reaction $NiO+Al_2O_3 \rightarrow NiAl_2O_4$. In the case of coatings on a Si–Al–O refractory brick (Figure 5b), all four main nickel aluminide phases are present. On the other hand, coatings on $MgAl_2O_4$ spinel substrate (Figure 4) show only NiAl and Ni_2Al_3 phases.

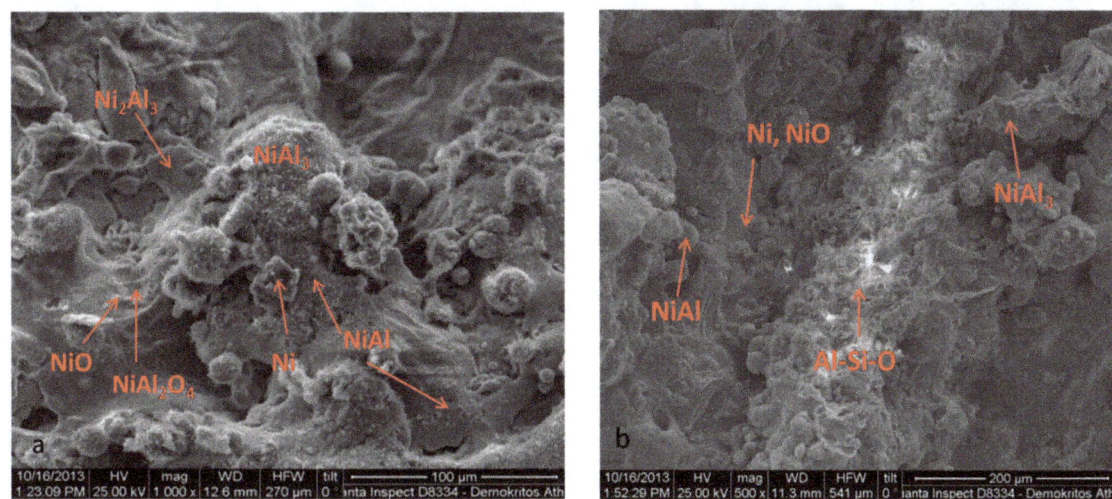

Figure 5. SEM/EDX analysis of catalytic coatings prepared by CAFSY method from initial powder mixture of 65.1% Ni + 34.9% Al (giving 4% nickel aluminides in the coating) using one gun pass on (**a**) spinel (Sp-Ni65-1) and (**b**) refractory brick (RB-Ni65-1) substrates.

The surface of the coating RB-Ni-65-1 (Figure 6a) shows the presence of the intermetallic phases NiAl and Ni_3Al and some traces of NiO. In addition, there are unreacted phases Ni and Al. Cross-section of the same coating (Figure 6b) shows the formation of intermetallic phases NiAl, Ni_3Al, and $NiAl_3$ and also some traces of NiO. Combining SEM/EDX analysis with the XRD results of the coatings indicates that the light-grey areas in the SEM micrographs are probably Ni-rich, the dark-grey areas are probably Al-rich, while the intermediate-grey areas are probably the intermetallic phases $NiAl_3$, NiAl, and Ni_3Al. The grey-white areas are probably oxides Al_2O_3 and NiO. The $NiAl_2O_4$ spinel phase indicated by XRD could not be found on any surface examined by SEM/EDX.

During CAFSY, the yield of nickel aluminides increases with increasing Ni concentration in the powder mixture. This is because increasing the amount of nickel in the powder mixture increases the temperature of the flame due to the exothermic combustion reaction between nickel and aluminium. This increases the reactivity of the powders in the flame, which increases the net yield of alloys.

The total porosity of the coatings also increases with increasing amount of nickel, because particles melt completely in the flame and tend to splash on the target surface. It is well known that fully molten particles often splash and scatter backwards instead of adhering to the surface, creating a rough surface with increased porosity (e.g., [32]). This also explains the fact that at higher concentration of nickel in the mixture, all of the aluminium has reacted with NiAl. Ni_3Al and $NiAl_3$ have low adiabatic temperatures and need high preheating temperatures of substrate [2] to react well. Melted aluminium reacts with NiAl on the substrate surface, with the subsequent appearance of $NiAl_3$ (Figure 6b). Increasing the substrate temperature increases $NiAl_3$ and Ni_3Al phases and decreases porosity [2].

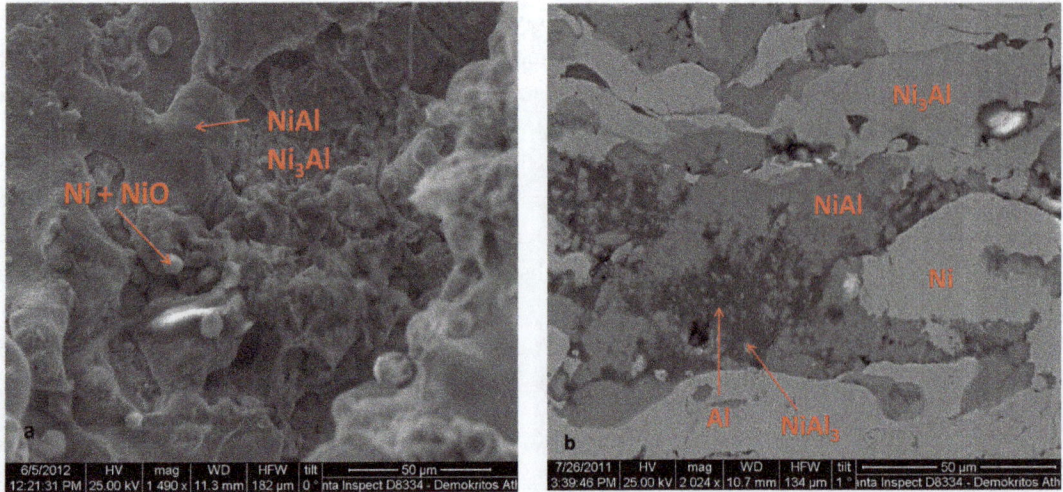

Figure 6. SEM/EDX analysis of the specimen RB-Ni65-1 of (**a**) surface of the catalytic coating and (**b**) a cross-section of the coating.

3.2. Catalytic Activity for Dry Reforming of Methane

Influence of Substrate Composition

As mentioned, three different substrates were used for thermal spraying of the nickel and aluminium mixtures by CAFSY. All the specimens listed in Table 1 were examined for catalytic activity at 900 °C for dry reforming of methane. The results of the most promising combinations are shown in Figure 7 compared with the catalytic activity of the three substrates alone.

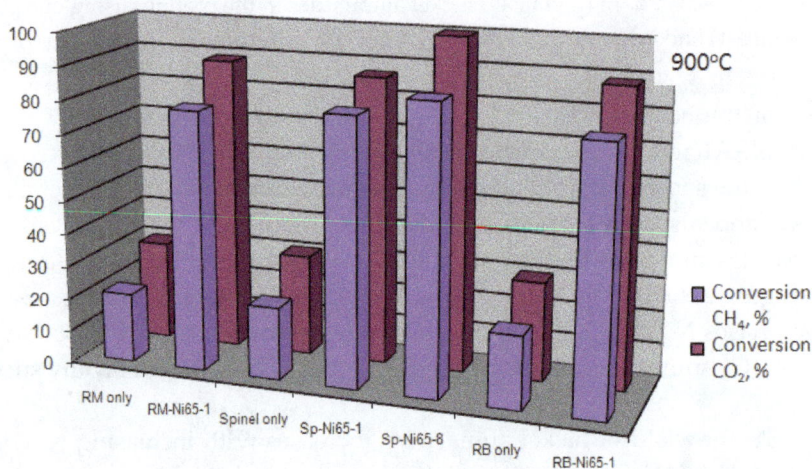

Figure 7. The catalytic activity of the most promising coatings produced by CAFSY for the process of dry reforming with regard to methane and carbon dioxide conversion.

Figure 7 shows that the substrate materials used also display catalytic activity for dry reforming of methane (comparison with data received for substrates without catalytic coatings). The conversion of methane measured on a catalyst containing 4 wt.% nickel aluminide coating on a standard spinel substrate (Sp-Ni65-1) is higher than that for the same coating on the refractory mats (RM-Ni65-1) and the refractory bricks (RB-Ni-65-1) substrates. On the other hand, carbon dioxide conversion was higher on RM-Ni-65-1 and RB-Ni-65-1 coatings than on the RB-Ni-65-1 coatings, where an increase of total content of nickel aluminides (from 4% (Sp-Ni65-1) to 70% (Sp-Ni65-8)) on the substrate leads to an increase of carbon dioxide conversion by 10%.

More detailed information about the influence of the number of layers (gun passes) on the conversion of methane and carbon dioxide and on the ratio H_2/CO in the outlet gases is presented in Figure 8.

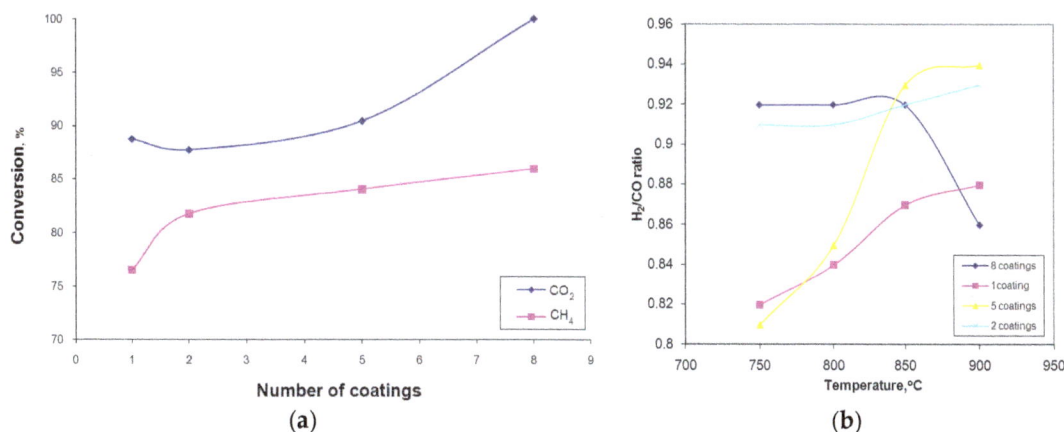

Figure 8. Influence of (**a**) number of coatings (gun passes) during CAFSY on the conversion of CH_4 and CO_2; and (**b**) temperature of catalysis on the H_2/CO ratio in the outlet gases.

Figure 8a shows that increasing the number of catalytic layers from 2 to 8 does not change methane conversion significantly (only by about 3%) but does increase CO_2 conversion by up to 13%. This means that a catalytic coating with just 4% nickel aluminide (specimen Sp-Ni65-1) content (this concentration is very often used in industrial catalysis) is very active for carbon dioxide dry reforming of methane.

Ratio of H_2 to CO in the outlet gas produced by the catalytic reaction depends on the number of coatings as well as the reaction temperature. At high temperatures (Figure 8b, 850 °C) all catalysts produce gases with a ratio of 0.92–0.94, which is suitable for industrial synthesis of alcohols, aldehydes, olefins, acetic acid, and so on. At lower temperatures, 750–800 °C, the number of catalytic layers plays a more significant role regarding the efficiency of the catalysts. It is worth noting that at 900 °C, the activity of the 70% coated catalyst (Sp-Ni65-8) decreases because of coke formation. This could be connected to the increase of contact time on the surface, which increases the concentration of active centres and desorption becomes the limiting factor of catalysis.

The amount of nickel in the initial powder mixture also plays a significant role in the catalyst activity (Figure 9). Significant changes in catalytic activity occur at 42.1%–59.3% Ni (specimens Sp-Ni42-2 and Sp-Ni59-2), where the main phases in the catalytic coating are: NiAl, Ni_2Al_3, Ni_3Al.

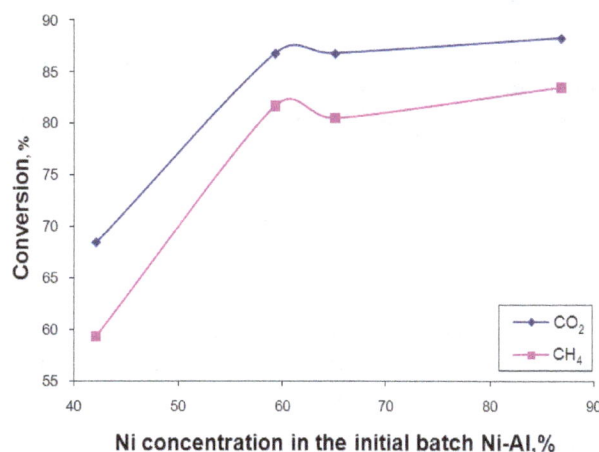

Figure 9. Influence of amount of nickel in the initial mixture on the catalytic activity.

Reaction temperature, type of substrate, and the amount of nickel aluminides in the coatings affect the H_2/CO ratio (Figure 10). Such influences could be connected with structural changes during coating.

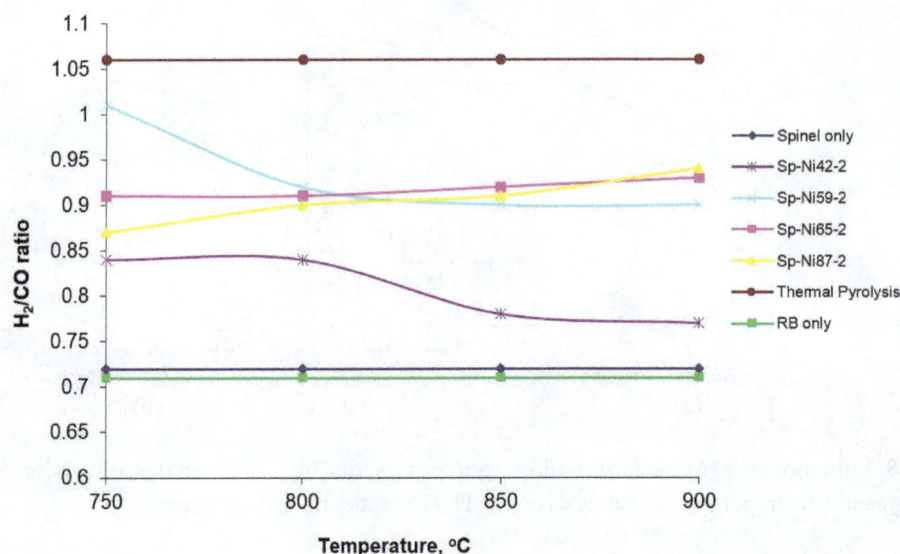

Figure 10. Influence of reaction temperature, type of substrate, and Ni concentration in the initial Ni + Al mixture on the H_2/CO ratio.

4. Conclusions

The work presented here demonstrated that the CAFSY coating method is able to rapidly produce low-cost nickel aluminide coatings on substrates with good catalytic properties. Studies of the catalytic activity of such CAFSY nickel aluminide coatings for dry reforming of methane shows that very high conversion rates are possible. By varying the processing parameters, it is possible to obtain a range of materials with varying ratios of the three main catalytic phases (NiAl, Ni_3Al, and Ni_2Al_3) on the substrates, with a wide range of catalytic properties. The coating substrates (which serve as carriers for the catalysts) influence the catalytic activity of the nickel aluminide coatings. Such CAFSY coatings can therefore be recommended as a low-cost and convenient method of catalyst production for industrial applications, especially in areas where conversion of flowing gases is required, such as flue outlets.

Acknowledgments: The Authors thank Pyrogenesis S.A. for offering thermal coating equipment for the CAFSY experiments.

Author Contributions: Galina Xanthopoulou conceived and designed the experiments, analysed and discussed the data and co-authored the paper. Amalia Marinou performed the experiments for the thermal coatings, conducted SEM/EDS investigations and analysed and discussed many of the results. Konstantinos Karanasios performed some experiments on catalytic activity of the coatings and finally George Vekinis analysed SEM/EDX and XRD results and co-authored the paper.

Conflicts of Interest: The authors declare no conflict of interest.

References

1. Marinou, A.; Xanthopoulou, G.; Vekinis, G.; Lekatou, A.; Vardavoulias, M. Synthesis and heat treatment of sprayed high-temperature NiAl-Ni$_3$Al coatings by in-flight combustion synthesis (CAFSY). *Int. J. SHS* **2015**, *24*, 192–201. [CrossRef]

2. Xanthopoulou, G.; Marinou, A.; Vekinis, G.; Lekatou, A.; Vardavoulias, M. NiAl and NiO-Al composite coatings by combustion-assisted flame spraying. *Coatings* **2014**, *4*, 231–252. [CrossRef]

3.   Xanthopoulou, G. Self-propagating high-temperature synthesis as method of catalysts production. *Sci. Cent. Asia* **2010**, *4*, 35–56.

4.   Xanthopoulou, G.; Vekinis, G. Catalytic oxidation of CO over a Cu-Cr-oxide catalyst made by self-propagating high-temperature synthesis. *Appl. Catal. B Environ.* **1998**, *19*, 37–44. [CrossRef]

5.   Xanthopoulou, G. Oxide catalysts for pyrolysis of diesel fuel made by self-propagating high-temperature synthesis. Part I: Cobalt-modified Mg-Al spinel catalysts. *Appl. Catal. A Gen.* **1999**, *182*, 285–295. [CrossRef]

6.   Xanthopoulou, G. Oxide catalysts for pyrolysis of diesel fuel made by self-propagating high-temperature synthesis(SHS): Part II: Fe-Cr oxide catalysts based on chromite concentrates. *Appl. Catal. A Gen.* **1999**, *187*, 79–88. [CrossRef]

7.   Xanthopoulou, G. Oxidative dehydrodimerization of methane using lead and samarium based catalysts made by self-propagating high-temperature synthesis. *Appl. Catal. A Gen.* **1999**, *185*, L185–L192. [CrossRef]

8.   Xanthopoulou, G.; Vekinis, G. Deep oxidation of methane using catalysts and carriers produced by self-propagating high temperature synthesis. *Appl. Catal. A Gen.* **2000**, *199*, 227–238. [CrossRef]

9.   Xanthopoulou, G.; Vekinis, G. An overview of some environmental applications of the self-propagating high-temperature synthesis. *Adv. Environ. Res.* **2001**, *5*, 117–128. [CrossRef]

10. Xanthopoulou, G.; Vekinis, G. Catalytic pyrolysis of naphtha on the SHS catalysts. *Eurasian Chem. Technol. J.* **2010**, *12*, 17–21. [CrossRef]

11. Dinka, P.; Mukasyan, A.S. Solution combustion catalysts for steam reforming of JP-8 surrogate. *J. Power Sources* **2007**, *167*, 472–481. [CrossRef]

12. Hirano, T.; Purwanto, H.; Akiyama, T.W.T. Self-propagating high-temperature synthesis with post-heat treatment of $La_{1-x}Sr_xFeO_3$ ($x = 0–1$) perovskite as catalyst for soot combustion. *J. Alloy. Compd.* **2009**, *470*, 245–249. [CrossRef]

13. Xanthopoulou, G. Catalytic properties of the SHS products—Review. *Adv. Sci. Technol.* **2010**, *63*, 287–296. [CrossRef]

14. Karanasios, K.; Xanthopoulou, G.; Vekinis, G.; Zoumpoulakis, L. Co-Al oxide SHS catalysts for dry reforming of methane. *Int. J. Self-Propag. High-Temp. Synth.* **2014**, *23*, 222–231. [CrossRef]

15. Kim, W.Y.; Lee, Y.H.; Park, H.; Choi, Y.H.; Lee, M.H.; Lee, J.S. Coke tolerance of Ni/Al_2O_3 nanosheet catalyst for dry reforming of methane. *Catal. Sci. Technol.* **2016**, *6*, 2060–2064. [CrossRef]

16. Xu, S.; Wang, X. Highly active and coking resistant $Ni/CeO_2–ZrO_2$ catalyst for partial oxidation of methane. *J. Fuel* **2005**, *84*, 563–567. [CrossRef]

17. Pengpanich, S.; Meeyoo, V.; Riksomboon, T. Methane partial oxidation over $Ni/CeO_2–ZrO_2$ mixed oxide solid solution catalysts. *J. Catal. Today* **2004**, *93–95*, 95–105. [CrossRef]

18. Larimi, S.; Alavi, S.M. Partial oxidation of methane over $Ni/CeZrO_2$ mixed oxide solid solution catalysts. *Int. J. Chem. Eng. Appl.* **2012**, *3*, 6–9.

19. Soloviev, S.O.; Kapran, A.Y.; Orlyk, S.N.; Gubareni, E.V. Carbon dioxide reforming of methane on monolithic Ni/Al_2O_3-based catalysts. *J. Natur. Gas Chem.* **2011**, *20*, 184–190. [CrossRef]

20. Fidalgo, B.; Arenillas, A.; Menindez, J.A. Synergetic effect of a mixture of activated carbon + Ni/Al_2O_3 used as catalysts for the CO_2 reforming of CH_4. *Appl. Catal. Ser. A* **2010**, *390*, 78–83. [CrossRef]

21. Moniri, A.; Alavi, S.M.; Rezaei, M. Syngas production by combined carbon dioxide reforming and partial oxidation of methane over Ni/α-Al_2O_3 catalysts. *J. Nat. Gas Chem.* **2010**, *19*, 638–641. [CrossRef]

22. ALuna, E.C.; Iriarte, M.E. Carbon dioxide reforming of methane over a metal modified $Ni–Al_2O_3$ catalyst. *Appl. Catal. Ser. A* **2008**, *343*, 10–15.

23. Meshkani, F.; Rezaei, M. Nanocrystalline MgO supported nickel-based bimetallic catalysts for carbon dioxide reforming of methane. *Int. J. Hydrog. Energy* **2010**, *35*, 10295–10301. [CrossRef]

24. Xanthopoulou, G.; Varitis, S.; Karanasios, K.; Vekinis, G. SHS-Produced Ni-Co-Al-Mg-O catalysts for dry reforming of methane. *SHS J.* **2014**, *23*, 92–100. [CrossRef]

25. Sikka, S.K.; Nair, G.J.; Roy, F.; Kakodkar, A.; Chidambaram, R. The recent Indian nuclear tests: A seismic overview. *Curr. Sci.* **2000**, *79*, 1359–1366.

26. Itin, V.I.; Naiborodenko, Y.S. *High Temperature Synthesis of Intermetallic Compounds*; Tomsk University Publisher: Tomsk, Russia, 1989.

27. Morsi, K. Review: Reaction synthesis processing of Ni-Al intermetallic materials. *Mater. Sci. Eng. A* **2001**, *299*, 1–15. [CrossRef]

28. Pidria, M.; Merlone, E.; Rostagno, M.; Tabone, L.; Bechis, F.; Vallauri, D.; Deorsola, F.A.; Amato, I.; Rodriguez, M.A. SHS production, processing and evaluation of advanced materials for wear-resistant cutting tools. *Mater. Sci. Forum* **2003**, *426*, 4373–4378. [CrossRef]

29. Xanthopoulou, G.; Marinou, A.; Karanasios, K.; Vekinis, G. Catalytic Activity of NiAl Composite Coatings Produced by In-Flight SHS during Thermal Spraying. In Proceedings of the SHS 2013 International Symposium, South Padre Island, TX, USA, 21–24 October 2013; pp. 277–279.

30. Marinou, A. Synthesis of High Temperature Coatings by the New Method CAFSY. Ph.D. Thesis, University of Ioannina, Ioannina, Greece, 2015.

31. Naiborodenko, Y.S.; Itin, V.I.; Savitskii, K.V. Exothermic effects during sintering of a mixture of nickel and aluminum powders. *J. Sov. Phys.* **1968**, *11*, 89–93. [CrossRef]

32. *Handbook of Thermal Spray Technology*; ASM Thermal Spray Technology; ASM International: Materials Park, OH, USA, 2004.

17

A New Finite Element Formulation for Nonlinear Vibration Analysis of the Hard-Coating Cylindrical Shell

Yue Zhang [1], Wei Sun [1,*] and Jian Yang [2]

[1]   School of Mechanical Engineering and Automation, Northeastern University, Shenyang 110819, China; zyuecn@163.com
[2]   School of Mechanical Engineering and Automation, University of Science and Technology Liaoning, Anshan 114051, China; 15904920669@163.com
*   Correspondence: weisun@mail.neu.edu.cn

Academic Editor: Alessandro Lavacchi

Abstract: In this paper, a four-node composite cylindrical shell finite element model based on Love's first approximation theory is proposed to solve the nonlinear vibration of the hard-coating cylindrical shell efficiently. The developed model may have great significance for vibration reduction of the cylindrical shell structures of the aero engine or aircraft. The influence of the strain dependence of the coating material on the complex stiffness matrix is considered in this model. Nonlinear iterative solution formulas with a unified iterative method are theoretically derived for solving the resonant frequency and response of the composite cylindrical shell. Then, a cylindrical shell coated with a thin layer of NiCoCrAlY + yttria-stabilized zirconia (YSZ) is chosen to demonstrate the proposed formulation, and the rationality is validated by comparing with the finite element iteration method (FEIM). Results show that the developed finite element method is more efficient, and the hard-coating cylindrical shell has the characteristics of soft nonlinearity due to the strain dependence of the coating material.

Keywords: finite element formulation; hard-coating; cylindrical shell; nonlinear vibration analysis

---

1. Introduction

The cylindrical shell has been widely used in the aviation and aerospace fields, especially in the aero engine and aircraft, which have an urgent demand to restrain their excessive vibration [1]. It is well known that the hard-coating treatment has found application in many engineering applications, such as the blades of gas turbines, new turbofans, cylindrical shell structures of the aero engine and other structures. The hard-coating plays an important role in improving the performance of the thermal barrier, anti-friction and anti-erosion of composite structures [2–5]. Recently, both Ivancic and Palazotto [6] and Patsias et al. [7] found that the hard-coating with special microstructural features can remarkably enhance the additional damping for vibration reduction of structures. Results also revealed that the storage modulus and the loss factor of the coating material change regularly with the strain amplitude, called the strain dependence, which makes the nonlinear vibration analysis of the hard-coating structures become more challenging.

To clearly explain the strain dependence of hard-coating, researchers have conducted many experiments. Blackwell et al. [8] experimentally studied the damping effects of hard-coating coated on the titanium plate and found its significant nonlinear damping characterization. Several research works, e.g., [9,10], indicated that both the elastic and damping properties of hard-coatings exhibit obvious nonlinearity, as well. Furthermore, Torvik [11] successfully extracted the mechanical parameters of hard-coating, such as the storage modulus and the loss factor, which are dependent on the equivalent strain amplitude and mainly responsible for the nonlinearity of the hard-coating structures. The above indicates that the hard-coating has nonlinear characteristics due to the strain dependence, which makes the establishment of an analysis model of the hard-coating structures become very difficult and time consuming.

To better apply the hard-coating for vibration reduction and solve the vibration problems of the hard-coating structures, several analysis models have been developed. Sun and Liu [12] developed an analytical model for the resonant frequencies and response analysis of the hard-coating beam with considering the strain dependence of the coating material. Chen et al. [13] proposed an analytical approach to calculate the free vibration characteristics and damping effect of the hard-coating on blades based on the constitutive model of the complex modulus and Rayleigh–Ritz method. Sun et al. [14] built an analytical model to solve the free vibration of the hard-coating cantilever cylindrical shell by Love's first approximation theory and the Rayleigh–Ritz method, but did not consider the strain dependence of the hard-coating. In addition to the analytical method, the finite element method with practicality and effectiveness has been used to solve the nonlinear vibration of the hard-coating structures. Filippi and Torvik [15] proposed a finite element method to solve the nonlinear vibration response of the hard-coating blade. However, it is difficult to exactly determine the nonlinear characteristics by using only the linear analytic expressions. Then, Li et al. [16] developed a finite element iterative method (FEIM), which successfully solved the resonant frequencies and responses of the hard-coating plate by satisfactorily characterizing the nonlinear storage modulus and loss modulus of hard-coating with the polynomial method. The practicability and reliability of the FEIM have been verified through comparing with the experimental results. The above research works provide an important support for the nonlinear vibration analysis of the hard-coating structures, such as beams, plates, blades and shells. However, it is quite inadequate for the nonlinear vibration analysis of the hard-coating cylindrical shell implemented by the finite element method with considering the strain dependence of the coating material. Moreover, the nonlinear vibration analysis of the hard-coating cylindrical shell may have great significance for vibration reduction of the cylindrical shell structures of the aero engine or aircraft.

Generally, the available FEIM needs to apply the commercial FEM software to obtain the element matrices of stiffness, mass and damping, and accurate results can only be obtained when with very small meshes, which makes the FEIM inefficient. Moreover, two different iteration methods have been used to solve the resonant frequencies and responses, respectively, leading to more inefficiency. The problems become even serious when solving the nonlinear vibration characteristics of the hard-coating cylindrical shell, which have more degrees of freedom. Unfortunately, no applicable elements exist to solve the problems more efficiently. The vast majority of elements are only used to deal with geometric nonlinearity [17–20]. Although Ferreira et al. [21] and Masti and Sainsbury [22] presented the composite cylindrical shell elements, which can describe the nonlinear behaviors of the hyperelastic coating and the viscoelastic coating, respectively, the elements have no ability to deal with the strain dependence of hard-coating, especially the strain dependence of the loss factor, which determines the value of the global material damping matrix.

Thus, this paper presents a new finite element formulation and a unified iterative method to solve the nonlinear vibration of the hard-coating cylindrical shell with considering the strain dependence of the coating material more efficiently. In Section 2, the geometry and element of the hard-coating cylindrical shell and Love's first approximation theory are briefly introduced, and then, the finite element formulation is deduced in detail. Meanwhile, the strain-dependent storage modulus and loss factor of hard-coating are characterized by the polynomial method in Section 3, and the solution formulas of the nonlinear vibration of the hard-coating cylindrical shell in Section 4 are theoretically derived. Finally, in Section 5, a cylindrical shell coated with a thin layer of NiCoCrAlY + yttria-stabilized zirconia (YSZ), which is a common hard-coating material [3], is set as the study subject; both the accuracy and convergence of the developed finite element method are compared with the FEIM, which indicates the rationality of the proposed method. Moreover, the nonlinear vibration analysis of the cylindrical shell coated with a thin layer of NiCoCrAlY + YSZ is implemented.

2. Derivation of the Finite Element Formulation for the Hard-Coating Cylindrical Shell

2.1. Geometry and Element of the Hard-Coating Cylindrical Shell

The hard-coating cylindrical shell with thickness h, axial length L and mean radius R is shown in Figure 1a. An orthogonal curvilinear coordinate system is established at the middle surface of the hard-coating cylindrical shell along the x (axial), θ (circumferential) and z (radial) directions, respectively. The finite element of the hard-coating cylindrical shell is shown in Figure 1b, which has four nodes and nine degrees of freedom at each node.

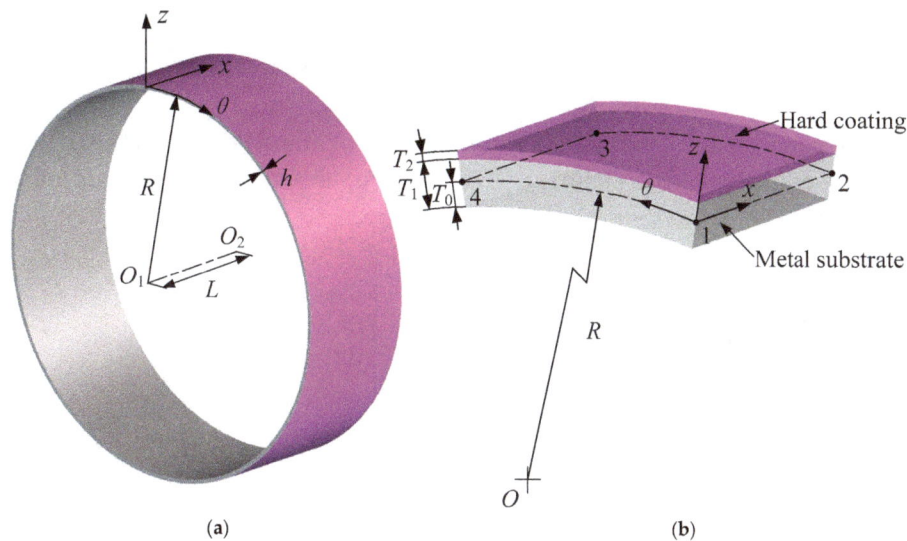

(a)	(b)

Figure 1. Geometry and element of the hard-coating cylindrical shell. (**a**) Shell geometry; (**b**) Shell element.

2.2. Love's First Approximation Theory

Based on Love's first approximation theory [23], the displacement fields u, v and w of the hard-coating cylindrical shell can be represented by the following relationships:

$$
\begin{aligned}
u(x,\theta,z) &= u^0(x,\theta) + z\psi_x(x,\theta) \\
v(x,\theta,z) &= v^0(x,\theta) + z\psi_\theta(x,\theta) \\
w(x,\theta,z) &= w^0(x,\theta)
\end{aligned}
\tag{1}
$$

where u^0, v^0 and w^0 are the orthogonal components of displacement of the mid-surface in the x, θ and z directions, respectively, and ψ_x and ψ_θ are the rotations of the mid-surface about the θ and x axes, respectively.

The magnitude of an arbitrary infinitesimal changing in the mid-surface of the hard-coating cylindrical shell ds is determined by:

$$
(\mathrm{d}s)^2 = g_x\mathrm{d}x^2 + g_\theta\mathrm{d}\theta^2 + g_z\mathrm{d}z^2
\tag{2}
$$

where g_x, g_θ, g_z, are Lamb constants [24]. For thin cylindrical shell ($h/R \ll 1$):

$$
g_x = 1, g_\theta = R, g_z = 1
\tag{3}
$$

Thus, ε_x^0, ε_θ^0, $\varepsilon_{x\theta}^0$ and κ_x, κ_θ, $\kappa_{x\theta}$, the strains and bending strains of the middle surface, can be given as follows [25]:

$$\varepsilon_x^0 = \frac{\partial u^0}{\partial x}, \varepsilon_\theta^0 = \frac{1}{R}\frac{\partial v^0}{\partial \theta} + \frac{w^0}{R}, \varepsilon_{x\theta}^0 = \frac{\partial v^0}{\partial x} + \frac{1}{R}\frac{\partial u^0}{\partial \theta}$$
$$\kappa_x = \frac{\partial^2 w^0}{\partial x^2}, \kappa_\theta = \frac{1}{R^2}\left(\frac{\partial^2 w^0}{\partial \theta^2} - \frac{\partial v^0}{\partial \theta}\right), \kappa_{x\theta} = \frac{2}{R}\left(\frac{\partial^2 w^0}{\partial x\partial \theta} - \frac{\partial v^0}{\partial x}\right)$$

(4)

The stress-strain equations of the hard-coating cylindrical shell can be expressed as:

$$\sigma = \left\{ \begin{array}{c} \sigma_x \\ \sigma_\theta \\ \sigma_{x\theta} \end{array} \right\} = \left[\begin{array}{ccc} Q_{11}^k & Q_{12}^k & 0 \\ Q_{21}^k & Q_{22}^k & 0 \\ 0 & 0 & Q_{66}^k \end{array} \right] \left\{ \begin{array}{c} \varepsilon_x \\ \varepsilon_\theta \\ \varepsilon_{x\theta} \end{array} \right\}$$

(5)

where Q_{ij}^k are the reduced stiffness coefficients of the k-th lamina, given by:

$$Q_{11}^k = Q_{22}^k = \frac{E_k}{1-v_k^2}, Q_{12}^k = Q_{21}^k = \frac{E_k v_k}{1-v_k^2}, Q_{66}^k = \frac{E_k}{2(1+v_k)}$$

(6)

The material parameters E_k and v_k are the Young's modulus and Poisson's ratio of the k-th lamina, respectively. For a double-layered cylindrical shell, the force and moment resultants are defined by:

$$\left\{ \begin{array}{c} N_x \\ N_\theta \\ N_{x\theta} \\ M_x \\ M_\theta \\ M_{x\theta} \end{array} \right\} = \sum_{k=1}^{2} \int_{z_{k-1}}^{z_k} \left\{ \begin{array}{c} \sigma_x \\ \sigma_\theta \\ \sigma_{x\theta} \\ z\sigma_x \\ z\sigma_\theta \\ z\sigma_{x\theta} \end{array} \right\} dz = S \left\{ \begin{array}{c} \varepsilon_x^0 \\ \varepsilon_\theta^0 \\ \varepsilon_{x\theta}^0 \\ \kappa_x \\ \kappa_\theta \\ \kappa_{x\theta} \end{array} \right\}$$

(7)

where:

$$S = \left[\begin{array}{cccccc} A_{11} & A_{12} & 0 & B_{11} & B_{12} & 0 \\ A_{21} & A_{22} & 0 & B_{21} & B_{22} & 0 \\ 0 & 0 & A_{66} & 0 & 0 & B_{66} \\ B_{11} & B_{12} & 0 & D_{11} & D_{12} & 0 \\ B_{21} & B_{22} & 0 & D_{21} & D_{22} & 0 \\ 0 & 0 & B_{66} & 0 & 0 & D_{66} \end{array} \right]$$

(8)

in which:

$$A_{ij} = \sum_{k=1}^{2} Q_{ij}^k (z_k - z_{k-1})$$
$$B_{ij} = \frac{1}{2}\sum_{k=1}^{2} Q_{ij}^k \left(z_k^2 - z_{k-1}^2\right)$$
$$D_{ij} = \frac{1}{3}\sum_{k=1}^{2} Q_{ij}^k \left(z_k^3 - z_{k-1}^3\right)$$

(9)

The three coordinates z_0, z_1 and z_2 are defined as follows:

$$z_0 = -T_0, z_1 = T_1 - T_0, z_2 = T_1 + T_2 - T_0$$

(10)

where T_1 and T_2 are the thickness of metal substrate and hard-coating, respectively (see Figure 1b). T_0 is the distance from the middle surface to the inner surface, defined by:

$$T_0 = \frac{E_1 T_1^2 + 2E_2 T_1 T_2 + E_2 T_2^2}{2(E_1 T_1 + E_2 T_2)}$$

(11)

2.3. Finite Element Formulation

The degrees of freedom at each node are defined as follows (see Figure 1b):

$$u^0, \; u^0_x, \; u^0_\theta, \; v^0, \; v^0_x, \; v^0_\theta, \; w^0, \; w^0_x, \; w^0_\theta \tag{12}$$

where:

$$u^0_x = \frac{\partial u^0}{\partial x}, \; u^0_\theta = \frac{\partial u^0}{\partial \theta}, \; v^0_x = \frac{\partial v^0}{\partial x}, \; v^0_\theta = \frac{\partial v^0}{\partial \theta}, \; w^0_x = \frac{\partial w^0}{\partial x}, \; w^0_\theta = \frac{\partial w^0}{\partial \theta} \tag{13}$$

The displacement field for u^0 in the coordinate directions x and θ is assumed to be of the form:

$$
\begin{aligned}
u^0 = \; & a_1 + a_2 x + a_3 \theta + a_4 x^2 + a_5 x\theta + a_6 \theta^2 + a_7 x^3 + a_8 x^2\theta + a_9 x\theta^2 + a_{10}\theta^3 \\
& + a_{11} x^3\theta + a_{12} x\theta^3
\end{aligned} \tag{14}
$$

The same form of expression is used for the displacements v^0 and w^0. The equations of the displacement field for u^0, u^0_x and u^0_θ can be written in matrix form:

$$\mathbf{U} = \mathbf{XA} \tag{15}$$

in which $\mathbf{U}$, $\mathbf{X}$ and $\mathbf{A}$ can be expressed as follows:

$$\mathbf{U} = \begin{bmatrix} \bar{\mathbf{U}}_1 & \bar{\mathbf{U}}_2 & \bar{\mathbf{U}}_3 & \bar{\mathbf{U}}_4 \end{bmatrix}^{\mathrm{T}} \tag{16}$$

$$\mathbf{X} = \begin{bmatrix} \bar{\mathbf{X}}_1 & \bar{\mathbf{X}}_1 & \bar{\mathbf{X}}_1 & \bar{\mathbf{X}}_1 \end{bmatrix}^{\mathrm{T}} \tag{17}$$

$$\mathbf{A} = \begin{pmatrix} a_1 & a_2 & a_3 & a_4 & a_5 & a_6 & a_7 & a_8 & a_9 & a_{10} & a_{11} & a_{12} \end{pmatrix}^{\mathrm{T}} \tag{18}$$

where:

$$\bar{\mathbf{U}}_i = \begin{bmatrix} u^0_i & u^0_{xi} & u^0_{\theta i} \end{bmatrix}^{\mathrm{T}} \tag{19}$$

$$\bar{\mathbf{X}}_i = \begin{bmatrix} 1 & x_i & \theta_i & x_i^2 & x_i\theta_i & \theta_i^2 & x_i^3 & x_i^2\theta_i & x_i\theta_i^2 & \theta_i^3 & x_i^3\theta_i & x_i\theta_i^3 \\ 0 & 1 & 0 & 2x_i & \theta_i & 0 & 3x_i^2 & 2x_i\theta_i & \theta_i^2 & 0 & 3x_i^2\theta_i & \theta_i^3 \\ 0 & 0 & 1 & 0 & x_i & 2\theta_i & 0 & x_i^2 & 2x_i\theta_i & 3\theta_i^2 & x_i^3 & 3x_i\theta_i^2 \end{bmatrix} \tag{20}$$

By applying inverse matrix method, the displacement interpolation functions can be calculated as follows:

$$\begin{pmatrix} N_1 & N_{x1} & N_{\theta 1} & N_2 & N_{x2} & N_{\theta 2} & N_3 & N_{x3} & N_{\theta 3} & N_4 & N_{x4} & N_{\theta 4} \end{pmatrix} = \mathbf{X}^{-1}\mathbf{U} \tag{21}$$

where N_i, N_{xi} and $N_{\theta i}$ are the displacement interpolation functions at node i.

Then, the shape function matrix $\mathbf{N}$ can be expressed as:

$$\mathbf{N} = \begin{bmatrix} \bar{\mathbf{N}}_1 & \bar{\mathbf{N}}_2 & \bar{\mathbf{N}}_3 & \bar{\mathbf{N}}_4 \end{bmatrix} \tag{22}$$

where $\bar{\mathbf{N}}_i$ is the interpolation functions corresponding to node i, given by:

$$\bar{\mathbf{N}}_i = \begin{bmatrix} N_i & N_{xi} & N_{\theta i} & 0 & 0 & 0 & 0 & 0 & 0 \\ 0 & 0 & 0 & N_i & N_{xi} & N_{\theta i} & 0 & 0 & 0 \\ 0 & 0 & 0 & 0 & 0 & 0 & N_i & N_{xi} & N_{\theta i} \end{bmatrix} \tag{23}$$

The general strain-displacement relations for the hard-coating cylindrical shell can be expressed as:

$$\varepsilon = \left\{ \begin{array}{c} \varepsilon_x^0 \\ \varepsilon_\theta^0 \\ \varepsilon_{x\theta}^0 \\ \kappa_x \\ \kappa_\theta \\ \kappa_{x\theta} \end{array} \right\} = \left\{ \begin{array}{c} \dfrac{\partial u^0}{\partial x} \\[2mm] \dfrac{1}{R}\dfrac{\partial v^0}{\partial \theta} + \dfrac{w^0}{R} \\[2mm] \dfrac{\partial v^0}{\partial x} + \dfrac{1}{R}\dfrac{\partial u^0}{\partial \theta} \\[2mm] \dfrac{\partial^2 w^0}{\partial x^2} \\[2mm] \dfrac{1}{R^2}\left(\dfrac{\partial^2 w^0}{\partial \theta^2} - \dfrac{\partial v^0}{\partial \theta}\right) \\[2mm] \dfrac{2}{R}\left(\dfrac{\partial^2 w^0}{\partial x\partial \theta} - \dfrac{\partial v^0}{\partial x}\right) \end{array} \right\} = \Gamma \left\{ \begin{array}{c} u^0 \\ v^0 \\ w^0 \end{array} \right\} = \Gamma \mathbf{N} d = \mathbf{B} x \qquad (24)$$

where the matrix differential operator Γ, the strain matrix $\mathbf{B}$ and the displacement vector x are defined, respectively, by:

$$\Gamma = \begin{bmatrix} \dfrac{\partial}{\partial x} & 0 & 0 \\[2mm] 0 & \dfrac{1}{R}\dfrac{\partial}{\partial \theta} & \dfrac{1}{R} \\[2mm] \dfrac{1}{R}\dfrac{\partial}{\partial \theta} & \dfrac{\partial}{\partial x} & 0 \\[2mm] 0 & 0 & \dfrac{\partial^2}{\partial x^2} \\[2mm] 0 & -\dfrac{1}{R^2}\dfrac{\partial}{\partial \theta} & \dfrac{1}{R^2}\dfrac{\partial^2}{\partial \theta^2} \\[2mm] 0 & -\dfrac{2}{R}\dfrac{\partial}{\partial x} & \dfrac{2}{R}\dfrac{\partial^2}{\partial x\partial \theta} \end{bmatrix} \qquad (25)$$

$$\mathbf{B} = \begin{bmatrix} \overline{\mathbf{B}}_1 & \overline{\mathbf{B}}_2 & \overline{\mathbf{B}}_3 & \overline{\mathbf{B}}_4 \end{bmatrix} \qquad (26)$$

$$x = \begin{pmatrix} \overline{x}_1 & \overline{x}_2 & \overline{x}_3 & \overline{x}_4 \end{pmatrix}^{\mathrm{T}} \qquad (27)$$

The matrix $\overline{\mathbf{B}}_i$ is:

$$\overline{\mathbf{B}}_i = \begin{bmatrix} N_i|_x & N_{xi}|_x & N_{\theta i}|_x & 0 & 0 & 0 & 0 & 0 & 0 \\[2mm] 0 & 0 & 0 & \dfrac{N_i|_\theta}{R} & \dfrac{N_{xi}|_\theta}{R} & \dfrac{N_{\theta i}|_\theta}{R} & \dfrac{N_i}{R} & \dfrac{N_{xi}}{R} & \dfrac{N_{\theta i}}{R} \\[2mm] \dfrac{N_i|_\theta}{R} & \dfrac{N_{xi}|_\theta}{R} & \dfrac{N_{\theta i}|_\theta}{R} & N_i|_x & N_{xi}|_x & N_{\theta i}|_x & 0 & 0 & 0 \\[2mm] 0 & 0 & 0 & 0 & 0 & 0 & N_i|_x^2 & N_{xi}|_x^2 & N_{\theta i}|_x^2 \\[2mm] 0 & 0 & 0 & \dfrac{-N_i|_\theta}{R^2} & \dfrac{-N_{xi}|_\theta}{R^2} & \dfrac{-N_{\theta i}|_\theta}{R^2} & N_i| & \dfrac{N_{xi}|_\theta^2}{R^2} & \dfrac{N_{\theta i}|_\theta^2}{R^2} \\[2mm] 0 & 0 & 0 & \dfrac{-2N_i|_x}{R} & \dfrac{-2N_{xi}|_x}{R} & \dfrac{-2N_{\theta i}|_x}{R} & \dfrac{2N_i|_{x\theta}^2}{R} & \dfrac{2N_{xi}|_{x\theta}^2}{R} & \dfrac{2N_{\theta i}|_\theta^2}{R} \end{bmatrix} \qquad (28)$$

where:

$$|_x = \frac{\partial}{\partial x}, \; |_x^2 = \frac{\partial^2}{\partial x^2}, \; |_\theta = \frac{\partial}{\partial \theta}, \; |_\theta^2 = \frac{\partial^2}{\partial \theta^2}, \; |_{x\theta}^2 = \frac{\partial^2}{\partial x\partial \theta} \qquad (29)$$

The displacement vector $\overline{x}_i$ corresponding to node i is defined as:

$$\overline{x}_i = \begin{pmatrix} u_i & u_{xi} & u_{\theta i} & v_i & v_{xi} & v_{\theta i} & w_i & w_{xi} & w_{\theta i} \end{pmatrix}^{\mathrm{T}} \qquad (30)$$

Thus, the element stiffness matrix $\mathbf{K}^{\mathrm{el}}$, the element mass matrix $\mathbf{M}^{\mathrm{el}}$ and the external load vector $\mathbf{F}^{\mathrm{el}}$ introduced by the base excitation can be expressed, respectively, by:

$$\mathbf{K}^{\mathrm{el}} = \frac{1}{2}\iiint \mathbf{B}^{\mathrm{T}}\mathbf{S}\mathbf{B}(R+z)\mathrm{d}z\mathrm{d}\theta\mathrm{d}x \qquad (31)$$

$$\mathbf{M}^{\text{el}} = \frac{1}{2} \iiint \rho \mathbf{N}^{\text{T}} \mathbf{N} (R + z) \mathrm{d}z \mathrm{d}\theta \mathrm{d}x \qquad (32)$$

$$\mathbf{F}^{\text{el}} = \frac{1}{2} \iiint \mathbf{N}^{\text{T}} f (R + z) \mathrm{d}z \mathrm{d}\theta \mathrm{d}x \qquad (33)$$

where the mass density per unit area ρ is defined by [22]:

$$\rho = \rho_1 T_1 + \rho_2 T_2 \qquad (34)$$

where ρ_1 and ρ_2 are the densities of metal substrate and hard-coating, respectively.

The f is a 3×1 order vector composed of the x, θ and z components of the base excitation per unit volume, defined by:

$$f = \rho a \{\mu_x, \mu_\theta, \mu_z\}^{\text{T}} \qquad (35)$$

in which a is the absolute acceleration of the base and μ_x, μ_θ and μ_z are the influence coefficients of the base excitation on the x, θ and z directions of the hard-coating cylindrical shell.

The global stiffness matrix $\mathbf{K}$, the global mass matrix $\mathbf{M}$ and the global external load vector $\mathbf{F}$ of the hard-coating cylindrical shell can be assembled by $\mathbf{K}^{\text{el}}$, $\mathbf{M}^{\text{el}}$ and $\mathbf{F}^{\text{el}}$ in the corresponding sequence, respectively.

3. Characterizing the Strain Dependence Using the High Order Polynomial

To describe the strain dependence of the hard-coating cylindrical shell, the complex modulus $E_2{}^*$ of the hard-coating is considered, given by:

$$E_2^* = E_2 + i\eta_2 \qquad (36)$$

where $*$ denotes the complex value, $i = \sqrt{-1}$, and E_2 (namely the E_2 in Equation (6)) and η_2 are the storage modulus and the loss factor of hard-coating, respectively, which can be expressed by the p-order polynomials,

$$E_2 = E_{20} + E_{21}\varepsilon_e + E_{22}\varepsilon_e^2 + \cdots + E_{2p}\varepsilon_e^p \qquad (37)$$

$$\eta_2 = \eta_{20} + \eta_{21}\varepsilon_e + \eta_{22}\varepsilon_e^2 + \cdots + \eta_{2p}\varepsilon_e^p \qquad (38)$$

in which E_{2j} and η_{2j} ($j = 0, 1, \ldots, p$) are specific j-order coefficients of the storage modulus and the loss factor, respectively. For convenience, one can define:

$$E_2^* = (E_{20} + i\eta_{20}) + E_{2,\text{inc}}^* \qquad (39)$$

where $E_{2,\text{inc}}^*$ is the increment of the complex modulus caused by the equivalent strain ε_e, given by:

$$E_{2,\text{inc}}^* = E_{21}^* \varepsilon_e + E_{22}^* \varepsilon_e^2 + \cdots + E_{2p}^* \varepsilon_e^p \qquad (40)$$

$$E_{21}^* = E_{21}(1 + i\eta_{21}),\ E_{22}^* = E_{22}(1 + i\eta_{22}), \cdots, E_{2p}^* = E_{2p}\left(1 + i\eta_{2p}\right) \qquad (41)$$

According to the principle of equal strain energy density [20], the equivalent strain ε_e of an element can be calculated as:

$$\varepsilon_e = \sqrt{x^{\text{T}} \frac{\mathbf{K}_{20}^{\text{el}}}{E_{20} V_2^{\text{el}}} x} \qquad (42)$$

where $\mathbf{K}_{20}^{\text{el}}$ and E_{20} are the element stiffness matrix and the storage modulus of the hard-coating at zero strain, respectively. V_2^{el} is the element volume of hard-coating. In addition, the displacement vector x is defined as:

$$x = \varphi_0 q \qquad (43)$$

in which q is the response vector in the normal coordinate and φ_0 is the normal mode shape matrix at zero strain without damping. Submitting Equation (43) to Equation (42), one can obtain:

$$\varepsilon_e = \sqrt{\frac{q^T \varphi_0^T K_{20}^{el} \varphi_0 q}{E_{20} V_2^{el}}} = \sqrt{\frac{q^T \Lambda_{el} q}{V_2^{el}}} \tag{44}$$

where Λ is the normalized element stiffness matrix per unit storage modulus of the hard-coating.

Usually, the storage modulus and the loss factor of the hard-coating can be identified by the relevant experiments. For the NiCoCrAlY + YSZ hard-coating considered here, the expressions of the strain-dependent storage modulus E_2 and loss factor η_2 obtained from the experiment [26] are:

$$E_2 = 54.494 - 0.06402\varepsilon_e + 8.09137 \times 10^{-5}\varepsilon_e^2 - 6.1681 \times 10^{-8}\varepsilon_e^3 \tag{45}$$

$$\eta_2 = 0.0212 - 0.43521\varepsilon_e - 0.37997\varepsilon_e^2 - 0.5441\varepsilon_e^3 \tag{46}$$

4. Solution of the Nonlinear Vibration of the Hard-Coating Cylindrical Shell

The dynamic equation of the hard-coating cylindrical shell in the normal coordinate is defined as:

$$\left(K_{\varepsilon N}^* - \omega^2 M_N \right) q = F_N \tag{47}$$

in which:

$$\mathbf{K}_{\varepsilon N}^* = \varphi_0^T \mathbf{K}_\varepsilon^* \varphi_0, \mathbf{M}_N = \varphi_0^T \mathbf{M} \varphi_0, \mathbf{F}_N = \varphi_0^T \mathbf{F} \tag{48}$$

where ω is the angular frequency of excitation, $\mathbf{K}_\varepsilon^*$ is the global complex stiffness matrix at ε strain and $\mathbf{M}$ and $\mathbf{F}$ are the global mass matrix and external force vector, respectively. The normalized response vector $\mathbf{q}$ can be obtained from Equation (47) by applying the QR-method.

It is assumed that the global material damping matrix $\mathbf{D}$ is independent of the strain level. Neglecting the coupling effect between the layers of hard-coating and the metal substrate of the cylindrical shell element, the global complex stiffness matrix at ε strain can be defined by:

$$\mathbf{K}_\varepsilon^* \approx \mathbf{K}_0^* + \sum_{el} \frac{\mathbf{K}_{20}^{el}}{E_{20}} \left(i\eta_{20} + E_{2,inc}^* \right) \tag{49}$$

in which:

$$\mathbf{K}_0^* = \mathbf{K} + i\mathbf{D} \tag{50}$$

The global material damping matrix $\mathbf{D}$ of the hard-coating cylindrical shell can be calculated by the same method as the global stiffness matrix $\mathbf{K}$, which only requires multiplying Equation (6) by the zero-strain loss factor of the metal substrate and hard-coating η_{10} and η_{20}, respectively

$$Q_{11}^1 = Q_{22}^1 = \frac{\eta_{10} E_1}{1 - v_1^2}, Q_{12}^1 = Q_{21}^1 = \frac{\eta_{10} E_1 v_1}{1 - v_1^2}, Q_{66}^1 = \frac{\eta_{10} E_1}{2(1 + v_1)} \tag{51}$$

$$Q_{11}^2 = Q_{22}^2 = \frac{\eta_{20} E_2}{1 - v_2^2}, Q_{12}^2 = Q_{21}^2 = \frac{\eta_{20} E_2 v_2}{1 - v_2^2}, Q_{66}^2 = \frac{\eta_{20} E_2}{2(1 + v_2)} \tag{52}$$

It should be noted that to improve the solution precision, the zero-strain loss factor of metal substrate η_{10} is considered.

The Newton-Raphson solution method [27,28] is employed to solve the nonlinear Equation (47). Further, the corresponding residual vector r can be expressed as:

$$r = \left(K_{\varepsilon N}^* - \omega^2 M_N \right) q - F_N \tag{53}$$

Submitting Equation (49) to Equation (53), one can obtain:

$$r = J_N q - F_N + \sum_{el} \frac{K_{20}^{el}}{E_{\varepsilon 0}^c} \left(i\eta_{20} + E_{2,\text{inc}}^* \right) q \tag{54}$$

in which:

$$J_N = K_{0N}^* - \omega^2 M_N \tag{55}$$

Since the normalized response vector q is complex-valued, it is necessary to separate the real and imaginary parts of q in the iterative formula derived by the Newton–Raphson solution method. The separated iterative formula can be expressed as:

$$\left\{ \begin{array}{c} R(q_{\text{new}}) \\ S(q_{\text{new}}) \end{array} \right\} = \left\{ \begin{array}{c} R(q_{\text{old}}) \\ S(q_{\text{old}}) \end{array} \right\} - \left[\begin{array}{cc} R(\partial r/\partial q_R) & R(\partial r/\partial q_S) \\ S(\partial r/\partial q_R) & S(\partial r/\partial q_S) \end{array} \right]^{-1} \left\{ \begin{array}{c} R(r) \\ S(r) \end{array} \right\} \tag{56}$$

where $R(\ )$ and $S(\ )$ are the functions of extracting the real and imaginary parts of the bracketed expressions, respectively. The derivatives of the residual vector r with respect to the real and imaginary parts, q_R and q_S, are:

$$\frac{\partial r}{\partial q_R} = J_N + \frac{K_{0N}^c}{E_{20}} \left[\frac{q_R^T \left(\Lambda_{el}^T + \Lambda_{el} \right) q}{V_2^{el}} \left(\frac{E_{21}^*}{2\varepsilon_e} + E_{22}^* + \frac{3E_{23}^* \varepsilon_e}{2} \right) + E_{2,\text{inc}}^* \right] \tag{57}$$

$$\frac{\partial r}{\partial q_S} = iJ_N + \frac{K_{0N}^c}{E_{20}} \left[\frac{q_S^T \left(\Lambda_{el}^T + \Lambda_{el} \right) q}{V_2^{el}} \left(\frac{E_{21}^*}{2\varepsilon_e} + E_{22}^* + \frac{3E_{23}^* \varepsilon_e}{2} \right) + iE_{2,\text{inc}}^* \right] \tag{58}$$

The termination of the Newton–Raphson solution method can be defined by:

$$\|r\|_2 \leq \zeta \tag{59}$$

where $\|r\|_2$ is the two-norm of the residual vector r and ζ is the solution precision. When the residual vector r of Equation (56) satisfies the condition of the solution precision, the newest global complex stiffness matrix at ε strain $K_{\varepsilon,\text{new}}^*$ (in Equation (49)) and the normalized response vector q_{new} can be output. Then, the n-order nonlinear resonant circular frequency ω_n and the nonlinear harmonic response x can be obtained, respectively, by:

$$\left| R\left(K_{\varepsilon,\text{new}}^* \right) - \omega_n^2 M \right| = 0 \tag{60}$$

$$x = \varphi_0^T q_{\text{new}} \tag{61}$$

The whole solution procedure of the nonlinear vibration of the hard-coating cylindrical shell is shown in Figure 2.

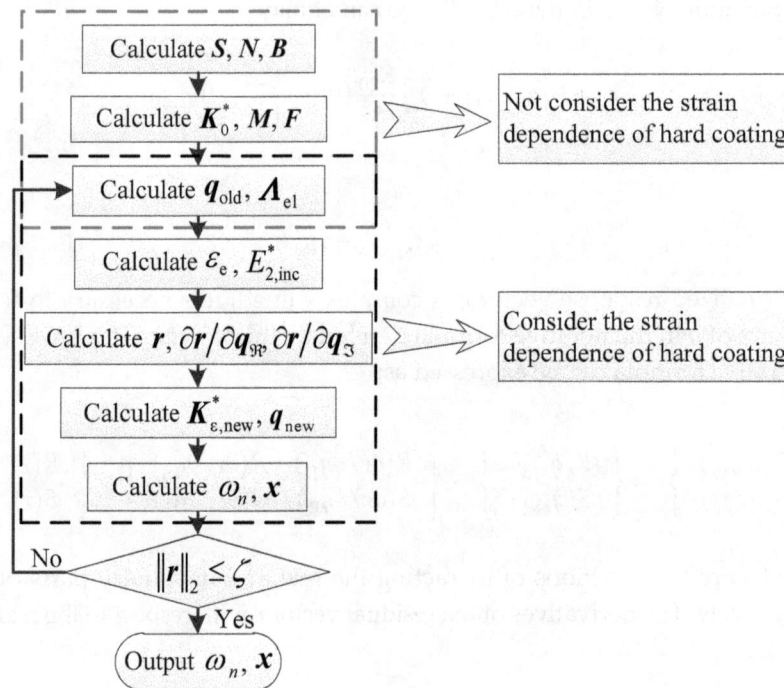

Figure 2. Solution procedure of the nonlinear vibration of the hard-coating cylindrical shell.

5. Case Study

For this study, the NiCoCrAlY + YSZ hard-coating with strain dependence consists of NiCrAlY and yttria-stabilized zirconia (YSZ), which is the most widely used in the hottest turbine sections of the aero engine and aircraft due to its optimal performance in high temperature and vibration damping applications [3]. The relevant geometrical parameters and material parameters of the hard-coating cylindrical shell are listed in Tables 1 and 2. It should be noted that E_2 and η_2 in Table 2 are functions of the equivalent strain ε_e (see Equations (45) and (46)).

Table 1. Geometrical parameters of the hard-coating cylindrical shell.

Parameters	L (mm)	R (mm)	T_1 (mm)	T_2 (mm)
Value	95	142	2	0.31

Table 2. Material parameters of the hard-coating cylindrical shell.

Lamina	Material	Young's Modulus (GPa)	Loss Factor	Density (kg/m³)	Poisson's Ratio
Metal substrate	Ti-6Al-4V	110.32	0.0007	4420	0.3
Hard coating	NiCoCrAlY + YSZ	E_2	η_2	5600	0.3

To simulate the vibration during the aero engine or aircraft working, the hard-coating cylindrical shell is forced in the horizontal direction by a sinusoidal base excitation. Assume that the absolute acceleration of the base is along the z direction in the Cartesian coordinate system (see Figure 3). Converting it into the orthogonal curvilinear coordinate system, one can obtain the values of the influence coefficients μ_x, μ_θ and μ_z in different quadrants. The μ_x is equal to zero; the values of μ_θ and μ_z are listed in Table 3.

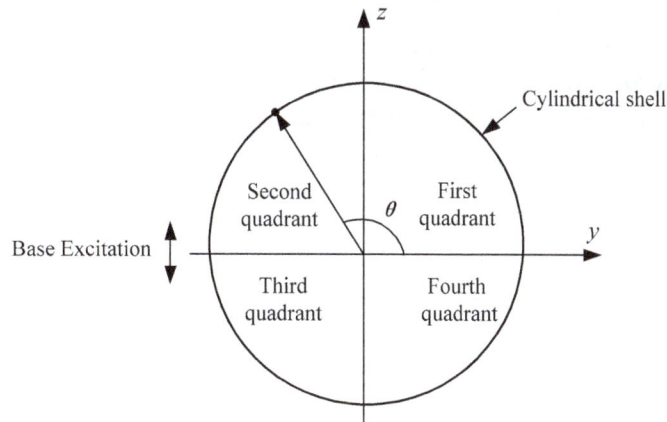

Figure 3. The quadrant distribution of the hard-coating cylindrical shell.

Table 3. Values of the influence coefficients in different quadrants.

Quadrant	μ_θ	μ_z
First	$\cos\theta$	$\sin\theta$
Second	$\sin\theta$	$-\cos\theta$
Third	$-\cos\theta$	$-\sin\theta$
Fourth	$\cos\theta$	$-\sin\theta$

5.1. Validation Analysis

The finite element models with element size 5 mm corresponding to the proposed method and the FEIM are shown in Figure 4, which contains 3496 and 6992 elements in total, respectively. The initial base excitation level is set to 1 g. Besides, all nodes at the bottom of the hard-coating cylindrical shell are constrained in all degrees of freedom to represent the clamped-free boundary condition. It should be noted that the finite element model of the FEIM contains two layers of the element, which are coupled together to describe the connection of the hard-coating and metal substrate [16].

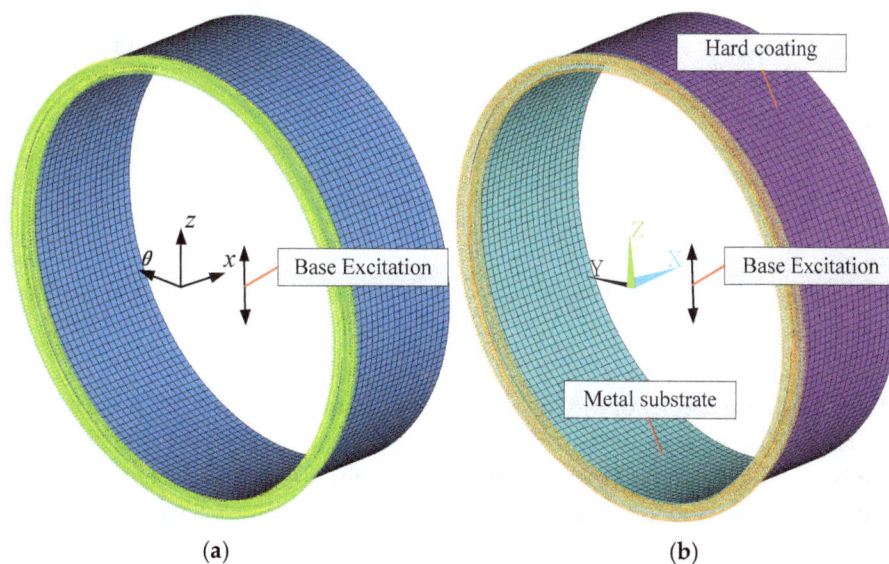

(a) (b)

Figure 4. Two finite element models of the hard-coating cylindrical shell: (**a**) Present model; (**b**) FEIM model.

Since the classical shell element (Shell281) applied in the FEIM contains eight nodes and six degrees of freedom at each node and the proposed cylindrical shell element contains four nodes and nine degrees of freedom at each node, the total degrees of freedom of the finite element model of the FEIM is about 2.67-times that of the present model under the same element size, which indicates the advantage of the lower computing cost of the present model briefly.

Based on the new finite element formulation, the whole process of nonlinear vibration analysis of the hard-coating cylindrical shell is implemented by a MATLAB (MathWorks Inc., Natick, MA, USA) self-made finite element program. The numerical results of the hard-coating cylindrical shell will now be compared with the FEIM, which applied the Shell281 element of ANSYS (ANSYS Inc., Canonsburg, PA, USA) to calculate the element matrices of stiffness, mass and damping. Numerical experiences indicate that the appropriate results can only be obtained with fine enough meshing. Thus, different element sizes are used to solve the nonlinear resonant frequency of the hard-coating cylindrical shell by the FEIM and the present method, respectively. The comparison of the convergences of the first order nonlinear resonant frequencies under different element sizes is shown in Figure 5, which indicates that the developed finite element method is less affected by the element size and has lower computing cost. The advantages become more obvious when solving the nonlinear resonant response spectrums of the hard-coating cylindrical shell, because its calculation process can be regarded as a combination of a number of nonlinear resonant frequency calculation processes, which have a basically consistent iterative method.

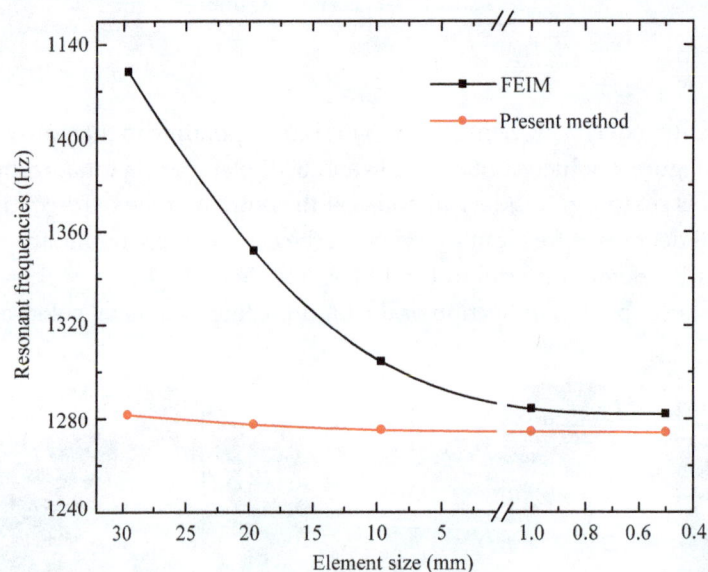

Figure 5. Convergences of nonlinear resonant frequencies of the finite element iteration method (FEIM) and present method under different element sizes.

In addition, the comparisons of the first six order nonlinear resonant frequencies and responses of the hard-coating cylindrical shell with the FEIM under element size of 1 mm are listed in Table 4. The difference represents the relative error between present method and the FEIM. As can be seen from Table 4, the nonlinear resonant frequencies and responses calculated by the present method and the FEIM show a good agreement, and the differences are within 0.873% and 4.940%, respectively, which indicate the rationality of the developed finite element method.

Table 4. Comparisons of the first 6 order nonlinear resonant frequencies and responses with the FEIM.

Modal Order	Nonlinear Resonant Frequencies			Nonlinear Resonant Responses		
	Present Method A (Hz)	FEIM B (Hz)	Difference $|A - B|/A$ (%)	Present Method C (10^{-2} mm)	FEIM D (10^{-2} mm)	Difference $|C - D|/C$ (%)
1	1274.561	1281.912	0.577	1.786	1.738	2.732
2	1283.778	1294.985	0.873	3.199	3.048	4.940
3	1443.842	1446.837	0.207	1.753	1.705	2.793
4	1513.250	1526.021	0.844	1.409	1.345	4.753
5	1738.061	1736.589	0.085	1.172	1.164	0.702
6	1990.964	2002.791	0.594	1.239	1.200	3.259

5.2. Nonlinear Vibration Analysis and Results Discussion

The finite element model for nonlinear vibration analysis of the hard-coating cylindrical shell is shown in Figure 4a, and the element size is changed to 1 mm. To analyze the nonlinear vibration characteristics of the hard-coating cylindrical shell, the linear resonant frequencies and responses without considering the strain dependence of the hard-coating (the equivalent strain $\varepsilon_e = 0$) need to be calculated, which are taken as the reference.

The first six order linear and nonlinear resonant frequencies and responses of the hard-coating cylindrical shell under a 5 g excitation level are listed in Tables 5 and 6. Then, the three-order linear and nonlinear resonant frequencies and responses of the hard-coating cylindrical shell under different excitation levels are listed in Tables 7 and 8. Moreover, the three-order frequency response spectrums of the hard-coating cylindrical shell are shown in Figure 6. It should be noted that the legend L represents the linear vibration response, and the legend Nl represents the nonlinear vibration response.

Table 5. The first 6 order linear and nonlinear resonant frequencies of the hard-coating cylindrical shell under a 5 g excitation level.

Modal Order	Linear (Hz) E	Nonlinear (Hz) F	Descent (Hz) $E - F$
1	1274.899	1274.561	0.338
2	1283.922	1283.778	0.144
3	1444.074	1443.842	0.232
4	1513.410	1513.25	0.160
5	1738.293	1738.061	0.232
6	1991.071	1990.964	0.107

Table 6. The first 6 order linear and nonlinear resonant responses of the hard-coating cylindrical shell under a 5 g excitation level.

Modal Order	Linear (10^{-2} mm) G	Nonlinear (10^{-2} mm) H	Descent (10^{-2} mm) $G - H$
1	8.926	8.529	0.397
2	16.003	15.472	0.531
3	8.740	8.444	0.296
4	7.067	6.794	0.273
5	5.869	5.714	0.155
6	6.195	6.067	0.128

Table 7. The 3 order linear and nonlinear resonant frequencies of the hard-coating cylindrical shell under different excitation levels.

Excitation Level (g)	Linear (Hz) E	Nonlinear (Hz) F	Descent (Hz) $E - F$
1	1444.074	1444.026	0.048
3	1444.074	1443.932	0.142
5	1444.074	1443.842	0.232
7	1444.074	1443.755	0.319
9	1444.074	1443.672	0.402

Table 8. The 3 order linear and nonlinear resonant responses of the hard-coating cylindrical shell under different excitation levels.

Excitation Level (g)	Linear (10^{-2} mm) G	Nonlinear (10^{-2} mm) H	Descent (10^{-2} mm) $G - H$
1	1.752	1.741	0.011
3	5.246	5.136	0.110
5	8.740	8.444	0.296
7	12.234	11.641	0.593
9	15.738	14.728	1.010

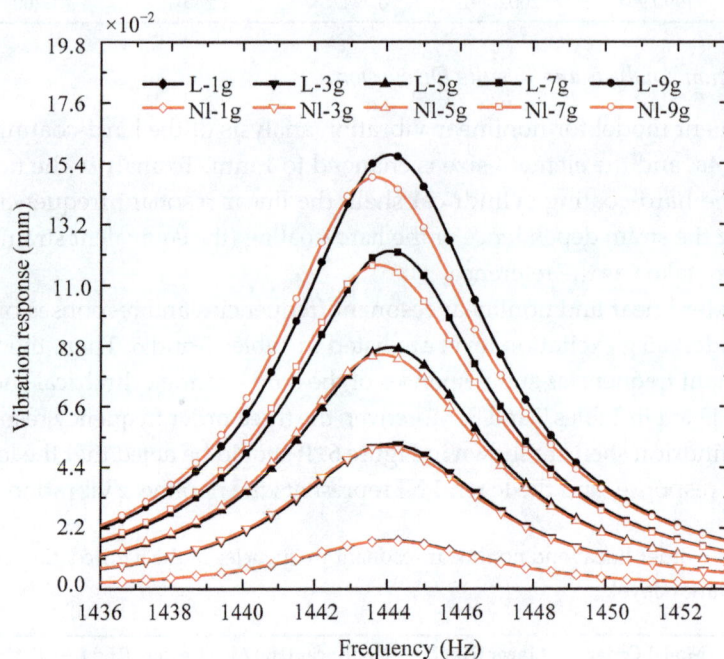

Figure 6. The three order linear and nonlinear frequency response spectrums of the hard-coating cylindrical shell under different excitation levels.

According to the results listed in Tables 5 and 6, it can be found that the nonlinear resonant frequencies and responses decrease to a certain degree compared with the linear calculation results, which indicate that the strain dependence of the hard-coating would make the resonant frequencies and responses present a certain degree of decline. As can be seen from Tables 7 and 8, the descents of the nonlinear resonant frequencies and responses of the hard-coating cylindrical shell increase continually compared with the linear calculation results, while the excitation level increases from 1 to 9 g; that is, the increase of the excitation level would make the strain dependence of the hard-coating more remarkable, which reveals the characteristics of the soft stiffness nonlinearity or "strain softening". This phenomenon may be more obvious and prominent in the frequency response spectrums (see Figure 6), which may indicate that the strain dependence of the hard-coating is beneficial for vibration reduction.

6. Conclusions

With considering the influence of the strain dependence of the coating material on the complex stiffness matrix, the nonlinear modeling, validation example and analysis are studied, and the following conclusions can be drawn.

- Based on Love's first approximation theory, a four-node composite cylindrical shell finite element model is proposed. Then, the nonlinear iterative solution formulas with a unified iterative method are theoretically derived for solving the resonant frequency and response of the hard-coating cylindrical shell.

- A cylindrical shell coated with a thin layer of NiCoCrAlY + YSZ is chosen to demonstrate the proposed formulation. The nonlinear resonant frequencies and responses calculated by the present method and the FEIM show a good agreement, which indicates the rationality of the developed finite element method. Moreover, the developed finite element method is less affected by the element size and has lower computing cost.

- Moreover, the nonlinear vibration analysis of the cylindrical shell coated with a thin layer of NiCoCrAlY + YSZ is implemented. Compared with the linear calculation results, the nonlinear resonant frequencies and responses of each order decrease to a certain degree, and the descents increase continually with the increase of the excitation level; that is, the increase of the excitation level would make the strain dependence of the hard-coating more remarkable, which reveals the characteristics of the soft stiffness nonlinearity or "strain softening".

Acknowledgments: This work was supported by the National Natural Science Foundation of China (Grant No. 51375079).

Author Contributions: Yue Zhang and Jian Yang conceived and designed the paper; Jian Yang analyzed the data; Wei Sun contributed materials/analysis tools; Yue Zhang, Jian Yang and Wei Sun wrote the paper.

Conflicts of Interest: The authors declare no conflict of interest.

References

1. Jin, G.; Ye, T.; Chen, Y.; Su, Z.; Yan, Y. An exact solution for the free vibration analysis of laminated composite cylindrical shells with general elastic boundary conditions. *Compos. Struct.* **2013**, *106*, 114–127. [CrossRef]
2. Limarga, A.M.; Duong, T.L.; Gregori, G.; Clarke, D.R. High-temperature vibration damping of thermal barrier coating materials. *Surf. Coat. Technol.* **2007**, *202*, 693–697. [CrossRef]
3. Gregori, G.; Li, L.; Nychka, J.A.; Clarke, D.R. Vibration damping of superalloys and thermal barrier coatings at high-temperatures. *Mater. Sci. Eng. A* **2007**, *466*, 256–264. [CrossRef]
4. Rodríguez-Barrero, S.; Fernández-Larrinoa, J.; Azkona, I.; López de Lacalle, L.N.; Polvorosa, R. Enhanced performance of nanostructured coatings for drilling by droplet elimination. *Mater. Manuf. Process.* **2016**, *31*, 593–602. [CrossRef]
5. Fernández-Abia, A.I.; Barreiro, J.; López de Lacalle, L.N.; Martínez-Pellitero, S. Behavior of austenitic stainless steels at high speed turning using specific force coefficients. *Int. J. Adv. Manuf. Technol.* **2012**, *62*, 505–515. [CrossRef]
6. Ivancic, F.T.; Palazzotto, A.N. Experimental considerations for determining the damping coefficients of hard coatings. *J. Aerosp. Eng.* **2005**, *18*, 8–17. [CrossRef]
7. Patsias, S.; Saxton, C.G.; Shipton, M.K. Hard damping coatings: an experimental procedure for extraction of damping characteristics and modulus of elasticity. *Mater. Sci. Eng. A* **2004**, *370*, 412–416. [CrossRef]
8. Blackwell, C.W.; Palazzotto, A.N.; George, T.; Cross, C. The evaluation of the damping characteristics of a hard coating on titanium. *Shock Vib.* **2007**, *14*, 37–51. [CrossRef]
9. Tassini, N.; Patsias, S.; Lambrinou, K. Ceramic coatings: a phenomenological modeling for damping behavior related to microstructural features. *Mater. Sci. Eng. A* **2006**, *442*, 509–513. [CrossRef]
10. Reed, S.A.; Palazzotto, A.N.; Baker, W.P. An experimental technique for the evaluation of strain dependent material properties of hard coatings. *Shock Vib.* **2008**, *15*, 697–712. [CrossRef]
11. Torvik, P.J. Determination of mechanical properties of non-linear coatings from measurements with coated beams. *Int. J. Solids Struct.* **2009**, *46*, 1066–1077. [CrossRef]
12. Sun, W.; Liu, Y. Vibration analysis of hard-coated composite beam considering the strain dependent characteristic of coating material. *Acta Mech. Sin.* **2016**, *32*, 1–12. [CrossRef]
13. Chen, Y.G.; Zhai, J.Y.; Han, Q.K. Vibration and damping analysis of the bladed disk with damping hard coating on blades. *Aerosp. Sci. Technol.* **2016**, *58*, 248–257. [CrossRef]
14. Sun, W.; Zhu, M.W.; Wang, Z. Free vibration analysis of a hard-coating cantilever cylindrical shell with elastic constraints. *Aerosp. Sci. Technol.* **2017**, *63*, 232–244. [CrossRef]

15. Filippi, S.; Torvik, P.J. A methodology for predicting the response of blades with nonlinear coatings. *J. Eng. Gas Turbines Power* **2011**, *133*, 984–992. [CrossRef]

16. Li, H.; Liu, Y.; Sun, W. Analysis of nonlinear vibration of hard coating thin plate by finite element iteration method. *Shock Vib.* **2014**, *21*, 1–12. [CrossRef]

17. Wang, T.; Tang, W.; Zhang, S. Nonlinear dynamic response and buckling of laminated cylindrical shells with axial shallow groove based on a semi-analytical method. *J. Shanghai Univ. Engl. Ed.* **2007**, *11*, 223–228. [CrossRef]

18. Naidu, N.V.; Sinha, P.K. Nonlinear finite element analysis of laminated composite shells in hygrothermal environments. *Compos. Struct.* **2005**, *69*, 387–395.

19. Patel, B.; Nath, Y.; Shukla, K.K. Nonlinear thermo-elastic buckling characteristics of cross-ply laminated joined conical-cylindrical shells. *Int. J. Solids Struct.* **2006**, *43*, 4810–4829. [CrossRef]

20. Khoroshun, L.P.; Babich, D.V.; Shikula, E.N. Stability of cylindrical shells made of a particulate composite with nonlinear elastic inclusions and damageable matrix. *Int. Appl. Mech.* **2007**, *43*, 1123–1131. [CrossRef]

21. Ferreira, A.J.; Sa, J.M.; Marques, A.T. Nonlinear finite element analysis of rubber composite shells. *Strength Mater.* **2003**, *35*, 225–235. [CrossRef]

22. Masti, R.S.; Sainsbury, M. Vibration damping of cylindrical shells partially coated with a constrained viscoelastic treatment having a standoff layer. *Thin Walled Struct.* **2005**, *43*, 1355–1379. [CrossRef]

23. Love, A. A treatise on the mathematical theory of elasticity. *Nature* **1944**, *47*, 529–530.

24. Cho, M.; Kim, J. A postprocess method for laminated shells with a doubly curved nine-noded finite element. *Compos. Part B Eng.* **2000**, *31*, 65–74. [CrossRef]

25. Lam, K.Y.; Loy, C.T. Effects of boundary conditions on frequencies of a multi-layered cylindrical shell. *J. Sound Vib.* **1995**, *188*, 363–384. [CrossRef]

26. Sun, W.; Wang, Z.; Zhu, M.; Du, G. Identifying the mechanical parameters of hard coating with strain dependent characteristic by an inverse method. *Shock Vib.* **2015**, *2*, 1–15. [CrossRef]

27. Groll, G.V.; Ewins, D.J. The harmonic balance method with arc-length continuation in rotor/stator contact problems. *J. Sound Vib.* **2001**, *241*, 223–233. [CrossRef]

28. Petrov, E.P.; Ewins, D.J. Method for analysis of nonlinear multiharmonic vibrations of mistuned bladed disks with scatter of contact interface characteristics. *J. Turbomach.* **2005**, *127*, 128–136. [CrossRef]

18

Characterization of the Anti-Graffiti Properties of Powder Organic Coatings Applied in Train Field

Stefano Rossi *,†, **Michele Fedel** †, **Simone Petrolli** † and **Flavio Deflorian** †

Department of Industrial Engineering, University of Trento, Via Sommarive, 9, 38123 Trento, Italy; michele.fedel@unitn.it (M.F.); petrolli.simone@gmail.com (S.P.); flavio.deflorian@unitn.it (F.D.)
* Correspondence: Stefano.rossi@unitn.it
† These authors contributed equally to this work.

Academic Editor: Mariateresa Lettieri

Abstract: The widespread prevalence of the phenomenon of graffiti and the growth of the removal cost—in particular in public transport systems—has pushed the research for technical solutions to this problem. Suitable solutions to address graffiti-related concerns are needed in order to reduce the cleaning costs as well as the downtime of trains. Graffiti are a big problem for painted metal, because the protective coatings and graffiti have the same chemical nature (polymeric matter). A permanent coating is expected to be able to resist the highest possible number of cleanings of the graffiti without modifying its aesthetic and corrosion protection properties. The purpose of this study is to develop a methodological approach for the characterization of graffiti-resistant organic coatings. For this purpose, a critical review of the existing standards is carried out. The anti-graffiti properties of a polyurethane organic coating were investigated before and after accelerated weathering. In order to understand the behavior of the coatings during cleaning, the aging of the coating in contact with the remover was carried out. The effect on the corrosion protection properties was assessed during the accelerated aging. The resistance of the coating was proved to be strongly affected by the surface finishing. UV exposure modified surface properties and graffiti removal efficiency.

Keywords: organic coating; painted metal; graffiti; graffiti cleaning agents

---

1. Introduction

The problem of graffiti originated in the second half of the last century, causing a lot of damages to transport vehicle bodyworks as well as to buildings and public areas [1–3]. The cost of graffiti removal is very high. Therefore, the coating industry has developed permanent solutions for minimizing the cleaning costs. For the urban administration and transportation companies, the damage caused by graffiti presents considerable costs. In 2007, in Australia, the rolling stock graffiti removal cost was $12 million and $8 million for railway infrastructure [4]. In 2014, a study reported that the cost of cleaning Milano (Italy) of graffiti would have been 100 million euros [5]. Trenord (the railroad agency of the Lombardia region, Italy) evaluated the cost of vandalism including the graffiti removal to be approximately 8 million euros [6]. From a technological point of view, there are two kinds of anti-graffiti products [7]. The first one consists of a "sacrificial" layer (commonly wax-based coatings) which is removed during the cleaning. The removal is carried out using hot water jet [8,9]. The second technological solution is a permanent layer which is not solubilized during graffiti removal. This solution is also well considered, not only on metallic substrates, but also on stone and concreate [10,11]. The permanent coatings are designed to withstand frequent and repeated cleaning cycles [12–14]. The choice of a permanent coating involves a larger initial cost, repaid in the following years thanks to a faster and more effective cleaning. The selection between the first and the second solution is influenced by some aspects. One of the most important is the cost of the coating and the revamping of the anti-graffiti system. In the train field, the actual trend is to move to the application of a permanent layer due to the reduction of graffiti removal time, thus avoiding long standstill of the rolling stock.

Polyurethane resins show good mechanical properties and high chemical resistance. The latter is needed to guarantee a sufficient solvent resistance, and therefore, corrosion resistance [15,16]. As far as polyurethane-based formulations are concerned, UV radiation resistance is critical. Ultra-violet radiation is recognized to affect the molecular bonds of the polymeric matrix and additives, thus leading to modifications of properties of the coating [17]. In the last years, a lot of solutions were investigated adding some additives (e.g., UV-absorbers) to the blend. Widely-accepted procedures to evaluate the properties of anti-graffiti coatings have been developed and are described in ASTM as well as in UNI standards. These standards referring to anti-graffiti evaluation present different critical points. In particular, in ASTM D6578 [18], several graffiti marking materials and removal agents are indicated. Thus, the testing of all the possible combinations of marking materials and removers is possible after considerable time and effort. Moreover, no limit to the re-cleanability procedure is indicated. This aspect is believed to be very important due to the frequency of graffiti damage—particularly on railroad components and materials. Considering this aspect, the UNI 11246 [19] standard seems more useful, as it indicates in ten, the maximum number of graffiti production and removal. However, this standard is believed to provide a relatively restricted threshold to consider a graffiti-resistant coating as acceptable. Starting from the approach recommended in the cited standards, the aim of this paper is to individuate a suitable method to characterize the anti-graffiti properties of organic coatings. The idea is to overcome the limits of the current ASTM and UNI standards, suggesting a different testing approach in order to fully understand the effect of the cleaning mechanism. A review of the existing standard would help to get some insight into the anti-graffiti coating durability, thus leading to a better improvement of both coating and remover formulation. For this aim, a polyurethane powder coating which is widely used on transport and infrastructure as a protection system [15,20–22] is considered. This application method is used in industrial plants to cover unpainted metallic substrates. In the case of degraded painted surfaces, to guarantee a good adhesion between paint and substrate, it is necessary to remove the old organic coating before application. The polymeric matrix was modified, changing the additives to reach two different surface finishes. Considering the importance of the damage produced by UV exposure and chemical solvent contact, these aspects are the object of another paper [17].

2. Materials and Methods

A polyurethane powder coating was deposited on Aluminum alloy 1050 (min 99.50% Al) panels. Two different coating finishes—obtained by modifying the additives in the paint—were investigated: smooth (S, roughness R_a in the range of 1 μm) and wrinkled (W, roughness R_a in the range of 25 μm). The surface roughness was measured by profilometer Mahr mod. MarSurf PS1 on a stretch of length 5.6 mm. Aluminum was selected as substrate considering the growing interest for lightweight materials both in the transport of both constructions. The polymeric blend consisted of a hydroxylated polyester resin based on terephthalic acid and neopentylglycol. The hardener agent was a polyisocyanate adduct blocked with caprolactam and uretidione. The coating layer was applied by a powder deposition method. Before deposition, the raw aluminum surface was degreased with acetone for 10 min. The powder consisted of a pre-blended mixture of resins, hardener, pigments, and additives. The powder was electrostatically applied on the substrate prior to a thermal treatment, which promoted melting and curing [21,23,24]. In our case, the curing process was performed at 190 °C for 20 min. The color of this permanent paint was grey (RAL 7035). The paint was opaque and completely covered the metallic aspect of the aluminum surface.

First, the coating was characterized with an optical microscope to verify the presence of macro and micro defects. Figure 1 shows the surface of both samples: observe the different surface roughness.

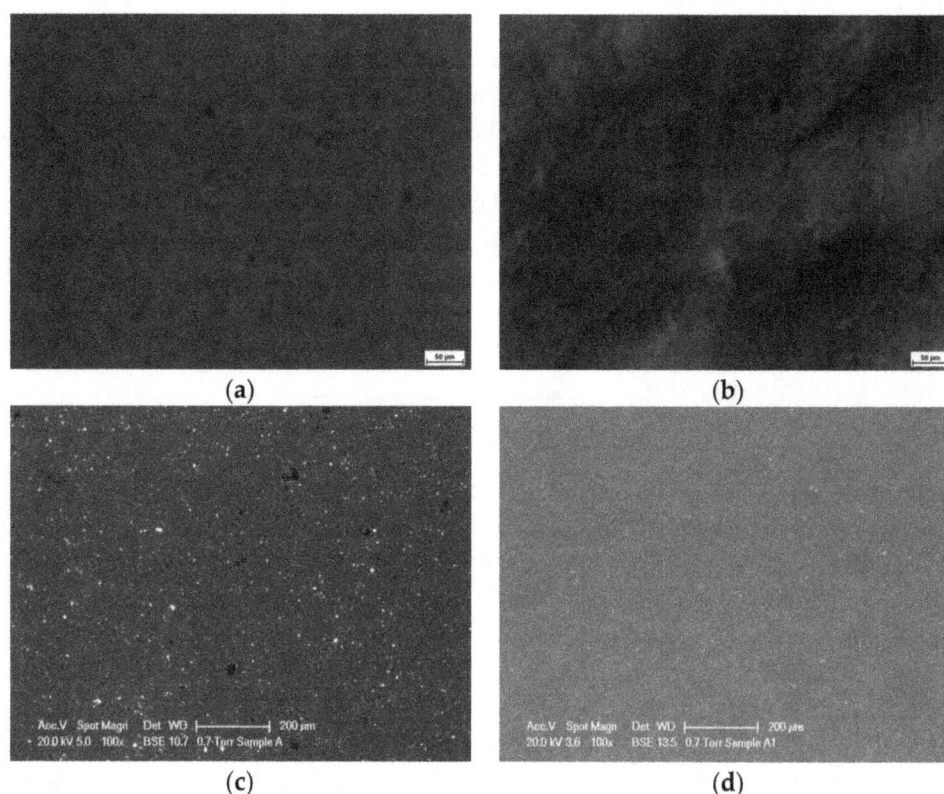

Figure 1. Optical and electronic pictures of surface of smooth sample (**a**,**c**) and wrinkled one (**b**,**d**).

The anti-graffiti behavior was evaluated considering some prescriptions of ASTM D6578 standard [18]. This standard indicates the application of different materials typically used as graffiti markings. The anti-graffiti behavior is evaluated considering the complete removal of graffiti by the mechanical action of a cotton cloth embedded with different cleaning agents (without producing any damage on the coating). To evaluate the clearness, visual observation or gloss and color change measurements are considered.

Considering the graffiti marking materials indicated in the standard, the graffiti were simulated using a red acrylic spray paint (AREXONS Acril color [25]), because sprayed coatings represent the most used damage methods in an urban environment. A 2.5 cm × 2.5 cm mask was employed in order to limit the graffiti areas. The surface was colored maintaining a 30 cm spraying distance in order to obtain a homogeneous layer of acrylic paint. According to the standard and the acrylic paint supplier recommendation, the samples were cured at room temperature. The graffiti removal was done using a white cotton cloth soaked in the remover at room temperature without water addiction. Several standard cleaning agents are indicated in the ASTM D6578 [18]. Nevertheless, some of them are too soft (for example, lint-free cotton cloth, mild detergent-based on sodium phosphate). The choice of removers to understand the cleaning mechanism was made considering chemical species such as MEK (methyl ethyl ketone, a polar aliphatic organic solvent), and xylene, an apolar aromatic solvent. The selection of this remover was made considering possible chemical interaction with the organic coating. The ideal solvent should remove the graffiti paint while only slightly affecting the properties of the underlying coating.

The cleaning was done as long as the surface would have been completely cleaned or a further cleaning action would have resulted ineffective. Following the guidelines of the cited standards, a visual evaluation of the completely cleaned surface was performed. After total graffiti removal, the cleaning action was interrupted in order to avoid to damage to the painted surface. If graffiti residuals visually remain on the surface, the removal was considered ineffective.

To have an objective evaluation of removal action of the solvents, color and gloss changes were measured as indicated in the ASTM D6578 [18] standard. A KonicaMinolta 2500 cc spectrophotometer (Tokyo, Japan) was used for color measurements, and an Erichsen NL3A digital glossmeter (Hemer, Germany) was used for gloss measurements. The gloss was measured as indicated in the ASTM D6578 standard, using a beam incident angle of 60° following ISO 2813 [26] and ASTM D523 [27] standards, containing information about specular gloss measurement on nonmetallic specimens considering test geometry and apparatus characteristic. The color measurement was obtained with the spectrophotometer according to the CIE $L^*a^*b^*$ system using the SCE (specular component excluded) considering the Cartesian coordinates of the color space: L^*, a^*, and b^*. The color difference ΔE_{ab}^* is calculated using the following equation [28–30]:

$$\Delta E_{ab}^* = \sqrt{(\Delta L^*)^2 + (\Delta a^*)^2 + (\Delta b^*)^2} \tag{1}$$

Considering ASTM D6578 standard, the criteria of acceptability for the anti-graffiti properties is a maximum gloss variation of 10% and a maximum color variation (ΔE) of two points. However, the total number of cleaning cycles are not indicated to consider an acceptable result. As far as the operative procedure is concerned, the UNI 11246 standard [19] is very similar to the ASTM standard. In this standard, no removal agents are specified, and the acceptable threshold is 5% for gloss reduction measured at 60° and one point in color variation (ΔE). Considering that the threshold related to the color variation is only slightly perceived by the human eye, UNI 11246 seems too restrictive. For this reason, the ASTM threshold will be considered throughout the paper. The color perception sensibility and the related acceptable thresholds are the objects of debate. Although a few research papers present in the literature propose different threshold values [31,32], to the best knowledge of the authors, no standard values are recognized. On the contrary, the UNI standard provides a threshold limit of ten maximum cleaning cycles, which was adopted as the limit value in this paper. The total standard deviation was calculated on 18 values obtained by measuring different spots from each cleaning area.

The ASTM D6578 standard indicates the possibility of evaluating the anti-graffiti behavior of materials subjected to outdoor or accelerated laboratory weathering. In fact, the aging of the organic coating due to UV irradiation would affect also its clean-ability and graffiti resistance. Considering the necessity to evaluate the properties in a reasonable time and to simultaneously produce a measurable damage, cyclic UV-B/humidity exposure was selected as the accelerated lab-scale test. The accelerated aging was performed with UV-B-313 lamps (where the number represents the characteristic nominal wavelength in nm of the peak of emission, Q-LAB, Westlake, OH, USA), according to ASTM G154 standard [33], which describes the basic principles for operating a fluorescent UV lamp and water apparatus. The cycle consisted of 8 h of UV exposure at 60 °C followed by 4 h of darkness and condensation at 50 °C for a total 1800 h. After different exposure times, the color, the gloss, and the roughness were measured. The standard indicates to check the anti-graffiti properties at the end of the accelerated exposure time. However, to follow the influence of the produced damage on anti-graffiti behavior, more interesting results are expected by monitoring the evaluation of the surface cleanability after each cycle. As previously explained, it was carried out considering the color and gloss changes. In order to understand the cleaning mechanism on the aged surface, contact angle measurements were carried out following ASTM D7334 [34] standard, using a 20 µL drop of distilled water. The contact angle was measured by digitally processing of the pictures, from optical stereomicroscope Stemi2000C (Carl Zeiss, Oberkochen, Germany).

The effect of the direct and prolonged exposure of the coatings to the solvent was investigated. The protective properties over solvent exposure time were evaluated by electrochemical impedance spectroscopy (EIS), which is a widely-used method to evaluate the corrosion behavior of painted metals [35,36]. The configuration of the test was inspired by ASTM D1308 standard [37], where the geometry test called "Spot Test, Open" is considered. At 23 °C and 50% relative humidity, a small portion of the reagent is placed on a horizontal surface (a specific standard procedure does not already exist for graffiti-resistant coating). A cylindrical PVC cell ensures the contact between solvent and coating surface for a fixed time. The amount of solvent was held constant for all the tests (25 mL). The same cell set-up was used for electrochemical impedance spectroscopy measurements, carried out in Harrison solution (3.5 wt % ammonium sulfate and 0.5 wt % sodium chloride), pouring the electrolyte after solvent removal. The tested area was 16.6 cm^2. A three-electrode configuration was used: the sample as working electrode, a platinum wire counter electrode, and an Ag/AgCl (+205 mV vs. standard hydrogen electrode (SHE)) reference electrode. A Parstat 2273 potentiostat (Princeton Applied Research Ametek, Berwyn, PA, USA) was used for data acquisition. A frequency range between 10^5 and 10^{-2} Hz with an amplitude of the signal of 30 mV was employed. The impedance modulus in the low-frequency range was exploited to provide a rough estimation of the protection level of the organic coating [38,39].

3. Results

The principal aim of the method is to better understand the effect of the interaction between the coating and the removers, not only focusing on the coating. Table 1 reports the total removal cycles possible on both surfaces finishing with the two removers (following the procedures described in ASTM D6578 standard).

Table 1. Cycles of the total removal of graffiti following ASTM D6578 standard.

Samples/Removers	MEK	Xylene
Smooth S	2	4
Wrinkled W	1	4

The xylene shows a higher efficiency on both surface finishes. The MEK very effectively removed the graffiti, but at the same time interacts strongly with the paint, reducing the gloss values. Considering the color change, the wrinkled surface shows noticeable changes, probably due to the higher contact surface with the solvent, due to the higher roughness.

The color change with increasing cleaning cycles is shown in Figure 2. In this example, it is possible to see that the polar solvent (MEK) cannot remove the graffiti over the fourth cycle without producing a color change outside the ASTM threshold.

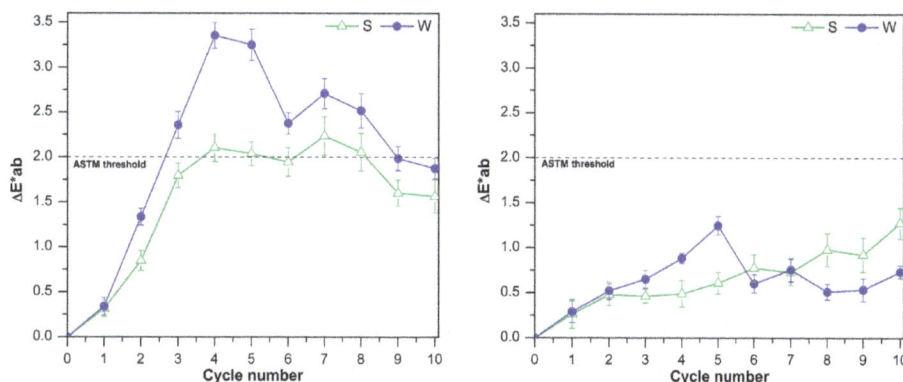

Figure 2. Color change as a function of the number of cleaning cycles using different removers: (**a**) Methyl ethyl ketone (MEK, polar); (**b**) Xylene (non-polar). Data extracted from [20].

The gloss variation was revealed to be more critical compared to the color variation (Figure 3). It is possible to observe that with the polar remover, the normalized gloss value exceeded the limit recommended by the standards after few cycles. In addition, the non-polar remover exceeded the limit value after only four removal actions. Considering the gloss changes with cycles, the sample with high roughness showed the worst behavior.

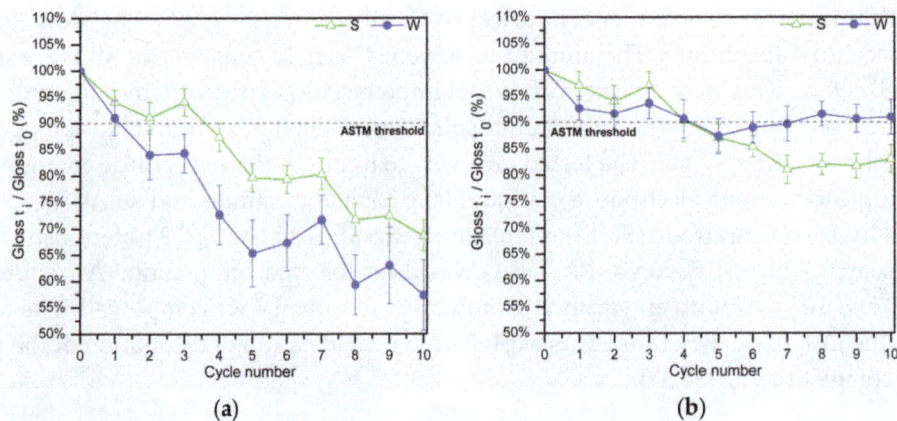

Figure 3. Cleaning gloss change as a function of the number of cleaning cycles using different removers: (**a**) MEK (polar); (**b**) Xylene (non-polar). Data extracted from [20].

The static contact angle was measured to verify the aging mechanism provided by the previous analysis. The measurements were repeated four times. During testing, the contact angle decreased for both samples (Figure 4), indicating an important modification of surface properties with an increase in hydrophilicity. To provide an easier comparison between the UV/humidity cycles and the contact angle measurements, "UV exposure time" and "total exposure time" are reported on the *x*-axis of Figure 4. Considering that a cycle consists of 8 h of UV exposure followed by 4 h of humidity (as reported in Section 2), the "UV exposure time" results equal to 2/3 of "total exposure time".

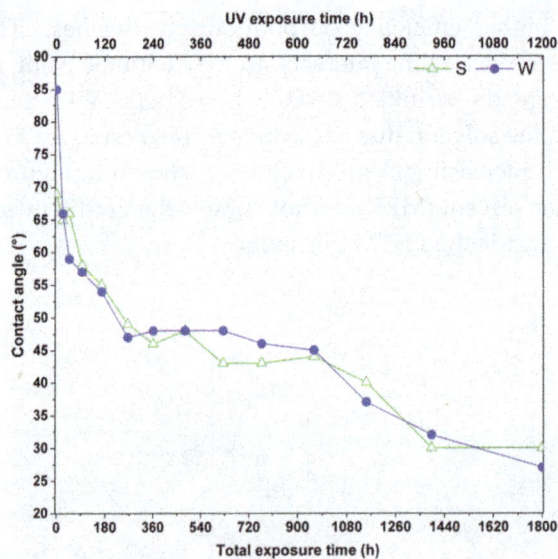

Figure 4. Contact angle as a function of the aging time.

The cleaning cycles analysis was carried out even on the aged surface to understand if the surface variation affects the graffiti cleanability. Table 2 shows the total UV-B cyclic exposure time, where the removal action is possible in accordance with the standard.

Table 2. Total UV-B exposure hours where the removal action results still effective.

Samples/Removers	MEK	Xylene
Smooth S	264	264
Wrinkled W	96	168

The chemical modification of the surface increased the graffiti adhesion. This caused a more difficult graffiti removal with both solvents. The polar one, having a stronger solubilization capability than the non-polar, better removed the graffiti, but at the same time produced higher gloss and color changes.

Electrochemical impedance spectroscopy was used to investigate the effect of the organic solvent exposure on the barrier properties of the organic coatings. During the solvent contact test, the organic solvent was not allowed to evaporate. After the solvent contact removal, the samples were conditioned at 20 °C to promote a partial evaporation from the soaked paint. Afterward, the samples were immersed in the electrolyte to carry out the EIS measurements. This procedure guarantees the partial inhibition of the polymer chains relocation after the organic solvent uptake. Therefore, it is believed to represent the coating behavior when an organic solvent is present (which is expected to lead to a barrier properties loss). The cycle aging test allows the solvent to partially evaporate before carrying out the electrochemical impedance spectroscopy measurements, and therefore the polymer chains have time to readapt, changing the original structure. This test is expected to highlight how much the coating can withstand contact–evaporation cycles before irreversible damages are formed.

To simplify the analysis, the impedance modulus in the low-frequency range was considered as an indication of the protection properties level of the organic coating. According to many reports [40–42], this property can be characterized by EIS measurements by examining the low-frequency limit of $|Z(\omega)|$ with an acceptable degree of approximation. In Figure 5, it is possible to observe how this value (total impedance $|Z|$ at 4×10^{-2} Hz) changes during contact time with MEK. Notice that the exposure promoted macroscopic phenomena due to the presence of water molecules and ions, such as swelling and then layer delamination. The test seems to be particularly tough for the samples with the wrinkled surface. After about 60 min of exposure, delamination of the paint from the substrate was observed, followed by the occurrence of cracks after about 200 min. These results suggest that the contact time with the organic solvent can be dramatically detrimental for the corrosion protection properties. From a practical point of view, it seems that particular care has to be taken to reduce the organic solvent contact time during graffiti removal operations—in particular if MEK is used on polyurethane paint with wrinkled surfaces.

On the other hand, xylene showed no changes of the coatings' impedance modulus in the low-frequency range (stable around 10^{11} Ω cm^2) for 360 min exposure time (not reported data). The noticeable differences observed compared to MEK are believed to rely on the apolar nature of xylene that probably does not affect the polyurethane intermolecular bonds.

It is important to highlight that the EIS measurements provide information on the reduction of corrosion protection properties of paints in a very short time (just a few hours).

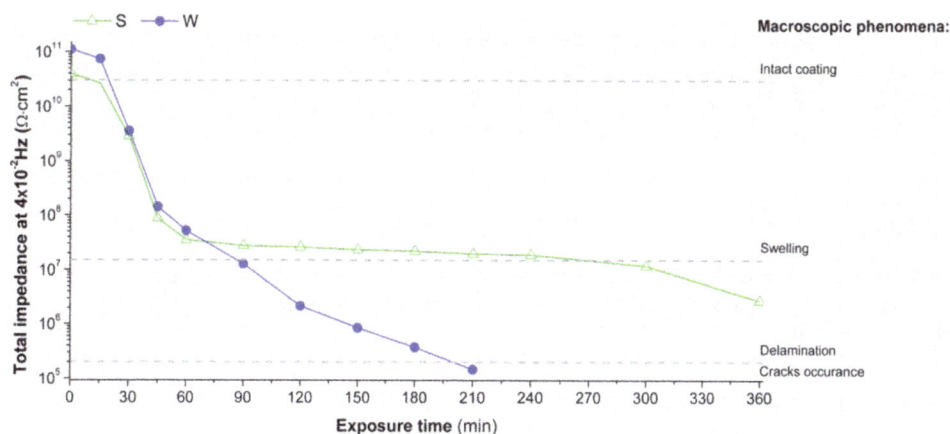

Figure 5. Impedance modulus at low frequency as a function of the aging time.

4. Conclusions

The field of powder coatings for anti-graffiti applications is relatively new, and therefore the performance of the products currently on the market can be optimized. In addition, a test protocol could be individuated to evaluate the efficiency of removers, and at the same time, the resistance of organic coatings to chemical remover.

In this work, a critical analysis of the commonly employed standards for anti-graffiti evaluation has been carried out. First of all, it can be stated that when quantifying the characteristics of an anti-graffiti paint it is not sufficient to consider only the visual changes, but it is necessary to obtain a quantification by measuring the change of gloss and color. The acceptance criteria specified in ASTM 6578 standard have been revealed to be more suitable to evaluate the coatings. On the other hand, UNI 11246 standard suggests a maximum limit of graffiti and removals.

Considering the graffiti removers, the standards recommend to test several graffiti-removers. Among them, the experimental data suggest that the use of only of two types of organic solvents—one polar and one non-polar—provides a complete evaluation of the susceptibility of the coating to damage by the removal agent.

As far as the studied formulations are concerned, the following conclusions can be drawn:

- UV-B/condensation weathering exposure leads to a wettability variation, toward an increased hydrophilic behavior of the surface with a consequently more difficult graffiti removal with both removers.
- The criticism of the polar remover is probably due to the chemical interaction between the coating and the remover. The apolar remover presents less interaction with the coating.
- The coating properties are much more affected when the painted sheet is continuously immersed in the solvent with respect to the cyclic aging (wet/dry cycles).
- Considering the ease of graffiti removal, the surface finish shows a significant influence. An increase in the roughness of the paint makes the cleaning action less effective.
- Electrochemical impedance spectroscopy measurements are a very useful method to evaluate the damage produced on the coatings due to polishing action and contact with removers. In addition, the information of residual protection properties is quantitative, and a short time is necessary for its acquisition. This electrochemical approach is frequently used to check the protection properties of the organic coating; nevertheless, it is not just used to evaluate the interaction of graffiti removals and coatings.
- Finally, with a simple modification of an electrochemical cell, EIS measurements can also be carried out in-field, resulting in a suitable method to check the damage level of existing structures and components without the necessity of a laboratory.

Based on the obtained data, permanent coatings seem to be a possible solution for train cars exposed to vandalism. A design made in symbiosis with the remover, which studies in detail the interaction between coating, fouling spray, and solvent, could further accelerate the development of this sector.

Acknowledgments: The authors are grateful to Roberto Paganica (Akzo Nobel Italy, Como Italy) for powder coated samples supply.

Author Contributions: Stefano Rossi conceived and designed the research and experiments, analyzed the data and wrote the paper; Michele Fedel designed the experiments, analyzed and discussed the data; Simone Petrolli performed the experiments, analyzed the data and wrote the paper; Flavio Deflorian analyzed and discussed the data.

Conflicts of Interest: The authors declare no conflict of interest.

References

1. Lettieri, M.; Masieri, M. Surface Characterization and Effectiveness Evaluation of Anti-Graffiti Coatings on Highly Porous Stone Materials. *Appl. Surf. Sci.* **2014**, *288*, 466–477. [CrossRef]
2. Watzlawik, M. The 'Art' of Identity Development—Graffiti Painters Moving Through Time and Space. *Culture Psychol.* **2014**, *20*, 404–415. [CrossRef]
3. Sanmartin, P.; Cappitelli, F.; Mitchell, R. Current Methods of Graffiti Removal: A Review. *Constr. Build. Mater.* **2014**, *71*, 363–374. [CrossRef]
4. Thompson, K.; Offler, N.; Hirsch, L.; Every, D.; Thomas, M. From broken windows to a renovated research agenda: A review of the literature on vandalism and graffiti in the rail industry. *Transp. Res. A Policy Pract.* **2012**, *46*, 1280–1290. [CrossRef]
5. Luongo, M. Quelli-che-dicono-no-ai-graffiti. *Corriere della Sera Newspaper*, 13 December 2014, p. 22.
6. Non Solo Graffiti Vandali Treni Costano Trenord Otto Milioni All'anno. Available online: http://milano.corriere.it/milano/notizie/cronaca/14_gennaio_28/test-yoodeal.shtml?refresh_ce-cp (accessed on 18 March 2017).
7. Garcia, O.; Malaga, K. Definition of the Procedure to Determine the Suitability and Durability of an Anti-Graffiti Product for Application on Cultural Heritage Porous Materials. *J. Cult. Herit.* **2012**, *13*, 77–82. [CrossRef]
8. Bengtsson, T. Waterborne Wax-Based Sacrificial Graffiti Protective Coatings: Demands and Experience. In Proceedings of the European Coatings Conference Antigraffiti Coatings, Berlin, Germany, 2–3 December 1999; pp. 169–189.
9. Black, R.H. Anti-Graffiti Coating Material and Method of Using Same. U.S. Patent 5,387,434, 7 February 1995.
10. Neto, E.; Magina, S.; Camões, A.; Begonha, A.; Evtuguin, D.V.; Cachim, P. Characterization of concrete surface in relation to graffiti protection coatings. *Constr. Build. Mater.* **2016**, *102*, 435–444. [CrossRef]
11. Moura, A.; Flores-Colen, I.; de Brito, J. Study of the effect of three anti-graffiti products on the physical properties of different substrates. *Constr. Build. Mater.* **2016**, *107*, 157–164. [CrossRef]
12. Manvi, G.N.; Singh, A.R.; Jagtap, R.N.; Kothar, D.C. Isocyanurate Based Fluorinated Polyurethane Dispersion for Anti-Graffiti Coatings. *Prog. Org. Coat.* **2012**, *75*, 139–146. [CrossRef]
13. Rabea, A.M.; Mohseni, M.; Mirabedini, S.M.; Tabatabaei, M.H. Surface Analysis and Anti-Graffiti Behavior of a Weathered Polyurethane-Based Coating Embedded With Hydrophobic Nano Silica. *Appl. Surf. Sci.* **2012**, *258*, 4391–4396. [CrossRef]
14. Licchelli, M.; Marzolla, S.J.; Poggi, A.; Zanchi, C. Crosslinked Fluorinated Polyurethanes for the Protection of Stone Surfaces from Graffiti. *J. Cult. Herit.* **2011**, *12*, 34–43. [CrossRef]
15. Rossi, S.; Fedel, M.; Deflorian, F.; Feriotti, A. Anti-graffiti properties of polyurethane powder coatings. *J. Test. Eval.* **2017**, in press. [CrossRef]
16. Chattopadhyay, D.K.; Raju, K.V.S.N. Structural Engineering of Polyurethane Coatings for High Performance Applications. *Prog. Polym. Sci.* **2007**, *32*, 352–418. [CrossRef]
17. Rossi, S.; Fedel, M.; Petrolli, S.; Deflorian, F. Accelerated weathering and chemical resistance of polyurethane powder coatings. *J. Coat. Technol.* **2016**, *13*, 427–437. [CrossRef]
18. *ASTM D6578-13 Standard Practice for Determination of Graffiti Resistance*; American Society for Testing and Materials: Philadelphia, PA, USA, 2013.
19. *UNI 11246 Alluminio e Leghe di Alluminio—Modalità di Valutazione di Prodotto Permanente Anti-Graffiti*; UNI, Enti Italiano di Normazione: Milano, Italy, 2007.
20. Papaj, E.A.; Mills, D.J.; Jamali, S.S. Complex shaped ZnO nano- and microstructure based polymer composites: mechanically stable and environmentally friendly coatings for potential antifouling applications. *Prog. Org. Coat.* **2014**, *77*, 2086–2090. [CrossRef]
21. Wicks, Z.W.; Jones, F.N.; Pappas, S.P. *Organic Coatings, Science and Technology*; John Wiley & Sons: New York, NY, USA, 1992.
22. Rossi, S.; Fedel, M.; Petrolli, S.; Deflorian, F. Behaviour of different removers on permanent anti-graffiti organic coatings. *J. Build. Eng.* **2016**, *5*, 104–113. [CrossRef]
23. *Powder Coating Technology Handbook*; Engineers India Research Institute: Delhi, India, 2014.
24. Ulrich, D.L. *User's Guide to Powder Coating*, 3rd ed.; Society of Manufacturing Engineers: Dearborn, MI, USA, 1993.

25. Acrilcolor. Available online: http://arexons.it/ (accessed on 15 June 2016).

26. *ISO 2813 Paints and varnishes—Determination of Gloss Value at 20 Degrees, 60 Degrees and 85 Degrees*; International Organization for Standardization: Geneva, Switzerland, 2014.

27. *ASTM D523-14 Standard Test Method for Specular Gloss*; American Society for Testing and Materials: Philadelphia, PA, USA, 2014.

28. *CIE 15 Technical Report: Colorimetry*, 3rd ed.; International Commission on Illumination, Bureau Central de la CIE: Paris, France, 2004.

29. Precise Color Communication. Konica Minolta Sensing. Available online: http://www.konicaminolta.com/instruments/about/network (accessed on 18 March 2017).

30. Gulrajani, M.L. *Colour Measurement: Principles, Advances and Industrial Applications*; Woodhead Publishing: Oxford, UK, 2016.

31. Sanmartín, P.; Silva, B.; Prieto, B. Effect of Surface Finish on Roughness, Color, and Gloss of Ornamental Granites. *J. Mater. Civ. Eng.* **2011**, *23*, 1239–1248. [CrossRef]

32. Prieto, B.; Sanmartín, P.; Pereira-Pardo, L.; Silva, B. Recovery of the traditional colours of painted woodwork in the Historical Centre of Lugo (NW Spain). *J. Cult. Herit.* **2011**, *12*, 279–286. [CrossRef]

33. *ASTM G154-16 Standard Practice for Operating Fluorescent Ultraviolet (UV) Lamp Apparatus for Exposure of Nonmetallic Materials*; American Society for Testing and Materials: Philadelphia, PA, USA, 2016.

34. *ASTM D7334–13 Standard Practice for Surface Wettability of Coatings, Substrates and Pigments by Advancing Contact Angle Measurement*; American Society for Testing and Materials: Philadelphia, PA, USA, 2013.

35. Macdonald, J.R. *Impedance Spectroscopy: Emphasizing Solid Materials and Systems*; Wiley: New York, NY, USA, 1987.

36. Scully, J.R.; Silverman, D.C.; Kendig, M.W. *Electrochemical Impedance: Analysis and Interpretation*; American Society for Testing and Materials: Philadelphia, PA, USA, 1993.

37. *ASTM D1308-02 Standard Test Method for Effect of Household Chemicals on Clear and Pigmented Organic Finishes*; American Society for Testing and Materials: Philadelphia, PA, USA, 2013.

38. Amirudin, A.; Thierry, D. Application of electrochemical impedance spectroscopy to the degradation of polymer-coated metals. *Prog. Org. Coat.* **1995**, *26*, 1–28. [CrossRef]

39. Rossi, S.; Deflorian, F.; Fontanari, L.; Cambruzzi, A.; Bonora, P.L. Electrochemical measurements to evaluate the damage due to abrasion on organic protective system. *Prog. Org. Coat.* **2005**, *52*, 288–297. [CrossRef]

40. Bierwagen, G.; Tallman, D.; Li, J.; He, L.; Jeffcoate, C. EIS studies of coated metals in accelerated exposure. *Prog. Org. Coat.* **2005**, *46*, 148–157. [CrossRef]

41. Hinderliter, B.R.; Croll, S.G.; Tallman, D.E.; Su, Q.; Bierwagen, G.P. Interpretation of EIS data from accelerated exposure of coated metals based on modeling of coating physical properties. *Electrochim. Acta* **2006**, *51*, 4505–4515. [CrossRef]

42. Olivier, M.G.; Romano, A.P.; Vandermiers, C.; Mathieu, X.; Poelman, M. Influence of the stress generated during an ageing cycle on the barrier properties of cataphoretic coatings. *Prog. Org. Coat.* **2008**, *63*, 323–329. [CrossRef]

Permissions

The contributors of this book come from diverse backgrounds, making this book a truly international effort. This book will bring forth new frontiers with its revolutionizing research information and detailed analysis of the nascent developments around the world.

We would like to thank all the contributing authors for lending their expertise to make the book truly unique. They have played a crucial role in the development of this book. Without their invaluable contributions this book wouldn't have been possible. They have made vital efforts to compile up to date information on the varied aspects of this subject to make this book a valuable addition to the collection of many professionals and students.

This book was conceptualized with the vision of imparting up-to-date information and advanced data in this field. To ensure the same, a matchless editorial board was set up. Every individual on the board went through rigorous rounds of assessment to prove their worth. After which they invested a large part of their time researching and compiling the most relevant data for our readers.

The editorial board has been involved in producing this book since its inception. They have spent rigorous hours researching and exploring the diverse topics which have resulted in the successful publishing of this book. They have passed on their knowledge of decades through this book. To expedite this challenging task, the publisher supported the team at every step. A small team of assistant editors was also appointed to further simplify the editing procedure and attain best results for the readers.

Apart from the editorial board, the designing team has also invested a significant amount of their time in understanding the subject and creating the most relevant covers. They scrutinized every image to scout for the most suitable representation of the subject and create an appropriate cover for the book.

The publishing team has been an ardent support to the editorial, designing and production team. Their endless efforts to recruit the best for this project, has resulted in the accomplishment of this book. They are a veteran in the field of academics and their pool of knowledge is as vast as their experience in printing. Their expertise and guidance has proved useful at every step. Their uncompromising quality standards have made this book an exceptional effort. Their encouragement from time to time has been an inspiration for everyone.

The publisher and the editorial board hope that this book will prove to be a valuable piece of knowledge for researchers, students, practitioners and scholars across the globe.

List of Contributors

Lei Shao, Yangyong Zhao and Yong Zhang
State Key Laboratory for Advanced Metals and Materials, University of Science and Technology Beijing

Alejandro Jiménez and Manuel Vázquez
Institute of Materials Science of Madrid, CSIC, 28049 Madrid, Spain

Łukasz Sadowski
Faculty of Civil Engineering, Wrocław University of Science and Technology, Wybrzeż e Wyspian´ skiego 27, 50-370 Wrocław, Poland

Mehdi Nikoo
Young Researchers and Elite Club, Ahvaz Branch, Islamic Azad University, Ahvaz, Iran

Mohammad Nikoo
SAMA Technical and Vocational Training College, Islamic Azad University, Ahvaz Branch, Ahvaz, Iran

Hongtao Cui, Lingling Sun, Fangyang Liu, Chang Yan and Xiaojing Hao
School of Photovoltaic and Renewable Energy Engineering, University of New South Wales, Sydney 2052

Xiaolei Liu
UCL Institute for Materials Discovery, University College London, London WC1E 7JE, UK

Wei Sun, Yue Zhang and Rong Liu
School of Mechanical Engineering & Automation, Northeastern University, Shenyang 110819, China

Jin-Ho Park, Jitendra Kumar Singh and Han-Seung Lee
Department of Architecture Engineering, Hanyang University, 55, Hanyangdaehak-ro, Sangrok-gu, Ansan-si, Gyeonggi-do 15588, Korea

Parnia Navabpour, Kevin Cooke and Hailin Sun
Miba Coating Group, Teer Coatings Ltd.,West Stone House, Berry Hill Industrial Estate,Droitwich WR9 9AS, UK

Kaushal Kumar
Department of Mechanical Engineering; Guru Jambheshwar University of Science & Technology, Hisar 125001, India

Satish Kumar, Gurprit Singh, Jatinder Pal Singh and Jashanpreet Singh
Department of Mechanical Engineering; Thapar University, Patiala 147004, India

Mustafa Çakır
Department of Metallurgical and Materials Engineering, Faculty of Technology, Marmara University, Goztepe 34722, Istanbul, Turkey

Emine Bakan, Georg Mauer, Yoo Jung Sohn and Robert Vaßen
Forschungszentrum Jülich GmbH, Institute of Energy and Climate Research (IEK-1), 52425 Jülich, Germany

Dietmar Koch
German Aerospace Center, Institute of Structures and Design, 70569 Stuttgart, Germany

Yasuyuki Kawaguchi, Fumihiro Miyazaki, Masafumi Yamasaki and Katsunori Muraoka
Plazwire Co., Ltd., 2-3-54 Higashi-naka, Hakata-ku, Fukuoka 812-0829, Japan

Yukihiko Yamagata and Nozomi Kobayashi
Interdisciplinary Graduate School of Engineering Sciences, Kyushu University, Kasuga-koen, Kasuga, Fukuoka 816-8580, Japan

Martina Pini, Paolo Neri and Anna Maria Ferrari
Department of Sciences and Engineering Methods, University of Modena and Reggio Emilia, Via Amendola, 2, 42100 Reggio Emilia, Italy

Cristina Siligardi
Department of Engineering "Enzo Ferrari", University of Modena and Reggio Emilia, Via Vignolese, 905/A, 41125 Modena, Italy

Erika Iveth Cedillo González
Department of Engineering "Enzo Ferrari", University of Modena and Reggio Emilia, Via Vignolese, 905/A, 41125 Modena, Italy
Facultad de Ciencias Químicas, Universidad Autónoma de Nuevo León, Guerrero y Progreso s/n Col. Treviño, Monterrey 64570, Mexico

Xuejie Liu, Congjie Kang, Haimao Qiao, Yuan Ren, Xin Tan and Shiyang Sun
School of Mechanical Engineering, Inner Mongolia University of Science & Technology, Baotou, Inner Mongolia 014010, China

Francisco Silva, Rui Martinho and Maria Andrade
ISEP—School of Engineering, Polytechnic of Porto, 4200-072 Porto, Portugal

António Baptista
INEGI—Instituto de Ciência e Inovação em Engenharia Mecânica e Engenharia Industrial, 4200-465 Porto

Ricardo Alexandre
TEandM—Tecnologia, Engenharia e Materiais; 3045-508 Taveiro, Portugal

Qiaoyan Ye and Karlheinz Pulli
Fraunhofer Institute for Manufacturing Engineering and Automation, Nobelstr. 12, Stuttgart 70569, Germany

Yung-I Chen, Tso-Shen Lu and Zhi-Ting Zheng
Institute of Materials Engineering, National Taiwan Ocean University, 2 Pei-Ning Road, Keelung 20224, Taiwan

Galina Xanthopoulou, Amalia Marinou, Konstantinos Karanasios and George Vekinis
Institute of Nanoscience and Nanotechnology, National Center for Scientific Research w"Demokritos", Aghia Paraskevi, Athens15310, Greece

Jian Yang
School of Mechanical Engineering and Automation, University of Science and Technology Liaoning, Anshan 114051, China

Stefano Rossi, Michele Fedel, Simone Petrolli and Flavio Deflorian
Department of Industrial Engineering, University of Trento, Via Sommarive, 9, 38123 Trento, Italy

Index